ANIMAL SURGERY

ANIMAL SURGERY

Editor

Andre Garrard

scitus
academics

Animal Surgery

Edited by **Andre Garrard**

ISBN: 9781681179322
Library of Congress Control Number: 2017948962

Notice

Printed in United States of America.

Preface

Veterinary surgery is surgery performed on animals by veterinarians, whereby the procedures fall into three broad categories: orthopaedics (bones, joints, muscles), soft tissue surgery (skin, body cavities, cardiovascular system, GI/urogenital/respiratory tracts), and neurosurgery. Veterinarian Surgeons tend to the healthcare needs of animals, including pets, livestock, and zoo and laboratory animals. Their primary role is to operate on animals – repairing physical normality's or injuries that occur. The aim of veterinary surgery may be quite different in pets and in farm animals. In the former, situation is a bit like in human beings, and more and more complex operations are performed, with sophisticated anaesthesia techniques. In the latter, the cost of the operation must not exceed the economic benefit in surgically treating the illness. Anesthesia in animals has many similarities to human anesthesia, but some differences as well. Local anesthesia is primarily used for wound closure and removal of small tumors. Lidocaine, mepivacaine, and bupivacaine are the most commonly used local anesthetics used in veterinary medicine.

This volume 'Animal Surgery' equips the readers with the latest advancements in the veterinary medicine and surgery that deals with prevention, diagnosis and treatment of disease, disorder and injury in non-human animals. It relates to the medical and surgical treatment of animals. It covers research studies that helps advance the theoretical and practical understanding of animal science and medicine, relevant to any area of the health and management of domestic animals, poultry, aquatic animals, wildlife and related fields including: infectious diseases, epidemiology, disease dynamics, animal medicine; and welfare, management and hygiene, nutrition and clinical nutrition, surgery, anesthesiology and diagnostic imaging, and experimental surgery. This book will be of invaluable to veterinarians and veterinary technicians provided with practical information with the goal of using current information from the literature to applied situations.

Table of Contents

CHAPTER 1

Twinning in Holstein-Friesian Dairy Cows: Proportion Carried to Term and Calf Sex Ratios

Peter D. Cockcroft [1,†,*] and Emma J. Sorrell [2]

[1]*School of Animal and Veterinary Sciences, University of Adelaide, Roseworthy Campus, Roseworthy, SA 5371, Australia*

[2]*The Brambles Veterinary Surgery, Churchdown, GL3 2HE, UK*

ABSTRACT

The purpose of this study was to investigate the proportion of twins carried to term and the sex ratio of twin calves at birth in Holstein-Friesian dairy cattle kept on commercial farms in Devon and Cornwall, England. Ten farms were used in the study. Fifty four cows with twin pregnancies were identified using trans-rectal ultra-sonographic examination between 30 and 70 days of gestation. The farm records were subsequently used to derive the number of calves born. Farm records of 66 additional sets of twin births with the sex of the calves recorded were also identified. Of the 54 cows diagnosed with twin pregnancies, 16 cows (29.6%) aborted or absorbed both fetuses, 11 cows (20.4%) carried one calf to term and 27 cows (50%) carried both calves to term. In the calf sex analysis of the additional 66 sets of twins: 13♂♂ calves (19.7%), 18 ♀♀ calves (27.3%) and 35 ♂♀ calves (53.0%). There was no statistically significant difference from an expected ratio of 1♂♂:2♂♀:1♀♀ (p = 0.61). This study provides bench marks for the expected abortion/absorption rates following the early ultra-sonographic diagnosis of twin pregnancies in comparable populations and supports earlier observations that the expected sex ratio for twinning approximates to1♂♂:2♂♀:1♀♀.

Keywords: dairy; cattle; twinning; sex ratio; abortion

1. INTRODUCTION

Twinning in dairy cattle is associated with higher milk yields [1], an increased incidence of abortion [1], dystocia [1], retained fetal membranes [1], metritis [1], Freemartinism [1,2] and an extended calving to conception interval [1,3]. Over 95% of female calves twinned to a male are sterile [4]. The additional cost of twinning has been reviewed [3] and estimated at £109 [5]. The incidence of twinning reported in dairy herds varies between studies, herds and parities but is an average of 4% has been reported in UK herds [5]. The rate of single and double embryonic/fetal losses in twin pregnancies in Holstein cows have been published [6,7]. Large scale studies of the twin calf sex ratios have been reported [8,9,10].

The purpose of this study was to investigate the proportion of twins carried to term and the calf sex ratio of twin calves in Holstein-Friesian dairy cattle kept on 10 commercial farms in Devon and Cornwall, England.

2. EXPERIMENTAL SECTION

Ten Friesian-Holstein dairy farms in Devon and Cornwall were used in this study. Herd size ranged from 70 to 310 animals with a mean of 168. None of the farms instigated a specifically designed management protocol to deal with cows identified as carrying twins. Fifty four cows with twin pregnancies were identified during routine trans-rectal ultra-sonographic examination for pregnancy. A 7.5 MHz linear probe was used. The operator was a large animal veterinarian experienced in ultra-sonographic pregnancy diagnosis. Twins were identified between 30 and 70 days of gestation according to the last recorded service date and a diagnosis of twinning was made on the observation of two embryos/fetuses. The farm records were subsequently used to derive the number of calves born following the ultra-sonographic diagnosis.

Farm records of an additional 66 sets of twin births with the sex of the calves recorded were identified. A chi-squared test was used to analyse the data for a random distribution giving a gender ratio of 1♂♂:2♂♀:1♀♀.

3. RESULTS

Of the 54 cows diagnosed with twin pregnancies between 30 and 70 days of gestation, 16 cows (29.6%) aborted or reabsorbed both fetuses, 11 cows (20.4%) carried one calf to term and 27 cows (50%) carried both to term. In the calf sex ratio analysis of the additional 66 sets of twins there were 13♂♂ calves (19.7%), 18 ♀♀ calves (27.3%) and 35 ♂♀ calves (53.0%). There was no statistically significant difference when the numbers of calves within each category were compared to an expected ratio of 1♂♂:2♂♀:1♀♀ ($p = 0.61$).

4. DISCUSSION

A comparative study of gestation losses in beef cattle at the US Meat Animal Research Centre found that 12.2% of cows with twin pregnancies aborted or reabsorbed both fetuses and 4.6% aborted or reabsorbed one fetus [11]. López-Gatius and Hunter (2005) [6] reported embryonic/fetal losses in 211 twin pregnancies in high yielding dairy cows in Spain. Loss of both pregnancies occurred in 73 cows (24.2%), with 13 cows (6.2%) maintaining a single pregnancy. Silva-del-Río *et al.* (2009) [7] reported outcomes for 98 twin pregnancies in Holstein cows in the USA where 13 cows (13.3%) lost both embryos/fetuses, and 11 cows (11.2%) lost a single embryo/fetus. The abortion/absorbtion rate in this study were higher and may be related to management or health status differences between the populations.

A sex ratio of 1♂♂:2♂♀:1♀♀ was reported in a large population of Holstein cows [12]. The calf sex ratio amongst 925 sets of twins in dairy cows was 24.9% ♂♂:48.7% ♂♀:26.4%♀♀ in the USA study [8]. An analysis of 96,069 twin births found a twin sex ratio of 30.1% ♂♂:43.6%♂♀:26.3%♀♀ in USA dairy cows [9]. Ghavi Hossein-Zadeh *et al.* (2008) reported a calf sex ratio in 4,045 twin births in Holstein cows in Iran of 28.2%♂♂:48.8%♂♀:22.9%♀♀ [10].

Within the power of this study the distribution of the sexes of the twins were consistent with this random distribution of the sexes. In a review of the literature of twinning in cattle 92.6% were classified as fraternal or dizygotic twins and 7.4% classified as monozygotic or identical twins [13,14]. Therefore the assumption of a random distribution of the sexes will be approximate rather than absolute.

5. CONCLUSION

This study provides bench marks for the expected abortion/absorption rates following the early ultra-sonographic diagnosis of twin pregnancies in comparable populations and supports earlier observations that the expected sex ratio for twinning approximates to 1♂♂:2♂♀:1♀♀.

ACKNOWLEDGMENTS

Rupert Kirkwood for participation in the data collection and for the trans-rectal ultra-sonographic pregnancy diagnosis. Fred Heath for statistical advice.

AUTHOR CONTRIBUTIONS

Peter Cockcroft was responsible for the experimental design and writing the manuscript. Emma Sorrell was responsible for the data collection and analysis.

CONFLICTS OF INTEREST

The authors declare no conflict of interest.

REFERENCES

1. Nielen, M.; Schukken, Y.H.; Scholl, D.T.; Wilbrink, H.J.; Brand, A. Twinning in dairy cattle: A study of risk factors and effects. Theriogenology 1989, 32, 845–862.
2. Kinsel, M.L.; Marsh, W.E.; Ruegg, P.L.; Etherington, W.G. Risk factors for twinning in dairy cows. J. Dairy Sci. 1997, 81, 989–993.
3. Eddy, R.G.; Davies, O.; David, C. An economic assessment of twin births in British dairy herds. Vet. Rec. 1991, 129, 526–529.

4. Gregory, K.E.; Echternkamp, S.E.; Cundiff, L.V. Effects of twinning on dystocia, calf survival, calf growth, carcass traits and cow productivity. J. Anim. Sci. 1996, 74, 1223–1233.
5. Esslemont, R.J.; Kossaibati, M. The Costs of Poor Fertility and Disease in UK Dairy Herds; 2002 DAISY Report No. 5; Intervet UK Ltd.: Milton Keynes, UK, 2002.
6. López-Gatius, F.; Hunter, R.H.F. Spontaneous reduction of advanced twin embryos: Its occurrence and clinical relevance in dairy cattle. Theriogenology 2005, 63, 118–125.
7. Silva-del-Río, N.J.D.; Colloton, P.M. Fricke Factors affecting pregnancy loss for single and twin pregnancies in a high-producing dairy herd. Theriogenology 2009, 71, 1462–1471.
8. Foote, R.H. Sex Ratios in Dairy Cattle under Various Conditions. Theriogenology 1977, 8, 349–356.
9. Silva del Río, N.; Stewart, S.; Rapnicki, P.J.; Chang, Y.M.; Fricke, P.M. An Observational Analysis of Twin Births, Calf Sex Ratio and Calf Mortality in Holstein Dairy Cattle. J. Dairy Sci. 2007, 90, 1255–1264.
10. Ghavi Hossein-Zadeh, N.; Nejati-Javaremi, A.; Miraei-Ashtiani, S.R.; Kohram, H. An Observational Analysis of Twin Births, Calf Stillbirth, Calf Sex Ratio, and Abortion in Iranian Holsteins. J. Dairy Sci. 2008, 91, 4198–4205.
11. Echternkamp, S.E.; Gregory, K.E. Reproductive, growth, feedlot and carcass traits of twin versus single births in cattle. J. Anim. Sci. 2002, 80 (E. Suppl. 2), E64–E73.
12. Cady, R.A.; Van Vleck, L.D. Factors effecting twinning and effects of twinning in Holstein dairy cattle. J. Anim. Sci.1978, 46, 950–956.
13. Gatica, R. Studies in micromanipulation of farm animal embryos. PhD Thesis, National University of Ireland, Galway, Ireland, 1988.
14. Gordon, I. Induction of Twin Births in Reproduction in Cattle and Buffalo, 1st ed.; CABI: Wallingford, UK, 1996; p. 375, Chapter 8.

CHAPTER 2

Comparative Aspects of Canine Melanoma

Adriana Tomoko Nishiya [1], Cristina Oliveira Massoco [1], Claudia Ronca Felizzola [1,2], Eduardo Perlmann [3], Karen Batschinski [1], Marcello Vannucci Tedardi [1], Jéssica Soares Garcia [1], Priscila Pedra Mendonça [3], Tarso Felipe Teixeira [4] and Maria Lucia Zaidan Dagli [1]

[1] *Department of Pathology, School of Veterinary Medicine and Animal Science, University of São Paulo, Av. Prof. Dr. Orlando Marques de Paiva, 87, CEP 05508-270, São Paulo, Brazil*

[2] *Surgical Oncology Rua Antônio Alves Magan, 124, CEP 01251-150, São Paulo, Brazil*

[3] *Department of Surgery, School of Veterinary Medicine and Animal Science, University of São Paulo, CEP 05508-270, São Paulo, Brazil*

[4] *Pathology Veterinary Medicine, FEPI Itajubá University Center, Dr. Antonio Braga Filho Street, 687, Itajubá, 37.501-002 Minas Gerais, Brazil*

ABSTRACT

Melanomas are malignant neoplasms originating from melanocytes. They occur in most animal species, but the dog is considered the best animal model for the disease. Melanomas in dogs are most frequently found in the buccal cavity, but the skin, eyes, and digits are other common locations for these neoplasms. The aim of this review is to report etiological, epidemiological, pathological, and molecular aspects of melanomas in dogs. Furthermore, the particular biological behaviors of these tumors in the different body locations are shown. Insights into the therapeutic approaches are described. Surgery, chemotherapy, radiotherapy, immunotherapy, and the outcomes after these treatments are presented. New therapeutic perspectives are also depicted. All efforts are geared toward better characterization and control of malignant melanomas in dogs, for the benefit of these companion animals, and also in an attempt to benefit the treatment of human melanomas.

Keywords: melanoma; canine; cancer; histopathology; therapy

1. INTRODUCTION

Melanoma is a malignant neoplasm originating from melanocytes. Melanocytic neoplasms have been described in men and in most domesticated animal species, including dogs, cats, horses, and also in wild terrestrial and marine animals [1].

Since dogs live in close proximity to humans, they are influenced by similar environmental factors that may lead to chronic diseases and cancer. In fact, the domestic dog (*Canis familiaris*) reportedly presents more naturally inherited chronic diseases than any other animal species. Hence, those aspects have to be considered in the selection of an animal model for a multifactorial human disease, and the dog appears to be the one with the best characteristics [2,3,4]. Dogs are considered recommended models due to their relatively high susceptibility to cancers, including melanomas, and many behavioral similarities [5].

While many genetic alterations of melanomas have recently been discovered and studied, control of this disease is still challenging in both humans and animals, particularly in dogs and cats. The aim of this review is to report a comparative approach of the etiological, epidemiological, pathological, and molecular aspects of melanomas in dogs.

2. CANINE MELANOMA EPIDEMIOLOGY

Melanoma is one of the most devastating types of cancer, killing around 50,000 people worldwide every year [6,7]. In 2015, according to the American Cancer Society, it is estimated that about 73,870 new melanoma cases will be diagnosed (about 42,670 in men and 31,200 in women), and 9940 people are expected to die of melanoma (about 6640 men and 3300 women).

Human malignant mucosal melanoma of the head and neck is increasing rapidly in incidence in the United States. It is an uncommon and fatal malignancy that does not appear to have the same etiologic basis as cutaneous melanoma. In 452 observed cases, the majority of patients had sino-nasal location (72.6%) and white women ages 70 or older were most affected [8]. This kind of human malignant melanoma is uncommon and has aggressive behavior with increasing rates of recurrence and regional and distant metastasis [9]. Regional lymphatic metastasis occurred in 57.9% to 21% of patients and recurrence rates were 57.9% to 26.3% for patients treated with surgery alone and for those treated with surgery and radiotherapy, respectively. Distant metastases occurred in 52.6% receiving surgery alone and in 47.3% receiving both therapies [10].

Canine oral malignant melanomas in dogs accounted for 0.99% of all (338/33826) canine pathology cases received from 1992 through 1999 at the University of Missouri—Veterinary. The most common sites are the gingiva and labial mucosa [11].

In a retrospective study of neoplasms in domestic animals (data collected between 1993 and 2002 at the School of Veterinary Medicine and Animal Science of the University of São Paulo, Brazil), of 1813 cases of neoplasms in dogs, 58 were melanocytic neoplasms, accounting for 3% of the total in this species [12]. In another retrospective study, 2154 neoplasms were diagnosed at the University of São Paulo between 2000 and 2006. A total of 193 cases (8.9%) were melanocytic neoplasms, of which 186 cases occurred in dogs (96.4%), and only seven in cats (3.6%). Male mixed-breed dogs with black hair coats, with ages varying from 8 to 11 years, were the most commonly affected animals [13].

Canine malignant melanomas are located at different anatomical sites, such as the mouth (Figure 1A,B), lips, skin (Figure 1C), eyes, and digits (Table 1). Studies are controversial about the most common location for this disease in dogs; however, most studies point to the oral cavity and skin as the most common sites [14,15,16,17]. In a survey of 384 melanocytic canine tumors, oral, cutaneous, lips/feet, and eye sites comprised 19%, 59%, 19%, and 3%, respectively [18]. In another study considering malignant melanomas, oral cavity and cutaneous localization were the most common in 40%–62% and 27%–31%, respectively [13,19].

Data obtained from historical studies indicated that the following breeds of dogs are considered at risk to develop melanocytic tumors: Airedale Terrier, Boston Terrier, Boxer, Chihuahua, Chow Chow, Cocker Spaniel, Doberman Pinscher, English Springer Spaniel, Golden Retriever, Irish Setter [20], Miniature Schnauzer, Scottish Terrier [19,20], Poodles, Beauce Shepherds, Rottweilers, and Labrador Retrievers [19]. Some breeds of dogs, specifically the Schnauzer family (miniature and standard), Scottish Terriers, and Irish Setters are at risk for developing digital malignant melanomas [21,22].

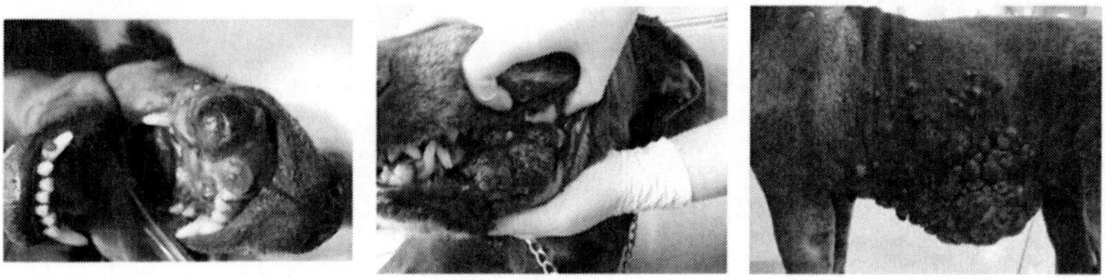

(A) (B) (C)

Figure 1. Clinical aspect of malignant oral melanomas (**A**) in maxilla; (**B**) in mandible. Clinical aspect of malignant cutaneous melanoma (**C**) in thoracic and abdominal wall (*Adriana Nihiya*).

Table 1. Common sites of malignant melanomas in dogs.

Sites/Total Number of Cases	*n* = 1652	*n* = 143	*n* = 209
Oral	1026	57	67
Cutaneous	448	44	88
Scrotum	-	4	-
Digit	94	21	-
Ungual	60	-	-
Ocular	24	17	-
Lips/Feet	-	-	54
Reference	Gillard *et al.*, 2014 [19]	Teixeira *et al.*, 2010 [13]	Spangler and Kass, 2006 [18]

Malignant melanomas typically occur in middle-aged to older dogs, as Miller *et al.* [23] found an average age of nine years, and with no gender predilections.

Teixeira *et al.* [24] studied 25 canine oral melanomas, including 16 melanotic and nine amelanotic specimens. In this study, 25% of the melanotic melanoma-bearing dogs were eight-year old animals, and 22.2% of amelanotic melanoma-bearing group were younger, with ages varying from four to eight years.

Most information on animal cancer epidemiology in the literature comes from laboratories or hospitals, usually from a unique center and using non-standard coding protocols. This may lead to epidemiological bias, and the data rarely generate reliable information useful to establish comparisons. Cancer registries are the most important source of epidemiological data for neoplasms [25]. There are some animal cancer registries around the world, situated mostly in Europe and the USA. In 2013, the São Paulo Animal Cancer Registry was created at the University of São Paulo, Brazil, serving as the first animal cancer registry in Latin America [25,26]. This system will allow for better and more accurate epidemiologic studies in domestic animals in this region.

3. CANINE MALIGNANT MELANOMA ETIOLOGY

In humans, malignant melanoma is the most aggressive, therapy-resistant, and deadly form of skin cancer [27]. The major risk factors for human melanoma include family history, skin and mucosal pigmentation characteristics, sun exposure, particularly to UVB light, and skin reactions to sun exposure. Another type of melanoma, oral malignant melanoma, is a rare aggressive neoplasm usually seen in middle-aged adults.

Several etiological factors are supposed to be involved in canine malignant melanomas, including consanguinity, trauma, chemical exposure, hormones, and genetic susceptibility [28]. However, there is no consensus regarding the etiology of malignant melanomas in dogs [29]. Sunlight may be involved in the development of this disease in the sun-exposed skin areas of the body, such as the face and pinnae; however, sunlight probably is not involved in mucosal melanomas, like the ones found in the canine buccal cavity. Other factors, like the presence of pigmented cells, trauma, chemical agents, or even the buccal microbiota, and inflammation may be associated with the etiology of these tumors [30].

Although melanomas can arise in any dog, the prevalence of this disease is higher in purebred dogs, especially Standard and Miniature Schnauzers, Doberman Pinschers, Scottish Terriers, Irish and Gordon Setters, and Golden Retrievers. This fact supports the theory that melanomas in dogs could have a genetic basis [28]. Genetic and epigenetic modifications in melanocytes that lead to alterations in expression or function of genes and proteins involved in cell cycle control and apoptosis are certainly involved in the development of melanomas.

4. CANINE MELANOMA PATHOLOGICAL ASPECTS AND PROGNOSIS

Melanocytic neoplasms can be classified as benign or malignant tumors [31]. Malignant melanomas are considered malignant neoplasms arising from melanocytes, while melanocytomas are the benign counterparts [31].

Melanocytes, which are dendritic cells derived from the neuroectoderm and melanoblasts of the neural crest, migrate during embryogenesis to the dermis and epidermis, mucous membranes, and eyes [29,32]. These dendritic cells in melanoma development have demonstrated altered expression of cell–cell adhesion molecules, such as decreased expression of E-cadherin and V-CAM 1 and increased expression of N-cadherin, Mel-CAM 1, ICAM 1, and α β integrins [33]. The development of malignant melanoma is generally characterized by a series of transitions that are outlined in [34], and arises from melanocytes that normally reside within the basal layer of epidermis [35].

According to Head *et al.* [36], a melanocytoma is a benign tumor composed of melanocytes in the basal region of the epithelium and extending into the submucosa (junctional melanocytoma) or in the sub-epithelial connective tissue under normal epithelium (dermal melanocytoma). Malignant melanoma can be subdivided into three patterns on the basis of cell shape [36]:

-epithelioid—round and polygonal cells (Figure 2)

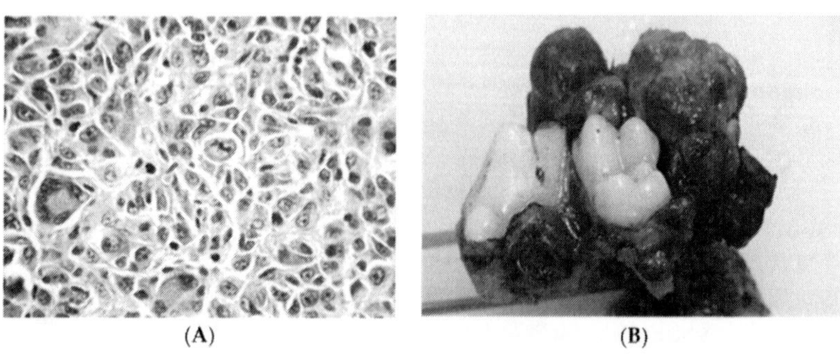

Figure 2. Photomicrograph of the most common histopathology type of malignant melanoma (epithelioid), showing formation of cell nests, with the presence of multinucleated giant cells and atypical mitotic figures, HE, Objective 40 × (**A**). Canine oral malignant melanoma, pigmented, exophytic mass protruding from the gingiva (**B**). Formalin-fixed tissue. Photographs are courtesy of Professor Dr. José Guilherme Xavier.

-spindle cell—tumor resembles fibroblasts

-mixed tumors—show both cell types

Malignant melanoma histologically can be the highly melanotic type or the highly anaplastic amelanotic type. Anaplastic melanocytes can be large with abundant cytoplasm with one or more oval or elongated nuclei, with obvious nucleoli. These are frequently characterized to form nests in the submucosa by a mixed structure of epithelial-like cells and fusiform cells and junctional infiltration between basal cells and in the submucosa [36] (Figure 3).

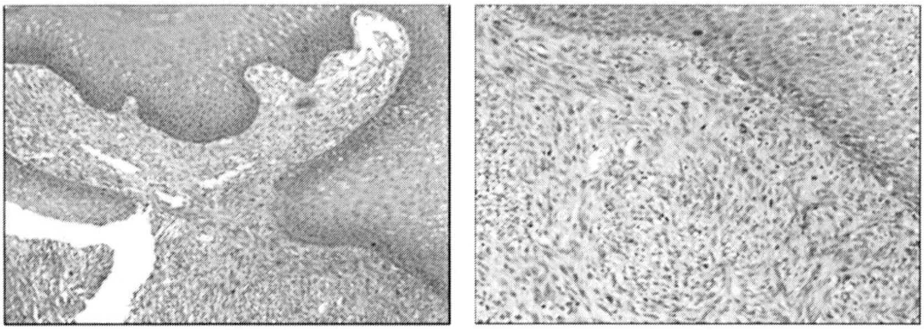

(A) (B)

Figure 3. Photomicrograph of oral amelanotic melanoma ((**A**), Objective 10 ×; (**B**), Objective 20 ×). Hematoxylin and eosin.

The most prominent biological property of melanoma cells is the ability to produce melanin [37]. The disruption of homeostasis of close association between melanocytes and basal keratinocytes may trigger a continuous proliferation of the melanocytes, which may lead to the development of a malignant melanoma [7]. Once malignant melanoma cells have escaped from the keratinocyte control, they become able to invade the tissue by the degradation of extracellular matrix through the action of metalloproteinases [38].

The histopathological diagnosis of melanoma may be difficult if the tumor does not contain melanin. Amelanotic malignant melanomas may represent one third of all melanoma cases in dogs.

Their histopathological aspect may resemble carcinomas, sarcomas, lymphomas, and osteogenic tumors. For this reason, the diagnosis of malignant melanoma should be made with the use of immunohistochemistry [11,39,40]. Teixeira *et al.* [13] used two different antibodies to confirm the diagnosis of malignant melanoma in amelanotic cases: HMB 45 (Mouse monoclonal anti-human HMB45, from Dako) and Melan A (Monoclonal mouse anti-human Melan-A, from Dako), and both were considered suitable to immunostain canine melanocytes. Ramos-Vara *et al.* [11] considered Melan A as a specific and sensitive marker for malignant melanomas. Smedley *et al.* [41] evaluated 49 amelanotic melanomas and created a cost-effective and efficient immunodiagnostic cocktail panel containing antibodies against PNL2, Melan-A, TRP-1, and TRP-2 that had 100% specificity and 93.9% sensitivity in identifying canine oral amelanotic melanocytic neoplasms. In addition, malignant melanomas are positive for vimentin and variably positive for S-100 protein and neuron-specific enolase (NSE) [42].

In order to use immunohistochemical techniques to study melanoma cell characteristics, it is frequently necessary to remove melanin from the histopathological specimen. There are several methods to do bleaching in melanoma cells, but the one described by Silva *et al.* [43] is one of the most effective, in which slides were immersed only in 10% hydrogen peroxide in pH 7.4 0.2 mol/L Tris-HCl buffer solution for 24 h in the dark at room temperature.

5. CANINE MELANOMA: MOLECULAR ASPECTS

Melanomas in humans seem to arise from normal melanocytes that suffer a number of molecular events that lead to their transformation. The progression from normal melanocytes to melanoma sometimes involves its transformation to melanocytic nevus. The genetic alterations involved in this progression are still under study. Several genes/genetic pathways were identified in human melanomas, and these are important for the diagnosis, establishment of prognosis, and drug targeting [44]. Some of these genes are described below:

-CDKN2A Locus—about 70% of melanomas harbor mutations or deletions in this locus on chromosome 9p21.

-Genes that are altered when normal melanocytes generate a nevus: BRAF, NRAS, and INK4a/ARF.

-Genes that are mutated or deregulated when a nevus is transformed to a melanoma: BRAF, NRAS, INK4a/ARF, PTEN, c-kit, NEDD9, and MITF.

Genetic alterations in canine malignant melanomas from mucosal or acral sites have not been fully described. Activating mutations in BRAF exon 15 have not been found [6,45]. Activating mutations of NRAS and c-kit appear to be absent in canine mucosal melanoma, in contrast to human mucosal melanoma, where these genes are mutated in 15% of tumors [44,46,47]. Further studies are necessary to understand the molecular basis of canine malignant melanomas.

Genetic alterations in canine malignant melanomas from human mucosal melanomas have not been fully studied (Table 2). Mutations in exon 15 of BRAF are not found [6], similar to human mucosal malignant melanoma [48]. Activating mutations of NRAS are presented in three of 77 cases [19] and c-kit appears absent in canine mucosal melanomas [49], in contrast to human mucosal melanoma, where these genes are mutated in 4% and 14% of tumors, respectively [50].

Table 2. Genes altered by mutations in canine and human mucosal malignant melanomas.

	Human	Dog
BRAF	4% [50], 2% [48], 6% [14]	0% [6,19]
C-KIT	12.5% [16], 4% [50]	0% [49]
NRAS	22% [15], 14% [50], 10% [14]	6% [19]
PTEN	48.1% [15]	4% [19]
GNAQ	0% [16]	0% [19]
CDK4	0% [17]	0% [19]

In 2014, Gillard *et al.* [19] compared the histopathology and genetic alterations of canine and human melanomas. They observed that dog malignant melanomas present cytological characteristics similar to human melanoma simulating “nevus” and “animal type,” suggesting a new melanoma classification. They also studied six genes in canine melanomas that are recurrent mutations in human melanomas, BRAF, NRAS, PTEN, KIT, GNAQ, and CDK4. In total, out of the 95 cases (77 oral and 18 cutaneous melanomas), no mutations were found in BRAF, KIT, GNAQ, and CDK4 genes; only three had a positive NRAS mutation and two had a PTEN gene mutation, of which one case had both changes [19]. BRAF mutations were seen in 57% of human cutaneous melanomas and NRAS was mutated in 17% of samples. Both changes were seen in 15.4% [51].

The low frequency of BRAF mutations associated with UV-independent carcinogenesis and oral anatomical distribution of canine melanomas supports the dog as a spontaneous model for investigation of non-UV-associated human melanomas [52].

Human and canine melanoma share differential gene expression patterns of MAPK, but similar sensitivity in cell culture to therapeutic inhibitors of MAPK (AZD6244) and PI3K/AKT (rapamycin) [6].

Dog malignant melanomas share numerous immunohistochemistry similarities (KIT, PTEN, and phosphorylated forms of AKT and ERK1/2) with human melanoma subtypes; particularly with mucosal, digital and ungual localizations that usually show high growth and aggressive behavior [5].

6. CANINE ORAL MELANOMAS

Melanoma is considered the most common oral malignancy in dogs [53], accounting for 14.4% to 45.5% of all oral tumors [54,55,56,57,58,59]. Some studies describe melanomas as the second most frequent oral cancer in dogs [54,56]. Oral malignant melanoma may also occur in cats and humans, but it is infrequent [60,61].

Malignant melanoma can be located in any portion of the oral cavity, but the gingival mucosa is the most common site [54,62]. The mandibular labial mucosa was the most common area, affected in 53% of cases, as described by Ramos-Vara *et al.* [11]. Although malignant melanoma rarely affects the tongue, it is considered the most common tumor type of this anatomic site as described by Dennis *et al.* [63], and large-breed dogs, particularly Chow Chow and Shar Pei, are at increased risk for malignant lingual melanoma [62,63].

Small-breed dogs and dogs with pigmented oral mucosa appear to be at increased risk for the development of oral malignant melanoma [61,62]. Cocker Spaniel, Poodle, Pekinese, Gordon Setter, Chow Chow, Golden Retriever, mixed-breed dogs, and Dachshund are the most commonly affected breeds [11,61]. Boxers and German Shepherds appear to be less affected by the disease, as described by Ramos-Vara *et al.* [11].

Dogs with oral malignant melanoma are usually 10.5 to 12 years old, with an average age of 11.4 years [11,54,61]. Males are supposed to be more predisposed to oral malignant melanomas, as verified by Todoroff *et al.* [54] and Vos *et al.* [55]; however, Ramos-Vara *et al.* [11], in a study of 328 dogs diagnosed with oral melanoma, and Teixeira *et al.* [13] found no such predisposition.

Oral malignant melanoma is a highly locally aggressive disease, with local invasion into bones occurring in 57% of lesions, with metastasis to regional lymph nodes (Figure 4) in 30.3% to 74% of cases, and to the lungs and other distant organs in 14% to 92% of affected dogs [11,61,64].

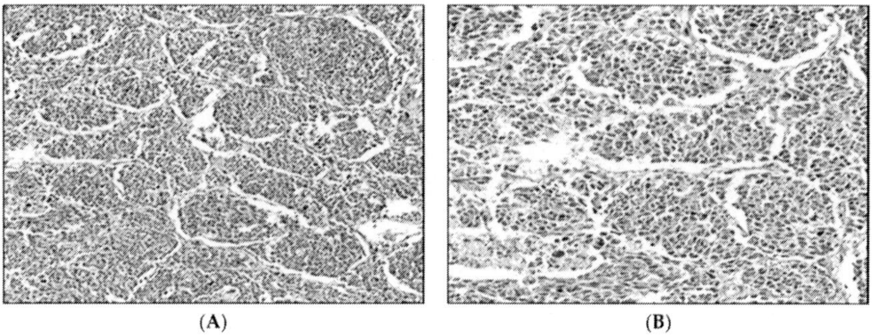

Figure 4. Photomicrograph of lymph node metastasis from oral melanocytic neoplasm ((**A**), Objective 10x; (**B**), Objective 20x). Hematoxylin and eosin.

Melanoma is one of the few neoplasms in animals for which location is an important prognostic indicator [65]. The statement that all oral melanocytic neoplasms should be considered malignant is commonly found throughout the literature and has become dogma [41].

The prognosis for oral melanoma varies from favorable to unfavorable, with an overall median survival time (MST) of less than 36 months, depending on the intrinsic characteristics of the tumor, and the stage of the disease [61,66,67,68,69].

Historically, many studies confirm that oral and lip melanomas have a poorer prognosis when compared to skin melanomas. The differences concerning the results of different studies of lip melanomas are directly related to the real site of the tumor, that is, either in the skin or in the mucocutaneous junction. In one study, oral melanomas had an MST of 147 days, while the MST of melanoma of the lips and digits were 676 days and 725 days, respectively [18]. Large and poorly or non-pigmented tumors, necrosis, ulceration, high rate of cell proliferation (PCNA expression), and p53 expression are considered poor prognostic factors [29]. Amelanotic melanomas showed a significantly higher amount of mitotic cells when compared to melanotic melanomas [24]. The PCNA (proliferating cell nuclear antigen) in pigmented and non-pigmented melanomas was also assessed. It was demonstrated that cells of all samples stained positive for PCNA, confirming the proliferative activity of this tumor.

The World Health Organization (WHO) staging scheme for dogs with oral melanoma is based on the size of the tumor and the presence of regional and distant metastasis (Table 3) [70]. These staging systems are old and size is not considered a prognostic factor for malignant melanoma [18,28,36], but they are still cited in veterinary practice.

Table 3. TNM classification of tumors in domestic animals, WHO, Geneva, 1980.

Clinical Staging System for Oral Tumor			
Primary Tumor (T)			
Tis	Tumor *in situ*		
T1	Tumor <2 cm in diameter at greatest dimension		
T1a	Without evidence of bone invasion		
T1b	With evidence of bone invasion		
T2	Tumor 2–4 cm in diameter at greatest dimension		
T2a	Without evidence of bone invasion		
T2b	With evidence of bone invasion		
T3	Tumor >4 cm in diameter at greatest dimension		
T3a	Without evidence of bone invasion		
T3b	With evidence of bone invasion		
Regional Lymph Nodes (N)			
N0	No regional lymph node metastasis		
N1	Movable ipsilateral lymph nodes		
N1a	No evidence of lymph mode metastasis		
N1b	Evidence of lymph mode metastasis		
N2	Movable contralateral lymph nodes		
N2a	No evidence of lymph mode metastasis		
N2b	Evidence of lymph mode metastasis		
N3	Fixed lymph nodes		
Metastasis (M)			
M0	No distance metastasis		
M1	Distance metastasis		
Stage Group	**Tumor (T)**	**Nodes (N)**	**Metastasis (M)**
I	T1	N0, N1a, N2a	M0
II	T2	N0, N1a, N2a	M0
III	T3	N0, N1a, N2a	M0
	Any T	N1b	M0
IV	Any T	N2b, N3	M0
	Any T	Any N	M0

Dogs with stage I melanomas have tumors smaller than 2 cm in diameter, while dogs with stage II and III have tumors measuring between 2 to 4 cm, and more than 4 cm, respectively. Dogs with lymph node metastasis are also classified as stage III disease, and dogs with distant metastasis are classified as stage IV disease. Median survival times for dogs with oral melanoma treated with surgery are approximately 17–18 months, 5–6 months, and 3 months for stage I, II, and III disease, respectively [67].

In a study of 70 dogs with oral melanoma, Tuohy *et al.* [71] observed that 11.4% of dogs had mandibular lymph node metastasis and 1.4% of dogs had lung metastasis at the time of diagnosis. After 821 days (mean time of follow-up), 8.6% of mandibular lymph node metastasis, 17.1% of lung metastasis, 5.7% of metastasis to both sites, and 4.2% of metastasis to lung and other locations were observed. Williams *et al.* [64] reported that 70% of canines with melanomas had metastasis when lymphadenomegaly was present, and that 40% had metastasis in the absence of lymphadenomegaly. Harvey *et al.* [72] observed that dogs with untreated oral melanomas have a median survival time of 65 days.

Gillard *et al.* [19] performed clinical and histopathological analyses of 153 malignant canine melanomas with a four-year follow-up. These samples were histopathological characterization by human classification and showed that most canine tumors are intradermal and homologous to human nevocytoid (72%) type, animal type (16.5%), composite type (6.5%), pleomorphic type (4.5%), and superficial spreading melanoma type (0.5%) [19]. Figure 5 shows a representative histology of canine oral malignant melanoma on which local invasion, sparsely pigmented cells, and pleomorphism with variation in shape and cell or nuclear size can be seen.

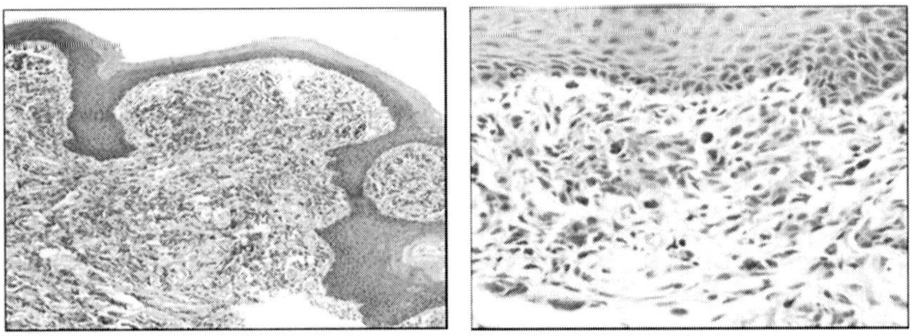

(A) (B)

Figure 5. Photomicrograph of oral melanocytic neoplasm ((**A**), Objective 4 ×; (**B**), Objective 40 ×); Hematoxylin and Eosin.

A recent study of over 382 canine oral melanomas did not find statistically significant differences in survival among different sites or mitotic indices [11].

7. CANINE OCULAR MELANOMAS

Ocular melanocytic tumors, benign or malignant, may occur in humans and domestic animals. These tumors may affect the eyelids, conjunctiva, orbit, limbo, and uvea [73,74]. Melanomas have different biological behaviors depending on the location in the eye [75]. In general, canine ocular melanomas are less aggressive than oral melanomas. Within the ocular melanoma group, the uveal melanoma is characterized by being more aggressive than the epibulbar melanoma [75].

Eyelid tumors in dogs are usually benign, such as sebaceous adenomas and sebaceous epitheliomas. Among all eyelid tumors, the melanocytic neoplasms account for 20%, and only 8% of these are melanomas [73]. Most, if not all, eyelid skin melanomas are benign [75]. Eyelid melanomas in dogs are usually pedunculated with nodular internal infiltration [73]. They are generally non-aggressive tumors, but occasionally some pigmented melanomas act aggressively and become locally invasive with the development of metastasis [73]. When compared to dogs, melanomas of the eyelid in humans carry a worse prognosis, especially when the tumor is located on the lid margin [76].

Conjunctival melanomas generally affect the nictitating membrane and the bulbar and/or eyelid conjunctiva [77]. Rottweilers and Cockers Spaniels are predisposed [74]. Lesions appear as dark, raised solid masses. Metastasis is rarely seen, but when it occurs it is typically associated with melanomas of the conjunctival eyelid. Complete surgical removal is warranted in early-diagnosed cases [77]. Due to the high risk of local recurrence, enucleation should be considered for invasive tumors [74]. Although conjunctival melanomas in humans are unusual, they have the potential to cause vision impairment and reduce life expectancy, with a mortality rate varying from 18% to 44% [74,78].

Canine and feline limbal melanomas have a benign biological behavior. They appear as raised, well-circumscribed heavily pigmented masses that arise from scleral and subconjunctival connective tissue melanocytes. Most of them develop slowly and are located closely to the superior aspect of the limbus. Labrador Retrievers are genetically predisposed [79].

Among the intraocular melanocytic tumors in the dog, melanocytomas are the most common tumors encountered, representing approximately

80% of the neoplasms. Melanomas account for only 20% of tumors [74]. It is difficult and unreliable to clinically differentiate between melanocytomas and melanomas, since melanocytomas can be locally aggressive and even invade the drainage angle and adjacent tissues [73,80]. Therefore, histopathological analysis is indispensable for a definitive diagnosis [81]. Despite the fact that melanocytomas are benign, when located in the iris, they can spontaneously cause necrosis and pigment dispersion, which may lead to obstruction of drainage and cause glaucoma in dogs and humans [82] (Figure 6).

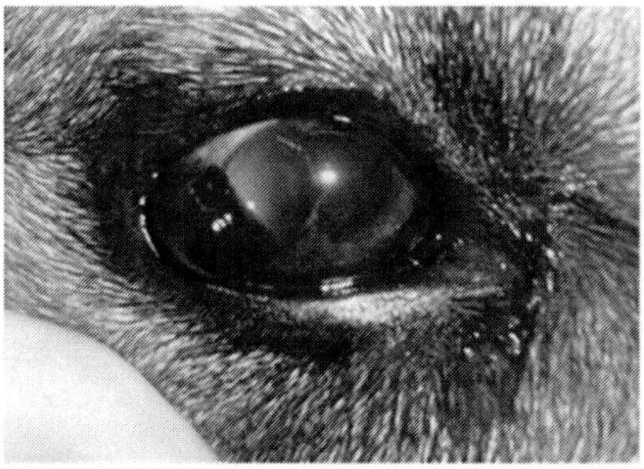

(A)

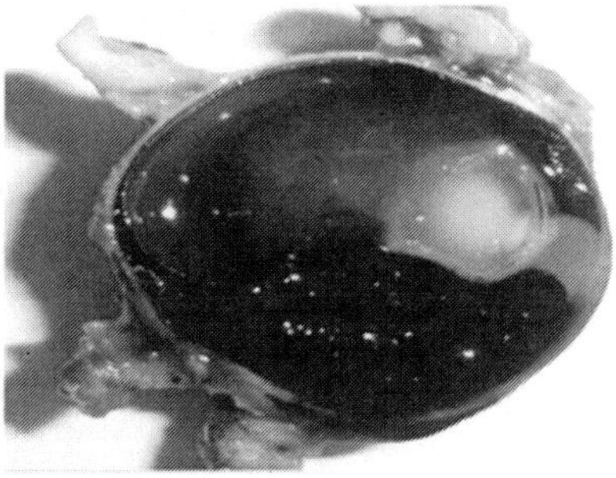

(B)

Figure 6. Photographs showing: blackened mass in the iris with extraocular extension (**A**) and sagittal section of the eyeball revealing melanocytic neoplasia of iris, ciliary body, and choroid (**B**) (*Eduardo Perlmann*).

Melanocytomas most commonly affect the anterior uvea in dogs, but rarely affect the choroid [29]. When located in the uvea, they may cause several changes, such as intraocular hemorrhage, uveitis, and secondary glaucoma. Despite the fact that human intraocular melanocytomas are commonly located at the level of the optical disk or adjacent tissues, this location was never described in dogs [74]. An early diagnosis and rapid surgical treatment increase the chances of a favorable prognosis [83]. Wilcock *et al.* [75] studied 91 canine uveal melanocytic neoplasms, and only three developed metastatic disease. These data emphasize the low metastatic rate of uveal melanoma in dogs. Giuliano *et al.* [84], however, showed that animals affected by uveal melanomas had a lower life expectancy when compared with patients diagnosed with uveal melanocytomas. Tumors located in the uvea most frequently are consistent with primary neoplasia; however, animals presenting with intraocular-pigmented tumors should have a complete diagnostic work-up and physical exam to rule out metastasis [85].

In general, when tumors are only confined to the intraocular space, the recommended treatment is enucleation. Exenteration is indicated when the tumor extends to the extraocular space [83]. Other treatment options include iridectomy or diode laser photocoagulation in small-pigmented tumors of the iris [86]. There is no evidence that chemotherapy is effective in the treatment of ocular melanomas [87], but long-term follow-up evaluation would be needed. Ocular malignant melanoma had significantly shorter survival time than benign melanocytomas [84].

8. CANINE DIGITAL MELANOMAS

Melanoma is considered the second most common type of cancer of the digits in dogs [22,29,88]. Local invasion is a common feature of digital melanomas and 5%–58% of dogs with this disease have evidence of bone lysis [22,88,89,90]. Tumors of the footpad and interdigital region are less likely to invade into the bone [22,90]. Besides possible local invasion, regional and distant metastasis via lymphatics is frequently noted. It is estimated that 30%–40% of dogs have metastasis at the time of presentation and that even after appropriate local treatment with surgery, subsequent metastasis is frequently seen [22,88].

Due to its aggressive behavior, canine digital melanomas often require digital amputation. The median survival time for dogs without metastasis at the time of diagnosis treated with digital amputation is approximately 12 months, with 42%–57% of dogs surviving one year, and 11%–13% surviving two years [22,88,90,91].

Canine Cutaneous Melanomas

In contrast to digital melanomas, cutaneous melanomas typically have benign behavior in dogs, with the exception of melanomas that develop on the mucocutaneous junctions [18,92]. They account for 0.8%–2% of all canine cutaneous tumors [64] and are more commonly seen in dogs with heavily pigmented skin. Predisposed breeds include Scottish Terrier, Poodle, Golden Retriever, Dachshund, Cocker Spaniel, Miniature Poodle, Chow Chow, and Gordon Setter [93]. Benign skin melanomas are usually solitary, small, pigmented, firm, and freely moveable over deeper structures. Malignant melanomas tend to be fast-growing tumors, and often are ulcerated, and pigmented (grey, brown, or black) [21].

The most common sites for benign cutaneous melanomas are the face (near the eyelids), trunk, and extremities. Malignant melanoma is found most frequently on the head, ventral abdomen, and scrotum. When metastases occur, lymph nodes, followed by lungs, are the main organs involved [29].

Brockley *et al.* [94] studied 63 dogs diagnosed with melanomas at different locations. Surgery was performed in 24 cutaneous melanomas, with 88% of dogs alive a year after treatment.

Most dogs are cured with complete surgical excision, although some characteristics of the tumor, such as a mitotic index of ≥3/10 HPF, the presence of ≥20% of nuclear atypia, a Ki67 index ≥15%, ulceration, lymphatic invasion, and tumors extending beyond dermis, may have a negative influence on prognosis [18,41,92].

9. TREATMENT MODALITIES FOR CANINE MELANOMAS

9.1. Surgery

Surgery is the most common and main treatment option for local management of all melanomas, including oral [69], ocular [74], cutaneous, and digital melanomas [90,92]. Due to its high metastatic potential, systemic therapy should also be considered [61,62].

Tuohy *et al.* [71] studied 70 cases of canine oral malignant melanomas treated with curative intent surgery. Surgical resections were done with wide margins, including 2 to 3 cm of bone margins, and 1 cm of soft tissue margins. Histopathological analysis showed that 72.9% of tumors were completely excised, with 10% of these patients developing local tumor recurrence. In this study, dogs that had surgery as their sole treatment option had a progression-free interval (PFI) longer than 567 days, and an MST of 874 days. In a survey performed by Boston *et al.* [69], among wide-margin surgeries performed, 79.3% (73/92) were completely excised based on histologic evaluation, and the recurrence rate was 8.3% (6/73) in this group. The MST of these dogs was 354 days. Esplin [65] studied 61 dogs diagnosed with well-differentiated oral melanomas that were surgically excised, and reported an MST of 34 months with 3.2% of local recurrence.

Regional lymph node resection has been studied in maxillofacial tumors. The analysis of sentinel lymph nodes provides essential information regarding the clinical stage of disease, and also helps determine the appropriate treatment plan [95,96]. As previously shown, in a survey of 100 dogs with oral malignant melanoma, 53% had cytological or histopathological evidence of mandibular lymph node metastasis, even in normal sized nodes [64]. Therefore, regional lymph node removal is warranted, especially in dogs with oral melanomas.

9.2. Radiation Therapy

Radiation therapy is effective for malignant melanomas in dogs and may be useful for the control of local disease and involved regional lymph nodes [97]. Response rates are generally best for smaller tumors rather than for salvage therapy against large tumor burdens late in their clinical progression [98]. Another prognostic factor that appears to affect therapy outcome includes the presence of bone lysis. Proulx *et al.* [97] found that dogs without evidence of radiologic bone destruction experienced longer disease-free intervals and overall survival when compared to dogs with bone invasion.

Although melanoma has traditionally been viewed as a "radio-insensitive" disease, many studies suggest a significant role for radiation therapy in the treatment of melanoma [99]. Optimal fractionation schedules have not yet been established for canine malignant melanoma; however, hypofractionated protocols have been utilized with some success [93]. Overall response rates range from 82% to 94.4%, with the complete and partial resolution of tumors in 51% to 69% and 25% to 31% of cases, respectively [91,92,93,94,95,96,97,98,99,100,101,102]. The reported median time to progression ranges from 3.6 to 7.9 months with a median survival time varying from 5.3 to 11.9 months [97,98,100,101,102,103].

Chemotherapy alone has generally offered little benefit in terms of clinical outcome for the control of local canine melanomas; however, a few published studies have shown that the association of radiation with chemotherapy may potentially slow local progression and/or improve overall survival. Cancedda *et al.* [102] compared the efficacy of radiation therapy (5 × 6 Gy) alone and radiation therapy with post-radiation temozolamide (60 mg/m^2 PO for five consecutive days every 28 days) in dogs with measurable malignant melanoma. Dogs treated with temozolamide had a significantly longer median time to progression (6.8 months) than dogs treated only with radiation (3.6 months). Freeman *et al.* [103] retrospectively looked at dogs with incompletely resected oral melanoma treated with either cisplatin (10–30 mg/m^2 IV) or carboplatin (90 mg/m^2 IV) given once weekly 1 h before receiving radiation therapy (6 × 6 Gy). The reported median survival time was 11.9 months, representing the longest survival time when compared to the survival time previously reported for dogs with incompletely resected oral malignant melanoma [103]. Other studies combining radiation with chemotherapy agents, such as carboplatin, cisplatin, or melphalan, did not have a significant impact on either time to progression or overall survival [97,101,104].

9.3. Chemotherapy

Despite treating dogs with appropriate local therapies, such as surgery and/or radiation therapy, the ultimate cause of death, especially in oral melanomas, is metastasis. Therefore, the use of systemic therapies is typically recommended. Many chemotherapy protocols have been published in the veterinary literature with minimal improvement in survival times when compared to local treatment alone.

Many chemotherapy protocols using platinum agents have been studied. Rassnick *et al.* [104] reported an overall response rate of 28% using carboplatin as a single agent (300 to 350mg/m^2/IV/ every 21 days) prior to surgery in dogs with oral (25/27) and cutaneous melanomas (2/27). Boria *et al.* [105] reported an overall response rate of 18% and a MST of 119 days using cisplatin (50 to 55 mg/m^2/IV/ every three weeks) in combination with piroxicam (0.3 mg/kg/PO/24 hs) in oral melanomas.

Brockley *et al.* 2013 [94] evaluated the effect of carboplatin on the survival of 63 canine patients diagnosed with oral, cutaneous, or digital malignant melanomas. In this survey, carboplatin was administered at 300 mg/m^2 IV every 21 days for a total of 4–6 cycles. The addition of carboplatin after local therapy did not lead to a significant increase in survival times, and the anatomic site was significantly associated with survival. The overall median survival time for patients with oral, digital, and cutaneous melanomas were 389 days, 1350 days, with a median follow-up of 776 days, respectively.

In another survey, Dank *et al.* [106] compared the outcome of 17 dogs treated with surgery, adjuvant carboplatin (150 to 300 mg/m^2/IV every three weeks), with or without radiation therapy. The MST was 387 days for the group treated with radiotherapy and 440 days for the group without the addition of radiotherapy.

In a comparative study reported by Tuohy *et al.* [71], surgery with or without any adjuvant therapy (carboplatin, cyclophosphamide as a metronomic therapy, radiotherapy, or xenogeneic canine melanoma vaccine) did not improve overall survival time in dogs with oral melanoma. The MST reported was 874 and 396 days for dogs that had surgery or surgery plus adjuvant therapy, respectively. Boston *et al.* [69] treated 151 dogs with oral melanoma and also concluded that there were no benefits in survival times among dogs that had surgery alone (MST of 335 days for 98 dogs) *versus* dogs that had surgery followed by systemic adjuvant therapy (MST of 352 for 53 dogs). Carboplatin, lomustine, dacarbazine, doxorubicin, metronomic chemotherapy, or commercial melanoma vaccine were the adjuvant therapies used in this study. Postoperative radiation therapy was associated with a longer survival time (MST of 1747 days), but was not significant in the multivariate analysis of this study.

9.4. Immunotherapy

Many researchers are currently using and developing therapies to activate the immune system of dogs. The immune responses by therapeutic vaccine represent a potential and powerful strategy for the treatment of melanomas [107]. There is wide evidence that the immune system could modulate the progression and metastasis of melanoma.

Many biological agents have the capacity to demonstrate antitumor activation, such as BCG, IFN-alpha, IL-2, and immunization activators, using an anticancer vaccine [108].

Bacillus Calmette–Guérin (BCG) is used intralesionally in melanomas to activate an immune response and induce tumor regression; however, there may be some complications, such as granulomas and hypersensitivity reactions [109].

Corynebacterium parvum is a bacterial substance that has antitumor activity in humans and canines with melanoma. MacEwen *et al.* [67] used *C. parvum* in dogs with stage II and III oral melanomas and compared the outcomes between surgical excision alone and surgery plus *C. parvum* immunotherapy. Animals who received *C. parvum* and surgery had a better prognosis than those with surgery only.

In rodents, dogs, and men, a liposome-encapsulad muramyl tripeptide phosphatidyl ethanolamine (L-MTP-PE), when used *in vivo*, actives monocytes and macrophages, resulting in antitumor activity, which can lead to melanoma metastasis tumor regression or at least stable disease. According to MacEwen *et al.* [108], when L-MTP-PE was used alone or in combination with GM-CSF the antitumor response was minimal, and only prolonged the survival time in dogs with stage I melanoma.

Bianco *et al.* [110], demonstrated that peripheral blood mononuclear cells with apoptotic melanoma cells significantly increased the antitumor immune response, leading to a reduction of the melanoma in three out of five dogs treated intralesionally with FasL-DNA.

Alexander *et al.* [111] used an allogeneic melanoma vaccine in combination with human a glycoprotein 100 (hgp 100), which is a xenogeneic melanoma protein. The vaccine was well tolerated in dogs, and the tumor response rate was 35% for six weeks (including complete response and stable disease).

Hogge *et al.* [112] used an autologous vaccine made of lysate cells transfected with GM-CSF in dogs with soft tissue sarcoma or melanoma. There was a large concentration of inflammatory cells at the site of vaccine administration, different from the ones that received a vaccine placebo only.

A xenogeneic human DNA-encoding tyrosinase protein vaccine (Oncept®) has been developed and used in dogs with stage II and III oral malignant melanoma [113]. This vaccine was generally well tolerated in a study of 111 dogs. No systemic reactions occurred, only local hematomas and pain at the injection site [114]. In another retrospective study, medical records from 45 dogs were reviewed, including 30 dogs with stage II and III disease, treated with the Oncept®vaccine after appropriate achievement of loco-regional cancer control. The review found that dogs that received the vaccine did not achieve a greater progression-free survival, disease-free interval, or median survival time than dogs that did not receive the vaccine [115].

The role of dendritic cells (DCs) as a biological adjuvant against malignant melanoma through the induction of an active immune response in the oncologic patient has been explored. Dendritic cells, as professional antigen-presenting cells with an ability to initiate the immune response, could therefore qualify for vaccination strategies to treat human or canine cancers. Compared with the human clinical trials using DC-based vaccines, reports of testing in dogs are scarce.

For humans, at least 13 clinical trials conducted in the United States have been registered using *ex vivo* autologous DCs [116] and, for dogs, only a few reports have been published to strengthen the rational use of DC vaccination for canine malignant melanoma [117,118]. Both protocols used bone marrow-derived DC of dogs with or without malignant melanoma. Gyorffy *et al.* [117] conducted a study using a few canine melanoma bearers and employed DC transduced with an adenovirus vector encoding a xenoantigen, human melanoma antigen gp100, as a coadjuvant of radiation therapy. One dog had an antigen-specific cytotoxic T lymphocyte (CTL) activity in peripheral blood lymphocytes, with no clinical signs, either locally or systemically, with melanoma recurrence 48 months after initial DC injection. However, another dog, which had a negative CTL activity, relapsed 22 months after vaccination. Tamura *et al.* [118], using an autologous DC pulsed with canine melanoma CMM2 cell lysate, efficiently elicited T cell-mediated immunity against CMM-2 cells *in vivo*, reinforcing the rationale to use DC as a therapeutic vaccine.

Although the immunotherapeutic approach using a DC vaccine to drive tumor-specific immune responses appears to be a promising strategy to control malignant melanoma, there are limitations, such as the cost of vaccine production for veterinary use as well as the very fragmentary designed strategies available.

9.5. Outcomes after Treatment

Table 4 describes the reported outcomes for canine oral malignant melanoma treated with different protocols.

Table 4. Outcomes for canine oral malignant melanoma after different therapeutic protocols.

Summary of Treatment Outcome for Canine Oral Malignant Melanoma
Radiation Therapy (RT)
- Response Rate = 82%–94.4% - Local Recurrence = 11%–27% - MST (Median Survival Time) = 5.3–11.9 months
References: Theon *et al.*, 1997 (98); Bateman *et al.*, 1994 (100); Murphy *et al.*, 2005 (101); Cancedda *et al.*, 2014 (102); Proulx *et al.*, 2003 (97); Freeman *et al.*, 2003 (103).
Surgery (SX) with Wide Margins
- Local Recurrence = 3.2%–10% - MST = 11.8–34 months
References: Tuohy *et al.*, 2014 (71); Boston *et al.*, 2014 (69); Esplin *et al.*, 2008 (65).
Chemotherapy (CT) for Macroscopic Disease
- Carboplatin response rate = 28% - Cisplatin + piroxicam response rate = 18%
References: Rassnick *et al.*, 2001 (104); Boria *et al.*, 2004 (105).
Adequate Local Treatment + Systemic Therapy
- SX ± RT + adjuvant carboplatin: MST = 14.6 months - SX ± RT + adjuvant carboplatin, metronomic therapy, or melanoma vaccine: MST = 13.2 months (adjuvant therapy did not improve overall MST) - SX + adjuvant carboplatin, lomustine, dacarbazine, doxorubicin, or melanoma vaccine: MST = 11.7 months (adjuvant therapy did not improve overall MST) - RT + post-RT temozolamide: CT did not improve MST, but was associated with a longer median time to progression when compared to dogs treated with RT alone (6.8 months *versus* 2.6 months) - SX + RT + cisplatin or carboplatin once weekly 1 h before RT: MST = 11.9 months
References: Dank *et al.*, 2012 (106); Tuohy *et al.*, 2014 (71); Boston *et al.*, 2014 (69); Cancedda *et al.*, 2014 (102); Freeman *et al.*, 2003 (103).

Abbreviations: RT = radiotherapy, SX = surgery, CT = chemotherapy, MST = median survival time.

9.6. New Perspectives for the Treatment of Canine Melanomas

The aggressive nature of melanomas in canines and humans, with short survival times and minimal clinical outcome benefits after surgery and adjuvant therapy, emphasizes the importance of new studies to find an effective therapy for treatment and prevention of melanomas metastasis.

A canine therapeutic vaccine has been commercially available since 2007. The Merial melanoma vaccine for dogs (Oncept®) consists of DNA encoding the gene for the human melanocyte protein tyrosinase but apparently seems not to cure all melanoma cases. Published studies evaluating the survival times in dogs with melanoma in the mouth undergoing surgery and treatment with the Merial melanoma vaccine report average survival times of 599 days (20 months) or longer [113,114]. These results, though not from randomized studies, are promising and have prompted many veterinary oncologists to recommend the vaccine for patients with stage II and III melanoma despite the lack of FDA-standard clinical trials. The vaccine is well tolerated and has not been reported to be associated with side effects requiring medical intervention. In patients with very large and aggressive melanoma, there may not be sufficient time for an immune response to be mounted against the tyrosinase protein to prevent progression of the melanoma. In fact, 15% of patients treated with the vaccine reportedly die within three months of initiation of treatment, presumably due to the aggressive nature of the melanoma and insufficient time for the vaccine to be effective [119].

Lupeol [120] is among the new therapies under study. Lupeol is a triterpene extracted from various fruits and vegetables that reportedly inhibits melanoma cell proliferation *in vitro* and *in vivo*. A total of 11 dogs bearing buccal melanomas (three, five, and three dogs diagnosed with clinical stage I, II, and III melanoma, respectively) were evaluated. Subcutaneous lupeol (10 mg/kg) was administered postoperatively at various time points. Of the 11 subjects, seven exhibited no local recurrence 180 days postoperatively and no severe adverse effects were observed in any of these cases. Furthermore, no metastases have been detected during the period of the trial.

The current and promising immunotherapy to further augment the options for treatment in malignant melanomas, at least in humans, has been based on checkpoint–blocking antibodies, which can restore anticancer immune responses. Immune checkpoint molecules regulate the immune response avoiding autoimmunity and deleterious effects of an inflammatory response, but when expressed on tumors cells they work as a mechanism of immune evasion. Good examples of immune checkpoint molecules include PD-1 immune checkpoint inhibitor antibody (Pembrolizumab- humanized IgG4; Nivolumab- human IgG4), which was the first drug to gain FDA approval to a phase I trial in patients with BRAF inhibitor-resistant advanced melanoma and BRAFV600 mutation-positive melanoma [121]; and a checkpoint inhibitor (ipilimumab) directed against the cytotoxic T lymphocyte Antigen (CTLA-4). Data from clinical trials has demonstrated the important role this therapeutic approach can play—even with the development of immune-related side effects—since it has been proved a meaningful clinical improvement and an impressive increase in overall survival in melanoma patients [122] when compared to conventional chemotherapy. In the veterinary medicine field, the perspective of this therapy was described by Maekawa *et al.* [123], who demonstrated that an antibovine PD-L1 monoclonal antibody effectively blocked the binding of recombinant PD-1 with PD-L1-expressing tumor cells in a dose-dependent manner, including canine melanoma cells.

Other therapy as future treatment strategy for canine malignant melanoma can be based on oncolytic virotherapy. Basically, the virus infects and kills tumor cells when replication, lysis, and release of viruses occur. Tumor oncolysis by reovirus infection that was related in Ras-transformed cells was studied in six canine malignant melanoma cell lines CMeC1, CMeC2, KMeC, LMeC, CMGD2, and CMGD5. All of them were susceptible to reovirus infection, although Ras activation was not related in these cells, and only one of them (CMGD2) demonstrated more than 50% of cell death [124].

All efforts are geared toward better characterization of malignant melanomas in dogs, for the benefit of these companion animals, and also in an attempt to benefit the treatment of melanoma in humans.

10. CONCLUSIONS

Melanomas are very aggressive tumors and dogs are considered good models to study this disease. All efforts are geared toward better characterization and control of malignant melanomas in dogs, for the benefit of these companion animals, and also in an attempt to benefit the treatment of human melanomas.

ACKNOWLEDGMENTS

Karen Batschinski is the recipient of a fellowship from Fundação de Amparo a Pesquisa do Estado de São Paulo, FAPESP, Proc. Number 2013/16423-8. The authors are grateful to Jose Guilherme Xavier for supplying photos of canine melanomas.

AUTHOR CONTRIBUTIONS

Adriana Tomoko Nishiya, Claudia Ronca Felizzola, Eduardo Perlmann, Claudia Ronca Felizzola, Priscila Pedra Mendonça, Jéssica Soares Garcia, Tarso Felipe Teixeira, Maria Lucia Zaidan Dagli and Maria Lucia drafted the manuscript, Karen Batschinski, Marcello Vannucci Tedardi and the others revised the manuscript. All authors read and approved the final manuscript.

CONFLICTS OF INTEREST

The authors declare no conflict of interest.

REFERENCES

1. Sweet, M.; Kirkham, N.; Bendall, M.; Currey, L.; Bythell, J.; Heupel, M. PLoS ONE 2012.
2. Starkey, M.P.; Scase, T.J.; Mellersh, C.S.; Murphy, S. Dogs really are man's best friend—Canine genomics has applications in veterinary and human medicine! Brief. Funct. Genomic. Proteomic. 2005, 4, 112–128.
3. Paoloni, M.; Khanna, C. Translation of new cancer treatments from pet dogs to humans. Nat. Rev. Cancer 2008, 8, 147–156.
4. Nowend, K.L.; Starr-Moss, A.N.; Murphy, K.E. The function of dog models in developing gene therapy strategies for human health. Mamm. Genome 2011, 22, 476–485.
5. Khanna, C.; Lindblad-Toh, K.; Vail, D.; London, C.; Bergman, P.; Barber, L.; Breen, M.; Kitchell, B.; McNeil, E.; Modiano, J.F.; et al. The dog as a cancer model. Nat. Biotechnol. 2006, 24, 1065–1066.
6. Fowles, J.S.; Denton, C.L.; Gustafson, D.L. Comparative analysis of MAPK and PI3K/AKT pathway activation and inhibition in human and canine melanoma. Vet. Comp. Oncol. 2015, 13, 288–304.
7. Haass, N.K.; Smalley, K.S.; Herlyn, M. The role of altered cell–cell communication in melanoma progression. J. Mol. Histol. 2004, 35, 309–318.
8. Marcus, D.M.; Marcus, P.M.; Prabhu, R.S.; Owonikoko, T.K.; Lawson, D.H.; Switchenko, J.; Beither, J. Rising incidence of mucosal melanoma of the head and neck in the United States. J. Skin Cancer 2012, 231693.

9. Gavriel, H.; McArthur, G.; Sizeland, A.; Henderson, M. Review: Mucosal melanoma of the head and neck. Melanoma Res. 2011, 21, 257–266.
10. Meleti, M.; Leemans, C.R.; De Bree, R.; Vescovi, P.; Sesenna, E.; van der Waal, I. Head and neck mucosal melanoma: Experience with 42 patients, with emphasis on the role of postoperative radiotherapy. Head Neck 2008, 30, 1543–1551.
11. Ramos-Vara, J.A.; Beissenherz, M.E.; Miller, M.A.; Johnson, G.C.; Pace, L.W.; Fard, A.; Kottler, S.J. Retrospective study of 338 canine oral melanomas with clinical, histologic, and immunohistochemical review of 129 cases. Vet. Pathol.2000, 37, 597–608.
12. Kimura, K.C.; Garate, A.P.; Dagli, M.L.Z. Retrospective study of neoplasms in domestic animals: A survey between 1993 and 2002 of the service of animal pathology, department of pathology, school of veterinary medicine and animal science, University of Sao Paulo, Southeast Brazil. Braz. J. Vet. Pathol. 2012, 5, 60–69.
13. Teixeira, T.F.; Silva, T.C.; Cogliati, B.; Nagamine, N.K.; Dagli, M.L.Z. Retrospective study of melanocytic neoplasms in dogs and cats. Braz. J. Vet. Pathol. 2010, 3, 100–104.
14. Omholt, K.; Grafström, E.; Kanter-Lewensohn, L.; Hansson, J.; Ragnarsson-Olding, B.K. KIT pathway alterations in mucosal melanomas of the vulva and other sites. Clin. Cancer Res. 2011, 17, 3933–3942.
15. Zanoni, M.T.; Medicina, D.; Lombardi, D.; Ungari, M.; Balzarini, P.; Rossini, C.; Pellegrini, W.; Battaglia, P.; Capella, C.; Castelnuovo, P.; et al. Sinonasal mucosal melanoma: Molecular profile and therapeutic implications from a series of 32 cases. Head Neck 2013, 35, 1066–1077.
16. Lyu, J.; Wu, Y.; Li, C.; Wang, R.; Song, H.; Ren, G.; Guo, W. Mutation scanning of BRAF, NRAS, KIT, and GNAQ/GNA11 in oral mucosal melanoma: A study of 57 cases. J. Oral Pathol. Med. 2015.
17. Curtin, J.A.; Fridlyand, J.; Kageshita, T.; Patel, H.N.; Busam, K.J.; Kutzner, H.; Cho, K.H.; Aiba, S.; Bröcker, E.B.; LeBoit, P.E.; et al. Distinct sets of genetic alterations in melanoma. N. Engl. J. Med. 2005, 353, 2135–2147.
18. Spangler, W.L.; Kass, P.H. The histologic and epidemiologic bases for prognostic consideration in canine melanocytic neoplasia. Vet. Pathol. 2006, 43, 136–149.
19. Gillard, M.; Cadieu, E.; De Brito, C.; Abadie, J.; Vergier, B.; Devauchelle, P.; Degorce, F.; Dreano, S.; Primot, A.; Dorso, L.; et al. Naturally occurring melanomas in dogs as models for non-UV pathways of human melanomas. Pigment Cell Melanoma Res. 2014, 27, 90–102.
20. Priester, W.A. Skin tumors in domesticated animals. Data from 12 United States and Canadian Colleges of Veterinary Medicine. J. Natl. Cancer Insit. 1973, 50, 457–466.
21. Goldschmidt, M.H. Pigmented lesions of the skin. Clin. Dermatol. 1994, 12, 507–514.
22. Marino, D.J.; Matthiesen, D.T.; Stefanacci, J.D.; Moroff, S.D. Evaluation of dogs with digit masses: 117 cases (1981–1991). J. Am. Vet. Med. Assoc. 1995, 207, 726–728.
23. Miller, W.H.; Griffin, I.; Campbell, K.L. Muller and Kirk's Small Animal Dermatology, 7th ed.; Elsevier Mosby: St. Louis, MO, USA, 2013; pp. 774–843.
24. Teixeira, T.F.; Gentile, L.B.; Silva, T.C.; Mennecier, G.; Chaible, L.M.; Cogliati, B.; Roman, M.A.; Gioso, M.A.; Dagli, M.L.Z. Cell proliferation and expression of connexins differ in melanotic and amelanotic canine oral melanomas. Vet. Res. Commun. 2014, 38, 29–38.
25. Tedardi, M.V.; Biondi, L.R.; Kimura, K.C.; Pedra Mendonça, P.; Grandi, F.; Latorre, A.M.; Dagli, M.L.Z. Comparative oncology in Sao Paulo, Brazil—What to learn from cancer registry. Source J. Vet. Sci. 2014, 1, 1–8.
26. Tedardi, M.V.; Veneziano, D.B.; Kimura, K.C.; Pedra Mendonça, P.; Biondi, L.R.; Grandi, F.; Latorre, A.M.; Dagli, M.L.Z. Sao Paulo animal cancer registry, the first in Latin America. Vet. Comp. Oncol. 2015, 13, 154–155.
27. Haass, N.K.; Ripperger, D.; Wladykowski, E.; Dawson, P.; Gimotty, P.A.; Blome, C.; Fischer, F.; Schmage, P.; Moll, I.; Brandner, J.M. Melanoma progression exhibits a significant impact on connexin expression patterns in the epidermal tumor microenvironment. Histochem. Cell Biol. 2010, 133, 113–124.

28. Modiano, J.F.; Ritt, M.G.; Wojcieszyn, J. The molecular basis of canine melanoma: Pathogenesis and trends in diagnosis and therapy. J. Vet. Intern. Med. 1999, 13, 163–174.
29. Smith, H.; Goldschmidt, H.; Mcmanus, P.M. A Comparative review of melanocytic neoplasms. Vet. Pathol. 2002, 39, 651–678.
30. Dzutsev, A.; Goldszmid, R.S.; Viaud, S.; Zivtogel, L.; Trinchieri, G. The role of the microbiota in inflammation, carcinogenesis, and cancer therapy. Eur. J. Immunol. 2015, 45, 17–31.
31. Goldschmidt, M.H.; Hendrick, M.J. Tumors of the skin and soft tissues. In Tumors in Domestic Animals, 4th ed.; Meuten, D.J., Ed.; Iowa State Press: Ames, IA, USA, 2002; pp. 78–82.
32. Scott, D.W.; Miller, W.H., Jr.; Griffin, C.E. Muller & Kirk's Small Animal Dermatology, 6th ed.; Elsevier Saunders: Philadelphia, PA, USA, 2001; pp. 1–70.
33. Li, G.; Herlyn, M. Dynamics of intercellular communication during melanoma development. Mol. Med. Today 2000, 6, 163–169.
34. Chin, L. The genetics of malignant melanoma: Lessons from mouse and man. Nat. Rev. Cancer 2003, 3, 559–570.
35. Gaggioli, C.; Sahai, E. Melanoma invasion—Current knowledge and future directions. Pigment Cell Res. 2007, 20, 161–172.
36. Head, K.W.; Else, R.W.; Dubielzig, R.R. Tumors of the Alimentary Tract. In Tumors in Domestic Animals, 4th ed.; Meuten, D.J., Ed.; Iowa State Press: Ames, IA, USA, 2002; pp. 426–430.
37. Tajima, S.; Ura-Ishiko, A.; Hayashi, A. Melanogenesis, biosynthetic phenotype of fibronectin and collagen, and migrating activity in cloned B16 mouse melanoma cells. J. Dermatol. Sci. 1996, 12, 24–30.
38. Simonetti, O.; Lucarini, G.; Brancorsini, D.; Nita, P.; Bernardini, M.L.; Biagini, G.; Offidani, A. Immunohistochemical expression of vascular endothelial growth factor, matrix metalloproteinase 2 and matrix metalloproteinase 9 in cutaneous melanocytic lesions. Cancer 2002, 95, 602–624.
39. Nakhleh, R.E.; Wick, M.R.; Rocamora, A.; Swanson, P.E.; Dehner, L.P. Morphologic diversity in malignant melanomas. Am. J. Clin. Pathol. 1990, 93, 731–740.
40. Chénier, S.; Doré, M. Oral Malignant Melanoma with Osteoid Formation in a Dog. Vet. Pathol. 1999, 36, 74–76.
41. Smedley, R.C.; Spangler, W.L.; Esplin, D.G.; Kitchell, B.E.; Bergman, P.J.; Ho, H.Y.; Bergin, I.L.; Kiupel, M. Prognostic markers for canine melanocytic neoplasms: A comparative review of the literature and goals for future investigation. Vet. Pathol. 2011, 48, 54–72.
42. Sandusky, G.E.; Carlton, W.W.; Wightman, K.A. Diagnostic immunohistochemistry of canine round cell tumors. Vet. Pathol. 1987, 24, 495–499.
43. Silva, A.P.; Silva, R.G.; Cogliati, B.; Dias, A.S.M.; Le Bas, A.E.; Hernandez-Blazquez, F.J. Bleaching of melanin in the epidermis of South American fur seal and its application on enzyme immunohistochemistry. Pesqui. Vet. Bras. 2011, 31, 267–270.
44. Fisher, D.E.; Kwong, L.N.; Chin, L. Melanoma. In DeVita VT, Hellman and Rosenberg's Cancer Principles & Practice of Oncology, 8th ed.; De Vita, V.T., Jr., Lawrence, T.S., Rosenberg, S.A., Eds.; Lippincott Williams & Wilkins: Philadelphia, PA, USA, 2008; pp. 1889–1951.
45. Shelly, S.; Chien, M.B.; Yip, B.; Kent, M.S.; Theon, A.P.; McCallan, J.L.; London, C.A. Exon 15 BRAF mutations are uncommon in canine oral malignant melanomas. Mamm. Genome 2004, 16, 211–217.
46. Chu, P.Y.; Pan, S.L.; Liu, C.H.; Lee, J.; Yeh, L.S.; Liao, A.T. KIT gene exon 11 mutations in canine malignant melanoma. Vet. J. 2013, 196, 226–230.
47. Curtin, J.A.; Busam, K.; Pinkel, D.; Bastian, B.C. Somatic activation of KIT in distinct subtypes of melanoma. J. Clin. Oncol. 2006, 24, 4340–4346.
48. Maldonado, J.L.; Fridlyand, J.; Patel, H.; Jain, A.N.; Busam, K.; Kageshita, T.; Ono, T.; Albertson, D.G.; Pinkel, D.; Bastian, B.C. Determinants of BRAF mutations in primary melanomas. J. Natl. Cancer Inst. 2003, 95, 1878–1890.

49. Murakami, A.; Mori, T.; Sakai, H.; Murakami, M.; Yanai, T.; Hoshino, Y.; Maruo, K. Analysis of KIT expression and KIT exon 11 mutations in canine oral malignant melanomas. Vet. Comp. Oncol. 2011, 9, 219–224.
50. Zebary, A.; Jangard, M.; Omholt, K.; Ragnarsson-Olding, B.; Hansson, J. KIT, NRAS and BRAF mutations in sinonasal melanoma. Br. J. Cancer 2013, 109, 559–564.
51. Goel, V.K.; Lazar, A.J.; Warneke, C.L.; Redston, M.S.; Haluska, F.G. Examination of mutations in BRAF, NRAS, and PTEN in primary cutaneous melanoma. J. Investig. Dermatol. 2006, 126, 154–160.
52. Mochizuki, H.; Kennedy, K.; Shapiro, S.G.; Breen, M. BRAF Mutations in Canine Cancers. PLoS ONE 2015.
53. Bergman, P.J. Canine Oral Melanoma. Clin. Tech. Small Anim. Pract. 2007, 22, 55–60.
54. Todoroff, R.J.; Brodey, R.S. Oral and pharyngeal neoplasia in the dog: A retrospective survey of 361 cases. J. Am. Vet. Med. Assoc. 1979, 175, 567–571.
55. Vos, J.H.; Gaag, I.V.D. Canine and feline oral-pharyngeal tumours. J. Vet. Med. 1987, 34, 420–427.
56. Guerra, J.L.; Dagli, M.L.; Godoy, L.F.; Birman, E.G. A retrospective study of oral neoplasms in dogs: A survey from 1941 to 1986. Rev. Fac. Med. Vet. Zootec. Univ. S. Paulo 1989, 26, 207–212.
57. Stopiglia, A.J.; Felizzola, C.R.; Dagli, M.L.Z.; Xavier, J.G.; Guerra, J.L.; Birman, E. Estudo clínico e anátomo-patológico das formações neoplásicas e não neoplásicas da cavidade bucal de cães. Comun. Cient. Fac. Med. Vet. Zootec. USP 1992, 16, 33–39.
58. Ghirelli, C.D.O. In Biblioteca Digital USP. Available online: http://www .teses.usp.br/teses/disponiveis/10/10137/tde-22072008-090727/pt-br.php (assecced on 15 November 2015).
59. Requicha, J.F.M.F. Repositório Universidade de Trás-os-Montes e Alto Douro. Available online: https://repositorio.utad.pt/bitstream/10348/711/1/ MsC_j fmfrequicha.pdf (accessed on 15 November 2015).
60. Bronden, L.B.; Eriksen, T.; Kristensen, A.T. Oral malignant melanomas and other head and neck neoplasms in Danish dogs-data from the Danish Veterinary Cancer Registry. Acta Vet. Scand. 2009, 51, 1–6.
61. Liptak, J.M.; Withrow, S.J. Cancer of gastrointestinal tract. In Small Animal Clinical Oncology, 5th ed.; Withrow, S.J., Vail, D.M., Page, R.L., Eds.; Elsevier Saunders: St. Louis, MO, USA, 2013; pp. 381–395.
62. McEntee, M.C. Clinical behavior of nonodontogenic tumors. In Oral and Maxillofacial Surgery in Dogs and Cats; Verstraete, F.J.M., Lommer, M.J., Eds.; Elsevier Saunders: Pennsylvania, PA, USA, 2012; pp. 387–402.
63. Dennis, M.; Ehrhart, N.; Duncan, C.; Barnes, A.; Ehrhart, E. Frequency of and risk factors associated with lingual lesions in dogs: 1196 cases (1995–2004). J. Am. Vet. Med. Assoc. 2006, 228, 1533–1537.
64. Williams, L.E.; Packer, R.A. Association between lymph node size and metastasis in dogs with oral malignant melanoma: 100 cases (1987–2001). J. Am. Vet. Med. Assoc. 2003, 222, 1234–1236.
65. Esplin, D.G. Survival of dogs following surgical excision of histologically well-differentiated melanocytic neoplasms of the mucous membranes of the lips and oral cavity. Vet. Pathol. 2008, 45, 889–896.
66. White, R.A.S.; Jefferies, A.R.; Freedman, L.S. Clinical staging for oropharyngeal malignancies in the dog. J. Small Anim. Pract. 1985, 26, 581–594.
67. MacEwen, E.G.; Patnaik, A.K.; Harvey, H.J.; Hayes, A.A.; Matus, R. Canine oral melanoma: Comparison of surgery versus surgery plus Corynebacterium parvum. Cancer Investig. 1986, 4, 397–402.
68. Hahn, K.A.; Denicola, D.B.; Richardson, R.C.; Hahn, E.A. Canine oral malignant melanoma: Prognostic utility of an alternative staging system. J. Small Anim. Pract. 1994, 35, 251–256.
69. Boston, S.E.; Lu, X.; Culp, W.T.N.; Montinaro, V.; Romanelli, G.; Dudley, R.M.; Liptak, J.M.; Mestrinho, L.A.; Buracco, P. Efficacy of systemic adjuvant therapies

administered to dogs after excision of oral malignant melanomas: 151 cases (2001–2012). J. Am. Vet. Med. Assoc. 2014, 245, 401–407.
70. Owen, L.N. TNM Classification of Tumours in Domestic Animals; World Health Organization: Geneve, Switzerland, 1980.
71. Tuohy, J.L.; Selmic, L.E.; Worley, D.R.; Ehrhart, N.P.; Withrow, S.J. Outcome following curative-intent surgery for oral melanoma in dogs: 70 cases (1998–2011). J. Am. Vet. Med. Assoc. 2014, 245, 1266–1273.
72. Harvey, H.J.; MacEwen, E.G.; Braun, D.; Patnaik, A.K.; Withrow, S.J.; Jongeward, S. Prognostic criteria for dogs with oral melanoma. J. Am. Vet. Med. Assoc. 1981, 178, 580–582.
73. Mould, J.R.B.; Petersen-Jones, S.M.; Peruccio, C.; Ratto, A.; Sassani, J.W.; Harbour, J.W. Uveal melanocytic tumors. In Ocular Tumors in Animals and Humans; Peiffer, R.L., Jr., Simons, K.B., Eds.; Iowa State Press: Ames, IA, USA, 2002; pp. 225–282.
74. Dubielzig, R.R.; Ketring, K.L.; Mclellan, G.J.; Albert, D.M. Veterinary Ocular Pathology: A Comparative Review; W.B. Saunders: Philadelphia, PA, USA, 2010; p. 472.
75. Wilcock, B.P.; Peiffer, J.R. Morphology and behavior of primary ocular melanomas in 91 dogs. Vet. Pathol. 1986, 23, 418–424.
76. Garner, A.; Koornneef, L.; Levene, A.; Collin, J.R. Malignant melanoma of the eyelid skin: Histopathology and behaviour. Br. J. Ophthalmol. 1985, 69, 180–186.
77. Conceição, L.F.; Ribeiro, A.P.; Piso, D.Y.T.; Laus, J.L. Considerations about ocular neoplasia of dogs and cats. Cienc. Rural 2010, 40, 2235–2242.
78. Lim, L.A.; Madigan, M.C.; Conway, R.M. Conjunctival melanoma: A review of conceptual and treatment advances. Clin. Ophthalmol. 2013, 7, 521–531.
79. Donaldson, D.; Sansom, J.; Scase, T.; Adams, V.; Mellersh, C. Canine Limbal melanoma: 30 cases (1992–2004). Part 1. Signalment, clinical and histological features and pedigree analysis. Vet. Ophthalmol. 2006, 9, 115–119.
80. Perlmann, E.; Barros, P.S.M.; Fernandez, L.; Di Cesare, S.; Bakalian, S.; Burnier, M.N., Jr. Histopathological evaluation of pigmented intraocular tumors in dogs. Investig. Ophthalmol. Vis. Sci. 2010, 5, 5159.
81. Esson, D.; Fahrer, C.; Zarfoss, M.K.; Dubielzig, R.R. Suspected uveal metastasis of a nail bed melanoma in a dog. Vet. Ophthalmol. 2007, 10, 262–266.
82. Fineman, M.S.; Eagle, R.C.; Shields, J.A.; Shields, C.L.; Potter, P. Melanocytomalytic glaucoma in eyes with necrotic iris melanocytoma. Ophthalmology 1998, 105, 492–496.
83. Ryan, A.M.; Diters, R.W. Clinical and pathologic features of canine ocular melanomas. J. Am. Med. Assoc. 1984, 184, 60–67.
84. Giuliano, E.A.; Chappell, R.; Fischer, B.; Dubielzig, R.R. A matched observational study of canine survival with primary intraocular melanocytic neoplasia. Vet. Ophthalmol. 1999, 2, 185–190.
85. Konrade, K.A.; Hoffman, A.; Schubert, C.S.; Dubielzig, R.R. A histologic series of secondary uveal melanomas in eleven dogs. In Proceedings of the 40th Annual Meeting of the American College of Veterinary Ophthalmologist, Chicago, IL, USA, 4–7 November 2009.
86. Willis, A.M.; Wilkie, D.A. Ocular oncology. Clin. Tech. Small Anim. Pract. 2001, 16, 77–85.
87. Cazalot, G.; Raymond-Letron, I.; Regnier, A. Choroidal melanoma presented as glaucoma in a dog: Case report and review of the literature. Rev. Med. Vet. 2008, 159, 74–78.
88. Henry, C.J.; Brewer, W.G.; Whitley, E.M.; Tyler, J.W.; Ogilvie, G.K.; Norris, A.; Fox, L.E.; Morrison, W.B.; Hammer, A.; Vail, D.M.; et al. Canine digital tumors: Veterinary cooperative oncology group retrospective study of 64 dogs. J. Vet. Intern. Med. 2005, 19, 720–724.
89. Aronsohn, M.G.; Carpenter, J.L. Distal extremity melanocytic nevi an malignant melanomas in dogs. J. Am. Anim. Hosp. Assoc. 1990, 26, 605–612.
90. Wobeser, B.K.; Kidney, B.A.; Powers, B.E.; Withrow, S.J.; Mayer, M.N.; Spinato, M.T.; Allen, A.L. Diagnosis and clinical outcomes associated with surgically amputated

canine digit submitted to multiple veterinary diagnostic laboratories. Vet. Pathol. 2007, 44, 355–361.

91. Manley, C.A.M.; Leibman, N.F.; Wolchok, J.D.; Riviere, I.C.M.; Bartido, S.; Craft, D.M.; Bergman, P.J. Xenogeneic Murine Tyrosinase DNA Vaccine for Malignant Melanoma of the Digit of Dogs. J. Vet. Intern. Med. 2011, 25, 94–99.
92. Bostock, D.E. Prognosis after surgical excision of canine melanoma. Vet. Pathol. 1979, 16, 32–40.
93. Bergman, P.J.; Kent, M.S.; Farese, J.P. Melanoma. In Small Animal Clinical Oncology, 5th ed.; Withrow, S.J., Vail, D.M., Page, R.L., Eds.; Elsevier Saunders: St. Louis, MO, USA, 2013; pp. 321–334.
94. Brockley, L.K.; Cooper, M.A.; Bennett, P.F. Malignant melanoma in 63 dogs (2001–2011): The effect of carboplatin chemotherapy on survival chemotherapy on survival. N. Z. Vet. J. 2013, 61, 25–31.
95. Smith, M.M. Surgical approach for lymph node staging of oral and maxillofacial neoplasms in dogs. J. Am. Anim. Hosp. Assoc. 1995, 31, 514–518.
96. Herring, E.S.; Smith, M.M.; Robertson, J.L. Lymph node staging of oral and maxillofacial neoplasms in 31 dogs and cats. J. Vet. Dent. 2002, 19, 122–126.
97. Proulx, D.R.; Ruslander, D.M.; Dodge, R.K.; Hauck, M.L.; Williams, L.E.; Horn, B.; Price, G.S.; Thrall, D.E. A retrospective analysis of 140 dogs with oral melanoma treated with external beam radiation. Vet. Radiol. Ultrasound.2003, 44, 352–359.
98. Théon, A.P.; Rodriguez, C.; Madewell, B.R. Analysis of prognostic factors and patterns of failure in dogs with malignant oral tumors treated with megavoltage irradiation. J. Am. Vet. Med. Assoc. 1997, 210, 778–784.
99. Khan, N.; Mohammad, K.K.; Almasan, A.; Singh, A.D.; Macklis, R. The evolving role of radiation therapy in the management of malignant melanoma. Int. J. Radiat. Oncol. Biol. Phys. 2011, 80, 645–654.
100. Bateman, K.E.; Catton, P.A.; Pennock, P.W.; Kruth, S.A. 0-7-21 radiation therapy for the treatment of canine oral melanoma. J. Vet. Intern. Med. 1994, 8, 267–272.
101. Murphy, S.; Hayes, A.M.; Blackwood, L.; Maglennon, G.; Pattinson, H.; Sparkes, A.H. Oral malignant melanoma: The effect of coarse fractionation alone or with adjuvant carboplatin therapy. Vet. Comp. Oncol. 2005, 3, 222–229.
102. Cancedda, S.; Bley, C.R.; Aresu, L.; Dacasto, M.; Leone, V.F.; Pizzoni, S.; Gracis, M.; Marconato, L. Efficacy and side effects of radiation therapy in comparison with radiation therapy and temozolomide in the treatment of measurable canine malignant melanoma. Vet. Comp. Oncol. 2014.
103. Freeman, K.P.; Hahn, K.A.; Harris, F.D.; King, G.K. Treatment of dogs with oral melanoma by hypofractionated radiation therapy and platinum-based chemotherapy (1987–1997). J. Vet. Intern. Med. 2003, 17, 96–101.
104. Rassnick, K.M.; Ruslender, D.M.; Cotter, S.M.; Al-Sarraf, R.; Bruyette, D.S.; Gamblin, R.M.; Meleo, K.A.; Moore, A.S. Use of carboplatin for treatment of dogs with malignant melanoma: 27 cases (1989–2000). J. Am. Vet. Med. Assoc. 2001, 218, 1444–1448.
105. Boria, P.A.; Murry, D.J.; Bennett, P.F.; Glickman, N.W.; Snyder, P.W.; Merkel, B.L.; Schlittler, D.L.; Mutsaers, A.J.; Thomas, R.M.; Knapp, D.W. Evaluation of cisplatin combined with piroxicam for the treatment of oral malignant melanoma and oral squamous cell carcinoma in dogs. J. Am. Vet. Med. Assoc. 2004, 224, 388–394.
106. Dank, G.; Rassnick, K.M.; Sokolovsky, Y.; Garrett, L.D.; Post, G.S.; Kitchell, B.E.; Sellon, R.K.; Kleiter, M.; Northrup, N.; Segev, G. Use of adjuvant carboplatin for treatment of dogs with oral malignant melanoma following surgical excision. Vet. Comp. Oncol. 2014, 12, 78–84.
107. Bergman, P.J.; Mcknight, J.; Novosad, A.; Charney, S.; Farrelly, J.; Craft, D.; Wulderk, M.; Jeffers, Y.; Sadelain, M.; Hohenhaus, A.E.; et al. Long-term survival of dogs with advanced malignant melanoma after DNA vaccination with xenogeneic human tyrosinase: A phase I trial. Clin. Cancer Res. 2003, 9, 1284–1290.
108. MacEwen, E.G.; Kurzman, I.D.; Vail, D.M.; Dubielzig, R.R.; Everlith, K.; Madewell, B.R.; Rodriguez, C.O., Jr.; Phillips, B.; Zwahlen, C.H.; Obradovich, J.; et al. Adjuvant therapy

for melanoma in dogs: Results of randomized clinical trials using surgery, liposome-encapsulated muramyl triptide, and granulocyte macrophage colony-stimulang factor. Clin. Cancer Res. 1999, 5, 4249–4258.

109. Moff, S.L.; Corey, G.R.; Gottfredsson, M. Distant cutaneous granulomas after bacille Calmette-Guérin immunotherapy for malignant melanoma: Case for direct infection. Clin. Infect. Dis. 1999, 29, 1569–1570.

110. Bianco, S.R.; Sun, J.; Fosmire, S.P.; Hance, K.; Padilla, M.L.; Ritt, M.G.; Getzy, D.M.; Duke, R.C.; Withrow, S.J.; Lana, S.; et al. Enhancing anti melanoma immune responses through apoptosis. Cancer Gene Ther. 2003, 10, 726–736.

111. Alexander, A.N.; Huelsmeyr, M.K.; Mitzey, A.; Dubielzig, R.R.; Kurzman, I.D.; MacEwen, E.G.; Vail, D.M. Development of an allogenic whole-cell tumor vaccine expressing xenogenic gp100 and implementation in phase II clinical trial in canine patients with malignant melanoma. Cancer Immunol. Immunother. 2006, 55, 433–442.

112. Hogge, G.S.; Burkholder, J.K.; Culp, J.; Albertini, M.R.; Dubielzig, R.R.; Keller, E.T.; Yang, N.S.; MacEwen, E.G. Development of human granulocyte-macrophage colony-stimulating factor- transfected tumor cell vaccines for treatment of spontaneous canine cancer. Hum. Gene Ther. 1998, 9, 1851–1861.

113. Bergman, P.J.; Wolchok, J.D. Of mice and men (and dogs): Development of a xenogeneic DNA vaccine for canine oral malignant melanoma. Cancer Ther. 2008, 6, 817–826.

114. Grosenbaugh, D.A.; Leard, A.T.; Bergman, P.J.; Klein, M.K.; Meleo, K.; Susaneck, S.; Hess, P.R.; Jankowski, M.K.; Jones, P.D.; Leibman, N.; et al. Safety and efficacy of a xenogeneic DNA vaccine encoding for human tyrosinase as adjunctive treatment for oral malignant melanoma in dogs following surgical excision of the primary tumor. Am. J. Vet. Res. 2011, 72, 1631–1638.

115. Ottnod, J.M.; Smedley, R.C.; Walshaw, R.; Hauptman, J.G.; Kiupel, M.; Obradovich, J.E. A retrospective analysis of the efficacy of Oncept vaccine for the adjunct treatment of canine oral malignant melanoma. Vet. Comp. Oncol. 2013, 11, 219–229.

116. Rotte, A.; Bhandaru, M.; Zhou, Y.; Mcelwee, K.J. Immunotherapy of melanoma: Present options and future promises. Cancer Metastasis Rev. 2015, 34, 115–128.

117. Gyorffy, S.; Rodriguez-Lecompte, J.C.; Woods, J.P.; Foley, R.; Kruth, S.; Liaw, P.C.; Gauldie, J. Bone marrow-derived dendritic cell vaccination of dogs with naturally occurring melanoma by using human gp100 antigen. J. Vet. Intern. Med. 2005, 19, 56–63.

118. Tamura, K.; Yamada, M.; Isotani, M., Arai, H.; Yagihara, H.; Ono, K.; Washizu, T.; Bonkobara, M. Induction of dendritic cell-mediated immune responses against canine malignant melanoma cells. Vet. J. 2008, 175, 126–129.

119. Elmslie, R. VRCC Veterinary Speciality & Emergency Hospital. Available online: http://www.vetcancerspecialists.com/blog/2013/1/31/the-controversy-surrounding-the-melanoma-vaccine-for-dogs/ (accesse on 15 November 2015).

120. Yokoe, I.; Azuma, K.; Hata, K.; Mukaiyama, T.; Goto, T.; Tsuka, T.; Imagawa, T.; Itoh, N.; Murahata, Y.; Osaki, T.; et al. Clinical systemic lupeol administration for canine oral malignant melanoma. Mol. Clin. Oncol. 2015, 3, 89–92.

121. Robert, C.; Ribas, A.; Wolchok, J.D.; Hodi, F.S.; Hamid, O.; Kefford, R.; Weber, J.S.; Joshua, A.M.; Hwu, W.J.; Gangadhar, T.C.; et al. Anti-programmed-death-receptor-1 treatment with pembrolizumab in ipilimumab-refractory advanced melanoma: A randomised dose-comparison cohort of a phase 1 trial. Lancet 2014, 384, 1109–1117.

122. Schadendorf, D.; Hodi, F.S.; Robert, C.; Weber, J.S.; Margolin, K.; Hamid, O.; Patt, D.; Chen, T.T.; Berman, D.M.; Wolchok, J.D. Pooled analysis of long-term survival data from phase II and phase III trials of ipilimumab in unresectable or metastatic melanoma. J. Clin. Oncol. 2015, 33, 1889–1894.

123. Maekawa, N.; Konnai, S.; Ikebuchi, R.; Okagawa, T.; Adachi, M.; Takagi, S.; Kagawa, Y.; Nakajima, C.; Suzuki, Y.; Murata, S.; et al. Expression of PD-L1 on canine tumor cells

and enhancement of IFN-c production from tumor infiltrating cells by PD-L1 blockade. PLoS ONE 2014.

124. Igase, M.; Hwang, C.C.; Coffey, M.; Okuda, M.; Noguchi, S.; Mizuno, T. The oncolytic effects of reovirus in canine solid tumor cell lines. J. Vet. Med. Sci. 2015, 77, 541–548.

CHAPTER 3

An Update on Drugs Used for Lumbosacral Epidural Anesthesia and Analgesia in Dogs

Paulo V. M. Steagall[1], Bradley T. Simon[2]*, Francisco J. Teixeira Neto[3] and Stelio P. L. Luna[3]

[1]*Faculty of Veterinary Medicine, Department of Clinical Sciences, Université de Montréal, Saint-Hyacinthe, QC, Canada*

[2]*Department of Small Animal Clinical Sciences, College of Veterinary Medicine and Biomedical Sciences, Texas A&M University, College Station, TX, USA*

[3]*Faculty of Veterinary Medicine and Animal Science, Department of Veterinary Surgery and Anaesthesiology, Universidade Estadual Paulista (UNESP), Botucatu, Brazil*

ABSTRACT

This review aims to report an update on drugs administered into the epidural space for anesthesia and analgesia in dogs, describing their potential advantages and disadvantages in the clinical setting. Databases searched include , , and CAB abstracts. Benefits of administering local anesthetics, opioids, and $alpha_2$ agonists into the epidural space include the use of lower doses of general anesthetics (anesthetic "sparing" effect), perioperative analgesia, and reduced side effects associated with systemic administration of drugs. However, the potential for cardiorespiratory compromise, neurotoxicity, and other adverse effects should be considered when using the epidural route of administration. When these variables are considered, the epidural technique is useful as a complementary method of anesthesia for preventive and postoperative analgesia and/or as part of a balanced anesthesia technique.

Keywords: epidural, canine, analgesia, anesthesia, opioids, local anesthetics, pain

INTRODUCTION

Epidural administration of drugs for pain management has been widely used in veterinary medicine (1–3). The advantage of this route is its proximity to the spinal cord receptors involved in the modulation and transmission of the nociceptive signal. The terms epidural and extradural refer to the space outside the dura mater. The terms intrathecal, subarachnoid, and spinal refer to the space between the pia mater and arachnoid membrane. *Epidural anesthesia* refers to the sensory, motor, and autonomic blockade produced by epidural administration of local anesthetics while *epidural analgesia* refers to epidural administration of analgesics, such as opioids.

The use of this technique may provide preemptive analgesia by inhibitingsensitization and modulating afferent signals to the dorsal horn, reducing pain and inhalant, and/or analgesic requirements during the perioperative period (4–6). In addition, epidural anesthesia suppresses the markers of stress response as represented by decreases in serum concentrations of cortisol and norepinephrine for up to 48 h after administration (7–9).

Absolute contraindications to epidural anesthesia include untreated hypovolemia, coagulation disorders, septicemia, bacteremia, skin trauma, neoplasia, and/or infection at the lumbosacral region (2, 10). Some neurological diseases, spinal cord trauma, low-dose heparin therapy, and anatomical changes including trauma of the pelvic area and obesity resulting in difficulty locating the lumbosacral space are considered relative contraindications (2). The use of ultrasonography to locate the epidural space may circumvent this latter issue (11).

The benefits of this technique often outweigh the risks; however, clinicians should be aware of potential complications such as technical failure, contamination, hematoma, accidental intrathecal injections, respiratory depression, sympathetic blockade, bradycardia, hypotension, neurological deficits, delayed hair growth, pruritus, and urinary retention (12–18). For example, an epidural abscess and lumbosacral discospondylitis has been reported in a dog most likely due to colonic perforation or fecal contamination (19). The technique was performed after multiple skin penetrations and subcutaneous needle redirections were done prior to achieving correct epidural placement. Epidural injections should only be performed after strict aseptic technique and manipulations should be gentle and cautious. In addition, epidural injections should be performed with preservative-free solutions to avoid the toxic effects produced by preservatives on the spinal cord. However, in research settings, epidural administration of morphine containing 0.1% of sodium metabisulfite did not cause neurotoxicity in dogs (20). In humans, side effects after intrathecal administration of morphine containing up to 0.004% sodium edetate and up to 0.04% of sodium metabisulfate appeared to be non-existent in 15 patients (21). In rats and sheep, chronic intrathecal administration of neostigmine containing methyl- and propylparaben did not produce any histologic evidence of neurotoxicity. Some institutions in North America and Europe use morphine containing 0.1% of sodium metabisulfite by the epidural route on a routine basis in dogs without reporting side effects (9).

Epidural administration of local anesthetics can be effective for a variety of surgical procedures such as cesarean section (22), orthopedic procedures in the hind limb, and soft tissue surgeries (8, 23). In these cases, minimal maternal and fetal depression (22) and less cardiopulmonary depression than general anesthesia (23) have been observed. Epidural anesthesia has been frequently used in combination with general anesthesia to reduce anesthetic and analgesic requirements, improve analgesia, and quality of anesthetic recovery (8, 24–26).

This review reports potential advantages and disadvantages of each class of drugs and their combinations used for epidural anesthesia and analgesia in dogs. The epidural technique *per se* is not the main focus of this article and has been reported elsewhere (10, 27).

Local Anesthetics

Local anesthetics diffuse across the dura mater to act on nerve roots and the spinal cord, and into the paravertebral area through the intervertebral foramina, producing multiple paravertebral nerve blocks after epidural administration (28).

Factors Influencing Onset and Duration of Autonomic, Sensory, and Motor Blockade

The extent of the sensory, motor, and autonomic blockade produced by a lumbosacral epidural injection will depend on the cranial distribution of the local anesthetic according to the anesthetic volume, concentration, velocity of injection, amount of the epidural fat, epidural space size, posture, and gravity (29–35). For example, the use of low concentration of local anesthetics (e.g., bupivacaine less than 0.25%) or low-dose bupivacaine (0.05 mL kg^{-1}) may be useful when sensory blockade is desired with minimum impairment of motor function (36–38). This property of "differential" or "selective" blockade may be attributed to the larger diameter and more profound myelination of Aα motor nerve fibers when compared with Aδ and C fibers (39). However, this is not always the case and motor function may be blocked even when low concentrations are administered (40).

It has been recommended that dogs be positioned in sternal recumbency and kept in this position following injection to provide bilateral analgesia or in lateral recumbency with the surgical site as the dependent side until the block has taken effect (32). One study using canine cadavers showed that if greater cranial migration of an epidurally administered drug is desired, placing the patient in lateral recumbency with the surgical site on the dependent side for 30–40 min may precede surgery (gravity effect) (32). In the authors' experience, placing the affected limb in a dependent position for the duration of the clipping and surgical preparation (approximately 30 min) is enough time to achieve this effect without delaying surgery.

When local anesthetics are administered epidurally using a Tuohy needle for epidural catheter placement, the needle bevel should face the head of the animal since it will facilitate the threading of the catheter to more cranial portions of the spinal canal (41). However, there is a lack of evidence to support the benefit of this technique when an epidural catheter is not being used, as the Tuohy needle bevel direction had limited clinical significance on preferential spread of 1.5% lidocaine after epidural administration in young (<40 years of age) human subjects (42). Another factor that may play a role in the cephalad migration of the local anesthetic in the spinal canal is the velocity of injection. However, a recent study showed that an injection velocity of 1 mL min^{-1} compared with 2 mL min^{-1} of 0.5% bupivacaine after epidural administration had limited effect on injectate distribution and sensory blockade in anesthetized dogs (33). The effect of epidural drug injection velocity on duration and quality of analgesia is yet to be determined (43). In the authors' routine, injectate volume is commonly administered over approximately 1 min. Peak cephalad migration of the injectate to the level of the cervical vertebra within the spinal canal is typically reached at doses of 0.4 mL kg^{-1} in dogs (40). Patients with high body mass index may have a smaller volume of epidural space due to accumulation of adipose tissue in the spinal canal (34, 44). This factor may contribute to a more cranial migration of the local anesthetic specifically with injections in the lumbosacral region (45). Lean body weight or spinal column length (occipital condyle to the first coccygeal vertebra) should be used when calculating the epidural dose in obese patients or those with abnormal spinal column conformations (35). An upward sloping body position of patients in sternal recumbency also did not affect injectate spread (44).

Side Effects after Epidural Administration of Local Anesthetics and Their Treatment

In general, small volumes (0.1 mL kg^{-1}) of 2% lidocaine or 0.5% bupivacaine produce blockade of the immediate lumbosacral area while larger volumes (0.2 mL kg^{-1}) produce complete desensitization of the caudal portions of the abdominal cavity, inguinal area, hind limbs, tail, and perineum (30, 46). Although volumes of local anesthetic exceeding 0.2–0.25 mL kg^{-1} may desensitize areas up to the thoracic and cervical vertebrae (31, 40), the use of greater volumes is not recommended because long-term blockade of sympathetic fibers may lead to severe cardiovascular depression, characterized by peripheral vasodilation, hypotension and bradycardia (47), and respiratory compromise as a result of blockade of the phrenic nerve (48). Hypovolemia should be treated before epidural anesthesia and cardiorespiratory monitoring is strongly recommended during injection otherwise serious complications may occur. Cardiac arrest due to excessive sympathetic blockade has been reported in a dog with uncorrected hypovolemia after lumbosacral administration of epidural 2% lidocaine (5.33 mg kg^{-1}) (49). Extensive cranial spread of epidural administration of 0.75% ropivacaine resulting in high sympathetic blockade (levels T1–T3) was likely to be the cause of cardiovascular depression and the Horner's syndrome in a dog (13). Horner's syndrome, ataxia, paraplegia, depression, stupor, and intermittent cough were also reported in dogs receiving epidural lidocaine (0.1–0.2 mL kg^{-1}) at the level of the seventh *thoracic* epidural space (50). The thoracic epidural space for the administration of anesthetics or analgesics has received little attention in small animal medicine, but may provide an alternative to the lumbosacral approach when analgesia for the thoracic or upper-abdominal region is required (50, 51). To avoid cephalad spread and adverse effects associated with the thoracic epidural technique, smaller volumes (0.1 mL kg^{-1}) should be considered when compared to the lumbosacral approach. In the authors' experience, complications are rare with lumbosacral epidural administration and avoided by following appropriate dosage regimens (Table 1) and technique.

Table 1. Commonly used and recommended epidural anesthetics and analgesics in dogs.

Drug	Dose (mg kg^{-1})	Final volume (mL kg^{-1}) injected into the LS epidural space	Onset time (min)	Duration of analgesia (h)	Comments	Reference
Local anesthetics						
2% Lidocaine with 1:200,000 epinephrine	5.0	0.25	4–6	1	Duration of motor blockade 60–120 min	(46, 60, 66, 69)
0.5% Bupivacaine	0.5–1.0	0.2–0.25	5–15	>2	Duration of complete motor blockade and ataxia was 65 and 240 min, respectively. May be prolonged with 0.75% bupivacaine. Complete motor blockade may not be observed at 0.25%	
0.5% Levobupivacaine	0.5–1.0	0.2	5–15	1–1.5	Duration of complete motor blockade and ataxia was 30 and 180 min, respectively. Complete motor blockade may not be observed at 0.25%	
0.75% Ropivacaine	1.65	0.22	7–15	1.5–2.5	Duration of motor blockade 90–150 min	
Opioids						
Morphine PF	0.1	0.1 for abdominal and pelvic procedures; 0.25 for thoracic procedures	45–90	12–24 for pelvic limb and abdominal procedures at 0.1 mL kg^{-1}; 5–6 for thoractomy procedures at 0.25 mL kg^{-1}	Reduced minimum alveolar concentration (MAC) by 30% and minimized CV depression from inhalant. Potential for urinary retention and pruritus	(4, 25, 60)
Buprenorphine	0.004	0.2	<60	Up to or greater than 24	Reduced risk for urinary retention	(63)
Local anesthetics and opioids						
Morphine PF and 0.5% bupivacaine	0.1 and 0.5–1.0	0.22	<15	Up to 24	67% return to normal motor function within 8 h. Potential for urinary retention	(125, 129)
Oxymorphone and 0.75% bupivacaine	0.1 and 1.0	0.2	<15	Up to 24	Decreases in heart rate. Transient hypotension. Systemic absorption of epidural oxymorphone is high	(24, 88)
Buprenorphine and 0.5% bupivacaine	0.004 and 1.0	0.2	<30	Up to 24	Low incidence of urinary retention	(130)
Alpha$_2$-Adrenoreceptor agonists						
Dexmedetomidine	0.003–0.006	0.25	<15	Up to 4.5	Dose-dependent MAC reduction up to 4.5 h. Bradycardia and elevated blood pressure may occur. Minimal effects on motor function	(110, 111)
Alpha$_2$-Adrenoreceptor agonists and local anesthetics						
Dexmedetomidine and 0.5% bupivacaine	0.004 and 1.0	0.22	<15	Up to 24	Less urinary retention when compared to opioid epidurals Prolonged motor blockade compared to local anesthetic and opioid epidurals	(128)

The prevalence of cardiovascular adverse effects has been reported in the literature (6). For example, mild cardiovascular depression developed in 20 of 219 dogs receiving a lumbosacral epidural injection of bupivacaine and morphine during inhalation anesthesia (6). Morphine does not cause sympathetic blockade and therefore it is likely that the cardiovascular depression reported in that study was associated with the blockade of sympathetic ganglia induced by the local anesthetic. Epidural administration of a large volume of 0.75% ropivacaine (1.65 mg kg^{-1} or 0.22 mL kg^{-1}) at the *thoracic* level was also shown to produce hypotension and bradycardia in isoflurane-anesthetized dogs (17, 52). Transient hypotension and bradycardia have also been reported in two obese dogs after lumbosacral epidural administration of local anesthetics (16). Although the volume injected into the epidural space was 0.2 mL kg^{-1} in both dogs, the total volume administered was relatively high (8 and 11 mL) (16). Cardiovascular depression may be observed due to the spread of local anesthetic to the upper four thoracic segments, low right-sided cardiac filling pressure, high epidural pressure raising cerebrospinal fluid (CSF) pressure resulting in increased vasomotor tone, and due to undetected hypovolemia and/or rapid and high volume injection of drugs into the epidural space (16). There is usually a recommendation by veterinarians without scientific support of limiting total injectate volume of local anesthetics to a maximum of 6 mL in dogs weighing more than 30 kg (10). However, larger volumes have been reported (8.6 mL) without complications (53). The authors have also used higher volumes (e.g., 8 mL) of epidural injectate in large canine breeds without complications but one should consider that smaller volume per kg of bodyweight (less than 0.2 mL kg^{-1}) is often enough for pelvic limb surgery.

Post-epidural hypotension during combined epidural/general anesthesia (systolic arterial blood pressure < 90 mmHg or mean arterial pressure < 60 mmHg) might be treated by increasing intravenous crystalloid administration and reducing vaporizer concentrations. In humans, post-epidural hypotension was found to be very frequent (55–90%) in patients undergoing cesarean section and spinal anesthesia (54). A crystalloid or colloid preload alone was ineffective in preventing hypotension following the administration of the local anesthetic (17, 54). Ephedrine (0.05 mg kg^{-1} IV), a non-selective agonist of beta and alpha receptors increases mean arterial blood pressure due to increases in systemic vascular resistance and cardiac output, may be effective in the treatment of post-epidural hypotension. Ephedrine has the advantages of being inexpensive and convenient to use as an intravenous (IV) bolus. Given the transient nature of hypotension observed after epidural anesthesia, the authors have found that a single dose is adequate in most cases. Ephedrine significantly increased cardiac index and stroke volume index but not arterial blood pressure in 5 out 6 dogs that had been treated with an epidural with morphine–bupivacaine (55) but one has to consider that this was a small clinical study where different anesthetic protocols had been administered. If ephedrine is ineffective to treat post-epidural hypotension, dopamine can be administered as a constant rate infusion (5–10 μg kg min^{-1}) as the rate can be adjusted according to individual blood pressure responses. If bradycardia-induced hypotension is detected, a reflex as a result of either low right heart filling pressures or from the generation of high extradural pressures following an epidural, it can be treated with the administration of an anticholinergic such as atropine (0.02–0.04 mg kg^{-1} IV).

Drugs

Table 1 shows the local anesthetics used for epidural anesthesia in dogs.

Lidocaine is an amide type local anesthetic that can be administered into the lumbosacral epidural space to produce a rapid desensitization with good muscle relaxation (46). Epidural administration of 0.25 mL kg^{-1} of 2% lidocaine with epinephrine (1:200,000) results in shorter latency period and shorter duration of desensitization, than 0.25 mL kg^{-1} of 0.5% bupivacaine with epinephrine (1:200,000) (46). Because the duration of sensory blockade is longer following epidural bupivacaine 0.5% alone (110 ± 14 min) or bupivacaine 0.5%–lidocaine 2% (94 ± 8 min) when compared with lidocaine alone (54 ± 5 min), the use of epidural bupivacaine either alone or combined with lidocaine may be preferred during prolonged surgeries (46). Onset of analgesia has been shown to be clinically similar following epidural administration of 2% lidocaine alone (5 ± 1 min) or in combination with 0.5% bupivacaine (7 ± 1 min) (46).

Bupivacaine is another amide type local anesthetic with a prolonged duration of effect compared with lidocaine (56) and may induce ventricular arrhythmias and cardiotoxicity at lower doses than lidocaine (57) but this is usually not a concern when doses and concentrations stated in this review are followed. At clinically used doses, epidural bupivacaine has limited effects on ventilation and oxygenation (24). Delayed hair growth following epidural bupivacaine has been reported in dogs (6, 14). However, this is observed following any epidural injection and not only with bupivacaine. Other factors such as season, housing, diet, or care of the coat may contribute to delayed hair growth following lumbosacral epidurals in dogs (14). The epidural administration of 0.2 mL kg^{-1} or 0.22 mL kg^{-1} of 0.25% or 0.75% bupivacaine results in analgesia to the level of the third or fourth lumbar vertebra, respectively (40, 58). A larger volume (0.31 mL kg^{-1}) is required to anesthetize up to the eleventh thoracic vertebra (59) but this volume may not be recommended for clinical use because sympathetic blockade followed by hypotension may result especially during the combine used of epidural anesthesia with inhalational anesthetics. The onset of motor blockade after lumbosacral administration of 0.5% bupivacaine (1.8 mg kg^{-1}) is approximately 2–4 min and lasts about 2–3 h (60).

Ropivacaine is a highly lipid soluble local anesthetic with the advantage of being less arrhythmogenic and toxic for the CNS and cardiovascular system than bupivacaine. Of interest, both drugs have similar pharmacokinetic profiles (61–64). It is believed that ropivacaine causes less motor blockade than bupivacaine in humans (65) and dogs (29, 66), but this has not been confirmed in clinical trials using dogs. To obtain a successful blockade (>80%) up to dermatomes T13–L1, a volume of 0.22 mL kg^{-1} using the concentration of 0.75% of bupivacaine or ropivacaine has been recommended (30).

Levobupivacaine is the S (-) isomer of bupivacaine and produced longer duration of motor block when compared with racemic bupivacaine in dogs (67). However, the dose-dependent analgesic and motor-blocking effects of epidurally administered levobupivacaine and racemic bupivacaine (0.25, 0.5, and 0.75%) were similar in conscious dogs (68). Like ropivacaine, levobupivacaine is less arrhythmogenic and cardiotoxic than bupivacaine (63).

Adjuvants Drugs without Intrinsic Analgesic Effect in Combination with Local Anesthetics

The duration and intensity of desensitization produced by the epidural technique will be influenced by the addition of adjuvant drugs without intrinsic analgesic effect. Epinephrine (1:200,000 or 5 μg mL^{-1}) aims to increase the duration of action and reduce the incidence of adverse effects by reducing systemic absorption of the local anesthetic from the injection site (69–71). In dogs, the epidural administration of mepivacaine with epinephrine (1:200,000) exerted considerable influence on the azygos venous blood uptake: peak concentrations of mepivacaine were reduced by two-thirds and systemic absorption of the drug was delayed. The azygos vein collects most of the blood draining from the internal vertebral venous plexus; therefore, the concentration of local anesthetics in this vein is an approximation of the total uptake by the epidural plexus (69). However, this effect may not be seen when epinephrine is added to long acting local anesthetics such as bupivacaine and ropivacaine, presumably because the greater solubility of these drugs causes them to bind avidly to tissues. In one study, the sensory and motor blockade produced by epidural administration of 0.75% bupivacaine or 1% ropivacaine in dogs was not further prolonged when epinephrine (1:200,000) was added to the local anesthetic solutions. However, epinephrine reduced the average anesthetic blood concentration observed in both treatment groups (62) possibly decreasing the likelihood of systemic toxicity. In a more recent study, dogs undergoing tibial plateau leveling osteotomy that received an epidural combination of ropivacaine, sufentanil, and epinephrine had longer lasting analgesia with less adverse effects when compared to epidural ropivacaine alone or combined with sufentanil (71). Therefore, the addition of epinephrine (1:200,000) to local anesthetics is an alternative to prolong duration of action of standard formulations of local anesthetics. However, in some cases, this may be a disadvantage especially if motor tone and normal ambulation are impaired hours after surgery.

Opioids

Opioids act by binding to opioid receptors that are located in thenervous system, spinal cord, and peripheral tissues. Epidural or spinal administration of opioids relieve somatic and visceral pain by selectively blocking nociceptive impulses without interfering with sensory and motor function or depressing the sympathetic nervous system (72). The pharmacodynamic profile is based on their receptor specificity, affinity, potency, and efficacy at the different opioid receptors and has been discussed elsewhere (73, 74).

Drugs

Highly hydrophilic drugs, such as morphine, produce analgesia 30–40 min after epidural administration and have long duration of effect due to the delayed systemic absorption and longer maintenance in the spinal cord (75, 76). Highly lipophilic drugs, such as fentanyl and meperidine have a shorter latency and duration of action (73, 77, 78). Lipophilicity can increase uptake into the adipose tissue of the epidural space, the degree of which will depend on the quantity of fat around the epidural space. Therefore, the tendency of drugs with high lipid partition coefficients to bind into the lipid phase leads to drug sequestration in lipid layers.

In dogs, epidural administration of 0.1 mg kg^{-1} of morphine reduced the minimum alveolar concentration (MAC) of halothane by 30% and minimized the hemodynamic depression produced by the inhalant agent (4, 79). Since morphine is more soluble in water than in fat, epidural administration of this opioid results in relatively high concentrations of the drug in the CSF (76). The tendency for epidural morphine to accumulate in the CSF rather than in fat results in delayed systemic absorption and facilitates the cranial spread of this opioid in the vertebral canal (76). These unique pharmacokinetic properties of epidural morphine (long residence time in the CSF and delayed systemic absorption) explain why lower doses of this opioid administered epidurally (0.1 mg kg^{-1}) will often result in more prolonged analgesic effects than those observed after systemic administration of higher doses and the extension of the analgesic effects to more cranial dermatomes, including the chest wall and thoracic limbs (75, 76, 80). In one study, epidural morphine (0.1 mg kg^{-1}) produced similar analgesia when compared with intercostal bupivacaine during the first 24 h after thoracotomy (75) and both techniques are often combined by the authors for this type of surgery. Epidural administration of morphine may not be used as a single analgesic technique to treat severe postoperative pain. In the two studies comparing the postoperative analgesic effects produced by epidural morphine (0.1–0.2 mg kg^{-1}) in combination with bupivacaine with those produced by epidural morphine alone (0.1–0.2 mg kg^{-1}) or epidural saline (placebo), the opioid-local anesthetic combination was always associated with longer lasting analgesia and improved pain scores than the other epidural treatments (8, 25). Epidural administration of morphine also provides analgesia for castration procedures (81). However, the authors do not recommend this technique for routine neutering (castration or ovariohysterectomy) since pain after neutering is well controlled with the administration of nonsteroidal anti-inflammatory drugs, parenteral opioids, and other local anesthetic blocks (intratesticular or intraperitoneal blocks).

Buprenorphine is a promising drug for epidural analgesia (82) with similar analgesic effect and duration of action when compared to morphine in dogs undergoing stifle surgery (83). Epidural buprenorphine may be useful for some animals because it does not appear to induce the urinary retention due to detrusor muscle relaxation seen with epidural morphine (84). It is apparent that in dogs and humans (83, 85), epidural buprenorphine may induce long lasting analgesia of similar duration of that provided by epidural morphine, but with fewer side effects. However, additional studies are necessary to confirm if the epidural administration of this opioid can replace epidural morphine in dogs.

Other opioids including different formulations (e.g., methadone, liposome-based sustained-release preparations of morphine, oxymorphone, butorphanol, and tramadol) have been administered by the epidural route in the experimental and clinical setting but do not provide consistent and additional information that would replace or change current practices (86–92).

Spinal (Intrathecal) versus Epidural Administration of Opioids

Spinal administration of opioids is not commonly used in small animals; however, it may provide some advantages over the epidural route such as lower intraoperative analgesic failure and faster return to normal motor function (93). A case report described the combination of spinal administration of fentanyl (0.77 μg kg^{-1}) and bupivacaine (0.31 mg kg^{-1}) and epidural administration of morphine (0.15 mg kg^{-1}) for tail resection and surgical exploration of the pelvic canal in a dog (94). In addition, a sparing effect of a low dose of spinal morphine (0.03 mg kg^{-1}) on intraoperative fentanyl requirements in dogs undergoing spinal surgery has been reported (95) but one should consider that dogs in the control group were intentionally given a higher loading dose and constant rate infusion of fentanyl when compared with the morphine group. The observer was also aware of the treatments administered (95). In both studies (94, 95), the authors reported good intra- and postoperative analgesia without neurological complications. However, spinal administration of opioids may not provide any clinical advantages over a single epidural injection of bupivacaine and/or morphine in most of the cases.

The anatomical location in which the opioid has been administered may also affect duration of analgesia. In one study, 0.1 mg kg^{-1} of morphine administered at T5–T6 provided significantly longer analgesia (9.9 ± 1.6 h) with minimal side effects when compared with its administration at L6–L7 (5.8 ± 0.8 h) following thoracotomy (80). This difference in duration of action may be attributed to its "local" effect when administered adjacent to the site of pain modulation. The use of topical epidural morphine administered at the site of injury *via* gelfoam, gelatin sponge, or "splash" technique resulted in varying effects on analgesia during hemilaminectomy in dogs (96–98). The administration of morphine (0.1 mg kg^{-1}) or oxymorphone (0.1 mg kg^{-1}) on a gelfoam pad then placed topically in the epidural space at the surgical site is used at the authors' institutions in dogs undergoing hemilaminectomy as part of a multimodal analgesic protocol.

Side Effects and Their Treatment

Side effects are not common but may occur following administration of opioids by the epidural route. Mild respiratory depression has been reported after epidural administration of 0.1 mg kg^{-1} of morphine in dogs (6) but this is not clinically relevant (38) if assisted or mechanical ventilation is promptly instituted. Pruritus has also been reported following epidural and intrathecal morphine administration (99–101). The use of a 5-hydroxytryptamine antagonists (e.g., ondansetron) has been described as treatment for neuraxial opioid-induced pruritus (102). Myoclonic activity upon recovery from anesthesia was also reported following intrathecal administration of morphine in dogs (15, 100) but this is not a clinical issue with epidural injections, providing that the maximum dose of 0.1 mg kg^{-1} is respected and accidental puncture of the subarachnoid space does not take place during needle insertion (103). Studies showed that approximately 3% of dogs receiving epidural morphine showed urinary retention postoperatively (6, 104). The prevalence of urinary retention increases when morphine is administered epidurally for postoperative pain control *via* indwelling epidural catheter. In this case, abnormal urinary patterns were observed in up to 39% of canine individuals (105). In practice, gentle bladder expression after surgery and before patient extubation avoids atony of the detrusor muscle and urinary retention (15).

Epidural administration of opioids seems to be a valuable tool for treatment of postoperative pain mainly as part of a balanced analgesia protocol and not as a sole technique. The dose, time to effect, and duration of action of epidural administration of commonly used opioids are described in Table 1.

Agonists of Alpha$_2$-Adrenergic Receptors

These drugs are potent sedatives and analgesics that bind to noradrenergic receptors of the spinal cord, inhibiting thetransmission of the afferent nociceptive impulses, by pre- and postsynaptic membrane hyperpolarization and inhibition of norepinephrine and substance P release (5, 106). The use of lower doses of epidural alpha$_2$ agonists minimizes the systemic side effects and produces a more effective antinociception when compared with their systemic use (5, 106).

Drugs

The MAC of isoflurane was reduced by 8, 22, and 33%, after epidural administration of 0.1, 0.2, and 0.4 mg kg^{-1} of xylazine, respectively (107). All doses of xylazine decreased the heart rate, but only the highest dose caused second-degree atrioventricular block (107). Epidural xylazine was also shown to provide longer postoperative analgesia when compared with epidural romifidine, dexmedetomidine, detomidine, and clonidine in dogs (108). This can be attributed to xylazine's local-like anesthetic properties and its low lipophilicity when compared with other $alpha_2$ agonists.

Medetomidine might produce dose-dependent antinociception when administered by the epidural route. Postoperative analgesia was observed for up to 90 min after epidural administration of 10 μg kg^{-1} of medetomidine in dogs (109) whereas 5 μg kg^{-1} of the drug administered by the epidural route did not provide analgesia when a towel clamp was applied to the tail (106). Epidural administration of medetomidine (15 μg kg^{-1}) also caused bradycardia, atrioventricular block, and transient hypertension (5). Epidural dexmedetomidine (1 μg kg^{-1}), the active isomer of medetomidine with a high $alpha_2$-selectivity and efficacy, produced an antinociceptive effect and reduced heart rate, without changes in carbon dioxide response, behavior, and motor function (110). Epidural administration of dexmedetomidine to isoflurane-anesthetized dogs decreased the MAC of the inhalant anesthetic in a dose-dependent fashion (111).

Epidural administration of these drugs will normally cause mild to moderate cardiovascular changes which may or may not be tolerated depending on the health status of the patient (112). In addition, there are concerns of using multi-vial drugs (risk of contamination and neurotoxicity) with preservatives for epidural administration. A study showed that epidural administration of preservative-free dexmedetomidine to rabbits may produce spinal neurotoxicity, characterized by demyelination of the oligodendrocytes in the white matter (113). The neurotoxic effect was attributed to vasoconstriction of the medullary spinal vessels and pH of dexmedetomidine that can cause demyelination (113). The prolonged duration of block may be also an issue after administration of these drugs in combination with local anesthetics (see below). The dose, time to effect, and duration of action of epidural administration of commonly used $alpha_2$-adrenoreceptor agonists are described in Table 1.

Dissociative Anesthetics

Ketamine is a non-competitive antagonist of N-methyl D-aspartate receptors involved in the transmission and modulation of nociceptive stimuli in the dorsal horn of the spinal cord, which plays a key role in chronic pain, reducing spinal cord hyperexcitability (114–116). Ketamine has intrinsic local anesthetic effects, producing motor, sensory, and sympathetic block when administered by the epidural route (115). Epidural administration of ketamine (2 mg kg^{-1}) caused minimal hemodynamic effects in dogs (114) and effectively reduced post-incisional hyperalgesia in the hind limbs for 12 h in dogs (117), but was more effective when administered preemptively (118). The advantages of epidural administration of ketamine over the systemic route in human medicine are controversial (119). Also, the efficacy of epidural racemic or ketamine S (+) has rarely been reported in veterinary medicine (117, 118). Preservative-free ketamine (S+) resulted in toxic damage to both the spinal cord and nerve roots in both rabbits and humans following repeated intrathecal administration (120, 121). Indeed, ketamine should not be administered by the epidural or intrathecal route until further studies report its neurotoxic effects in dogs.

Opioid and Local Anesthetic Combinations

The combination of opioids with local anesthetics may be beneficial because affinity of opioid drugs for their receptors in the spinal cord is increased by local anesthetics (122). The combination of opioids and local anesthetics decreased anesthetic requirements, postoperative analgesic requirements, and plasma cortisol concentrations and improved pain scores in dogs undergoing surgery when compared with dogs that received an opioid alone (6, 8, 24, 25, 123, 124). The combination morphine–bupivacaine provides similar analgesia to femoral-sciatic nerve blocks or constant rate infusions of analgesics in dogs undergoing orthopedic surgeries (125–128). However, the drug combination may result in more intraoperative hypotension and urinary retention than other forms of local regional techniques due to the effects of local anesthetics and morphine on vascular and bladder tone, respectively (125). Epidural administration of morphine (0.1 mg kg^{-1}) and 2% lidocaine (0.1 mL kg^{-1}) may be an option when shorter duration of motor blockade is desired (129). If urinary retention is a concern, the administration of epidural 0.5% bupivacaine (1 mg kg^{-1}) and buprenorphine (0.004 mg kg^{-1}) provided up to 24 h of analgesia in dogs undergoing stifle arthroplasty with a low incidence (1 out of 13) of urinary retention (130).

Other combinations using local anesthetics and different opioids have been reported (e.g., methadone–lidocaine, morphine–ropivacaine, tramadol–lidocaine, oxymorphone–bupivacaine) but it seems that none of these protocols provided clear benefits over the administration of morphine–bupivacaine (64, 88, 127, 131, 132). However, they may be useful when the latter two drugs are not available.

The administration of local anesthetics and opioids can be also performed *via* a constant rate infusion using an epidural catheter. In this case, bupivacaine should be diluted to a concentration less than or equal to 0.125% and will avoid motor weakness than when higher concentrations are used. When combined in a 50:50 mixture with morphine (1 mg mL^{-1}), 0.25% bupivacaine is diluted to 0.125% and when administered *via* an indwelling epidural catheter to deliver 0.3 mg kg^{-1} per day of morphine, the total amount of bupivacaine delivered to the patient is 0.75 mg kg^{-1} per day (12).

Alpha$_2$-Adrenoreceptor Agonist and Opioid or Local Anesthetic Combinations

The combination of alpha$_2$-adrenoreceptor agonists and opioids into the epidural space increases magnitude of analgesia. This synergism may be due to the activation of alpha$_2$-adrenoreceptors on opiate-containing interneurons, increasing the release of opioid peptides (133) and also due to the prolongation of the increased plasma morphine concentration produced by medetomidine (134). The duration of morphine analgesia was increased by almost twofold when morphine was administered with medetomidine in dogs (106). According to the authors, this may have been due to the waning alpha$_2$-adrenoreceptor-mediated vasoconstriction associated with medetomidine in epidural vasculature. Morphine present in the spinal cord, CSF, and epidural tissues may have had increased uptake into systemic circulation following the relative vasodilation of epidural vessels as the medetomidine effect subsided (134).

The addition of 0.05% dexmedetomidine (4 μg kg^{-1}) to 0.5% bupivacaine (1 mg kg^{-1}) provided adequate analgesia; however, prolonged duration of motor blockade was a clinical concern following pelvic orthopedic surgeries in dogs (126). Time to first urination was earlier but pain scores worsened at 1 h following surgery when compared with dogs receiving epidural bupivacaine–morphine for similar procedures (126). The combination of alpha$_2$ agonists with opioids or local anesthetics might be a good option for pain management in healthy dogs that are in severe pain. However, the same concerns described with the administration of agonists of alpha$_2$ receptors alone should be taken in consideration.

Dissociative Anesthetic and Opioid or Local Anesthetic Combinations

Ketamine might have an analgesic sparing effect when combined with local anesthetics and opioids (115). However, there was no potential advantage when opioids and ketamine were combined in humans (116) neither when ketamine (2.5 mg kg^{-1}) was administered with meperidine (2 mg kg^{-1}) by the epidural route of administration in dogs (135). On the other hand, levorotatory ketamine (ketamine S+) (0.5 or 1 mg kg^{-1}) combined with morphine (0.025 or 0.05 mg kg^{-1}) provided satisfactory intra- and postoperative analgesia with cardiorespiratory stability in dogs undergoing ovariohysterectomy (136). However, pain associated with ovariohysterectomy is moderate and the advantages of the use of epidural ketamine alone or combined with other drugs should be an objective of further studies.

Doses, injection volumes, onset period, and duration of epidural anesthesia or analgesia produced by combinations of drugs are described in Table 1.

Other drugs such as midazolam, ketorolac, and neostigmine have been administered epidurally in dogs in the experimental and clinical setting (137–141). None of these drugs have been shown to be promising safe and alternative analgesic techniques for clinical use.

CONCLUSION

The potential for cardiorespiratory compromise, neurotoxicity, and other side effects should be considered when drugs are to be administered into the epidural space. These drugs should be selected according to the potency, onset and duration of action, surgical procedure, cardiorespiratory changes, and local and systemic side effects. When these variables are considered, the epidural route is useful as a complementary method of anesthesia, preemptive and postoperative analgesia, and/or as part of a balanced anesthesia technique.

AUTHOR CONTRIBUTIONS

PS, BS, FN, and SL have contributed to gathering data and writing of the manuscript; have thoroughly evaluated the final product and checked all references; PS and BS assisted with the manuscript publication process.

CONFLICT OF INTEREST STATEMENT

The authors declare that the research was conducted in the absence of any commercial or financial relationships that could be construed as a potential conflict of interest.

FUNDING

The open access publishing fees for this article have been covered in part by the Texas A&M University Open Access to Knowledge Fund (OAKFund), supported by the University Libraries and the Office of the Vice President for Research.

REFERENCES

1. Ismail ZB. Epidural analgesia in cattle, buffalo, and camels. *Vet World* (2016) 9:1450–5.
2. Jones RS. Epidural analgesia in the dog and cat. *Vet J* (2001) 161:123–31.
3. Natalini CC. Spinal anesthetics and analgesics in the horse. *Vet Clin North Am Equine Pract* (2010) 26:551–64.
4. Valverde A, Dyson DH, Mc Donell WN. Epidural morphine reduces halothane MAC in the dog. *Can J Anaesth* (1989) 36:629–32.
5. Vesal N, Cribb PH, Frketic M. Postoperative analgesic and cardiopulmonary effects in dogs of oxymorphone administered epidurally and intramuscularly, and medetomidine administered epidurally: a comparative clinical study. *Vet Surg* (1996) 25:361–9.
6. Troncy E, Junot S, Keroack S, Sammut V, Pibarot P, Genevois JP, et al. Results of preemptive epidural administration of morphine with or without bupivacaine in dogs and cats undergoing surgery: 265 cases (1997-1999). *J Am Med Assoc* (2002) 221:666–72.
7. Stanek B, Schwartz M, Zimpfer M, Raberger G. Plasma concentrations of noradrenaline and adrenaline and plasma rennin activity during extradural blockade in dogs. *Brit J Anaesth* (1980) 52:305–11.
8. Hendrix PK, Raffe MR, Robinson EP, Felice LJ, Randall DA. Epidural administration of bupivacaine, morphine, or their combination for postoperative analgesia in dogs. *J Am Vet Med Assoc* (1996) 209:598–607.
9. Sibanda S, Hughes JM, Pawson PE, Kelly G, Bellenger CR. The effects of preoperative extradural bupivacaine and morphine on the stress response in dogs undergoing femoro-tibial joint surgery. *Vet Anaesth Analg* (2006) 33:246–57.
10. Torske KE, Dyson DH. Epidural analgesia and anesthesia. *Vet Clin North Am Small Anim Pract* (2000) 30:859–74.
11. Liotta A, Busoni V, Carrozzo MV, Sandersen C, Gabriel A, Bolen G. Feasibility of ultrasound-guided epidural access at the lumbo-sacral space in dogs. *Vet Radiol Ultrasound* (2015) 56:220–8.
12. Hansen BD. Epidural catheter analgesia in dogs and cats: technique and review of 182 cases (1991-1999). *J Vet Emerg Crit Care*(2001) 11:95–103.

13. Bosmans T, Schauvliege S, Gasthuys F, Marcilla MG, Polis I. Transient unilateral Horner's syndrome after epidural ropivacaine in a dog. *Vet Anaesth Analg* (2009) 36:401–6.
14. Kalchofner Guerrero KS, Guerrero TG, Schweizer-Kölliker M, Ringer SK, Hässig M, Bettschart-Wolfensberger R. Incidence of delayed hair re-growth, pruritus, and urinary retention after epidural anaesthesia in dogs. *Tierarztl Prax Ausg K Kleintiere Heimtiere* (2014) 42:94–100.
15. Kona-Boun JJ, Pibarot P, Quesnel A. Myoclonus and urinary retention following subarachnoid morphine injection in a dog. *Vet Anaesth Analg* (2003) 30:257–64.
16. Iff I, Moens Y. Two cases of bradyarrhythmia and hypotension after extradural injections in dogs. *Vet Anaesth Analg* (2008) 35:265–9.
17. Bosmans T, Schauvliege S, Gasthuys F, Duchateau L, Steblaj B, Gadeyne C, et al. Influence of a preload of hydroxyethylstarch 6% on the cardiovascular effects of epidural administration of ropivacaine 0.75% in anaesthetized dogs. *Vet Anaesth Analg* (2011) 38:494–504.
18. Naganobu K, Maeda N, Miyamoto T, Hagio M, Nakamura T, Takasaki M. Cardiorespiratory effects of epidural administration of morphine and fentanyl in dogs anesthetized with sevoflurane. *J Am Vet Med Assoc* (2004) 224:67–70.
19. Remedios AM, Wagner R, Caulkett NA, Duke T. Epidural abscess and discospondylitis in a dog after administration of a lumbosacral epidural analgesic. *Can Vet J* (1996) 37:106–7.
20. King FG, Baxter AD, Mathieson G. Tissue reaction of morphine applied to the epidural space in dogs. *Can Anaesth Soc J* (1984) 31:268–71.
21. Sjoberg M, Karlsson P, Nordborg C, Wallgren A, Nitescu P, Appelgren L, et al. Neuropathologic findings after long-term intrathecal infusion of morphine and bupivacaine for pain treatment in cancer patients. *Anesthesiology* (1992) 76:173–86.
22. Luna SP, Cassu RN, Castro GB, Teixeira Neto FJ, Silva Júnior JR, Lopes MD. Effects of four anaesthetic protocols on the neurological and cardiorespiratory variables of puppies born by caesarean section. *Vet Rec* (2004) 154:387–9.
23. Hewitt SA, Brisson BA, Sinclair MD, Sears WC. Comparison of cardiopulmonary responses during sedation with epidural and local anesthesia for laparoscopic-assisted jejunostomy feeding tube placement with cardiopulmonary responses during general anesthesia for laparoscopic-assisted or open surgical jejunostomy feeding tube placement in healthy dogs. *Am J Vet Res* (2007) 68:358–69.
24. Torske KE, Dyson DH, Pettifer G. End tidal halothane concentration and postoperative analgesia requirements in dogs: a comparison between intravenous oxymorphone and epidural bupivacaine alone and in combination with oxymorphone. *Can Vet J*(1998) 39:361–9.
25. Kona-Boun JJ, Cuvelliez S, Troncy E. Evaluation of epidural administration of morphine or morphine and bupivacaine for postoperative analgesia after premedication with an opioid analgesic and orthopedic surgery in dogs. *J Am Vet Med Assoc* (2006) 229:1103–12.
26. Romano M, Portela DA, Breghi G, Otero PE. Stress-related biomarkers in dogs administered regional anaesthesia or fentanyl for analgesia during stifle surgery. *Vet Anaesth Analg* (2016) 43:44–54.
27. Valverde A. Epidural analgesia and anesthesia in dogs and cats. *Vet Clin North Am Small Anim Pract* (2008) 38:1205–30.
28. Hogan Q. Distribution of solution in the epidural space: examination by cryomicrotome section. *Reg Anesth Pain Med* (2002) 27:150–6.
29. Feldman HS, Covino BG. Comparative motor-blocking effects of bupivacaine and ropivacaine, a new amino amide local anesthetic in the rat and dog. *Anesth Analg* (1988) 67:1047–52.
30. Duke T, Caulkett NA, Ball SD, Remedios AM. Comparative analgesic and cardiopulmonary effects of bupivacaine and ropivacaine in the epidural space of the conscious dog. *Vet Anaesth Analg* (2000) 27:13–21.

31. Lee I, Yamagishi N, Oboshi K, Yamada H. Distribution of new methylene blue injected into the lumbosacral epidural space in cats. *Vet Anaesth Analg* (2004) 31:190–4.
32. Gorgi AA, Hofmeister EH, Higginbotham MJ, Kent M. Effect of body position on cranial migration of epidurally injected methylene blue in recumbent dogs. *Am J Vet Res* (2006) 67:219–21.
33. Son WG, Jang M, Yoon J, Lee LY, Lee I. The effect of epidural injection speed onepidural pressure and distribution of solution in anesthetized dogs. *Vet Anaesth Analg* (2014) 41:526–33.
34. Reina MA, Franco CD, López A, Dé Andrés JA, van Zundert A. Clinical implications of epidural fat in the spinal canal. A scanning electron microscopic study. *Acta Anaesthesiol Belg* (2009) 60:7–17.
35. Otero P, Tarragona L, Ceballos M, Portela D. Epidural cephalic spread of a local anesthetic in dogs: a mathematical model using the column length. In: Axiak S, Trim C, editors. *Vet Anaesth Analg 2010 37: Proceedings of the 10th World Congress of Veterinary Anaesthesia*. Glasgow, Scotland (2009). 35 p.
36. Lacassie HJ, Habib AS, Lacassie HP, Columb MO. Motor blocking minimum local anesthetic concentrations of bupivacaine, levobpivacaine, and ropivacaine in labor. *Reg Anesth Pain Med* (2007) 32:323–9.
37. Anand GV, Kannan M, Palaramakrishnan D. Differential blockade to assess surgical repair by intraoperative active mobilization in knee injuries-beyond labour analgesia. *Indian J Anaesth* (2011) 55:181–3.
38. Abelson AL, Armitage-Chan E, Lindsey JC, Wetmore LA. A comparison of epidural morphine with low dose bupivacaine versus epidural morphine alone on motor and respiratory function in dogs following splenectomy. *Vet Anaesth Analg* (2011) 38:213–23.
39. Ford DJ, Raj PP, Singh P, Regan KM, Ohlweiler D. Differential peripheral nerve block by local anesthetics in the cat. *Anesthesiology* (1984) 60:28–33.
40. Freire CD, Torres ML, Fantoni DT, Cavalcanti RL, Noel-Morgan J. Bupivacaine 0.25% and methylene blue spread with epidural anesthesia in dog. *Vet Anaesth Analg* (2010) 37:63–9.
41. Visser WA, Lee RA, Gielen MJ. Factors affecting the distribution of neural blockade by local anesthetics in epidural anesthesia and a comparison of lumbar versus thoracic epidural anesthesia. *Anesth Analg* (2008) 107:708–21.
42. Park WY, Poon KC, Massengale MD, Macnamara TE. Direction of the needle bevel and epidural anesthetic spread. *Anesthesiology*(1982) 57:327–8.
43. Klumpner TT, Lange EM, Ahmed HS, Fitzgerald PC, Wong CA, Toledo P. An in vitro evaluation of the pressure generated during programmed intermittent epidural bolus injection at varying infusion delivery speeds. *J Clin Anesth* (2016) 34:632–7.
44. Son WG, Kim J, Seo JP, Yoon J, Choi M, Lee LY, et al. Cranial epidural spread of contrast medium and new methylene blue dye in sternally recumbent anaesthetized dogs. *Vet Anaesth Analg* (2011) 38:510–5.
45. Yokoyama M, Hanazaki M, Fujii H, Mizobuchi S, Nakatsuka H, Takahashi T, et al. Correlation between the distribution of contrast medium and the extent of blockade during epidural anesthesia. *Anesthesiology* (2004) 100:1504–10.
46. Cruz ML, Luna SPL, Clark RMO, Massone F, Castro GB. Epidural anaesthesia using lignocaine, bupivacaine or a mixture of lignocaine and bupivacaine in dogs. *Vet Anaesth Analg* (1997) 24:30–2.
47. Persson F. Epidural analgesia in dogs with special reference to intra-arterial blood pressure. *Acta Vet Scand* (1970) 11:186–96.
48. Warner DO, Brichant JF, Ritman EL, Rehder K. Epidural anesthesia and intrathoracic blood volume. *Anesth Analg* (1993) 77:135–40.
49. Savvas I, Anagnostou T, Papazoglou LG, Raptopoulos D. Successful resuscitation from cardiac arrest associated with extradural lidocaine in a dog. *Vet Anaesth Analg* (2006) 33:175–8.
50. Son WG, Jang M, Jo SM, Yoon J, Lee I. The volume effect of lidocaine on thoracic epidural anesthesia in conscious Beagle dogs. *Vet Anaesth Analg* (2015) 42:414–24.

51. Block BM, Liu SS, Rowlingson AJ, Cowan AR, Cowan JA Jr, Wu CL. Efficacy of postoperative epidural analgesia: a meta-analysis. *JAMA* (2003) 290:2455–63.
52. Bosmans T, Schauvliege S, Gasthuys F, Duchateau L, Marcilla MG, Gadeyne C, et al. Cardiovascular effects of epidural administration of methadone, ropivacaine 0.75% and their combination in isoflurane anaesthetized dogs. *Vet Anaesth Analg*(2011) 38:146–57.
53. Aarnes TK, Hubbell JA, Hildreth BE III. Use of sedation and ropivacaine-morphine epidural for femoral head and neck ostectomy in a dog. *J Small Anim Pract* (2014) 55:334–6.
54. Mercier FJ, Bonnet MP, De la Dorie A, Moufouki M, Banu F, Hanaf A, et al. Spinal anaesthesia for caesarean section: fluid loading, vasopressors and hypotension. *Ann Fr Anesth Reanim* (2007) 26:688–93.
55. Chen HC, Sinclair MD, Dyson D. Use of ephedrine and dopamine in dogs for the management of hypotension in routine clinical cases under isoflurane anesthesia. *Vet Anaesth Analg* (2007) 34:301–11.
56. Lawal FM, Adetunji A. A comparison of epidural anaesthesia with lignocaine, bupivacaine and a lignocaine-bupivacaine mixture in cats. *J S Afr Vet Assoc* (2009) 80:243–6.
57. Nath S, Haggmark S, Johansson G, Reiz S. Differential depressant and electrophysiologic cardiotoxicity of local anesthetics: an experimental study with special reference to lidocaine and bupivacaine. *Anesth Analg* (1986) 65:1263–70.
58. Heath RB. The practicality of lumbosacral epidural analgesia. *Semin Vet Med Surg (Small Anim)* (1986) 1:245–8.
59. Heath RB, Broadstone RV, Wright M. Using bupivacaine hydrochloride for lumbosacral epidural analgesia. *Compend Contin Educ Pract Vet* (1989) 11:50–5.
60. Franquelo C, Toledo A, Manubens J, Cristòfol C, Arboix M. Bupivacaine disposition and pharmacologic effects after intravenous and epidural administrations in dogs. *Am J Vet Res* (1995) 56:1087–91.
61. Arthur GR, Feldman HS, Covino BG. Comparative pharmacokinetics of bupivacaine and ropivacaine, a new amide local anesthetic. *Anesth Analg* (1998) 67:1053–8.
62. Hurley RJ, Feldman HS, Latka C, Arthur GR, Covino BG. The effects of epinephrine on the anesthetic and haemodynamic properties of ropivacaine and bupivacaine after epidural administration in the dog. *Reg Anesth Pain Manag* (1991) 16:303–8.
63. Leone S, Di Cianni S, Casati A, Fanelli G. Pharmacology, toxicology, and clinical use of new long acting local anesthetics, ropivacaine and levobupivacaine. *Acta Biomed* (2008) 79:92–105.
64. Albuquerque VB, Araújo MA, Ferreira GT, Fonseca MW, Arruda AM, ShiChen L, et al. Effects of ropivacaine combined with morphine at 0.15 and 0.2 mg kg(-1) in bitches undergoing epidural anesthesia. *Acta Cir Bras* (2015) 30:222–38.
65. De Negri P, Ivani G, Tirri T, Modano P, Reato C, Eksborg S, et al. A comparison of epidural bupivacaine, levobupivacaine, and ropivacaine on postoperative analgesia and motor blockade. *Anesth Analg* (2004) 99:45–8.
66. Feldman HS, Dvoskin S, Arthur GR, Doucette AM. Antinociceptive and motor-blocking efficacy of ropivacaine and bupivacaine after epidural administration in the dog. *Reg Anesth* (1996) 21:318–26.
67. Franquelo C, Toledo A, Manubens J, Valladares JE, Cristòfol C, Arboix M. Pharmacokinetics and pharmacologic effects of the s (-) isomer of bupivacaine after intravenous and epidural administration in dogs. *Am J Vet Res* (1999) 60:832–5.
68. Gomez de Segura IA, Menafro A, Garcia-Fernandez P, Murillo S, Parodi EM. Analgesic and motor-blocking action of epidurally administered levobupivacaine or bupivacaine in the conscious dog. *Vet Anaesth Analg* (2009) 36:485–94.
69. Burfoot MF, Bromage PR. The effects of epinephrine on mepivacaine absorption from the spinal epidural space. *Anesthesiology*(1971) 35:488–92.
70. Niemi G, Breivik H. The minimally effective concentration of adrenaline in a low-concentration thoracic epidural analgesic infusion of bupivacaine, fentanyl and

adrenaline after major surgery. A randomized, double-blind, dose-finding study. *Acta Anaesthesiol Scand* (2003) 47:439–50.

71. Adami C, Veres-Nyéki K, Spadavecchia C, Rytz U, Bergadano A. Evaluation of peri-operative epidural analgesia with ropivacaine, ropivacaine and sufentanil, and ropivacaine, sufentanil and epinephrine in isoflurane anesthetized dogs undergoing tibial plateau levelling osteotomy. *Vet J* (2012) 194:229–34.
72. Cousins MJ, Mather LE. Intrathecal and epidural administration of opioids. *Anesthesiology* (1984) 61:276–310.
73. Tung AS, Yaksh TL. The antinociceptive effects of epidural opiates in the cat: studies of the pharmacology and the effects of lipophilicity in spinal analgesia. *Pain* (1982) 12:343–56.
74. Pascoe PJ. Opioid analgesics. *Vet Clin North Am Small Anim Pract* (2000) 30:757–72.
75. Pascoe PJ, Dyson DH. Analgesia after lateral thoracotomy in dogs. Epidural morphine vs. intercostal bupivacaine. *Vet Surg* (1993) 22:141–7.
76. Valverde A, Conlon PD, Dyson DH, Burger JP. Cisternal CSF and serum concentrations of morphine following epidural administration in the dog. *J Vet Pharmacol Ther* (1993) 15:91–5.
77. Chrubasik J, Chrubasik S, Martin E. The ideal epidural opioid – fact or fantasy? *Eur J Anaesth* (1993) 10:79–100.
78. Saritas ZK, Saritas TB, Pamuk K, Korkmaz M, Demirkan I, Yaprakci MV, et al. Comparison of the effects of lidocaine and fentanyl in epidural anesthesia in dogs. *Bratisl Lek Listy* (2014) 115:508–13.
79. Valverde A, Dyson DH, Cockshutt JR, McDonell WN, Valliant AE. Comparison of the hemodynamic effects of halothane alone and halothane combined with epidurally administered morphine for anesthesia in ventilated dogs. *Am J Vet Res* (1991) 52:505–9.
80. Carregaro AB, Freitas GC, Lopes C, Lukarsewski R, Tamiozzo FS, Santos RR. Evaluation of analgesic and physiologic effects of epidural morphine administered at a thoracic or lumbar level in dogs undergoing thoracotomy. *Vet Anaesth Analg* (2014) 41:205–11.
81. Perez TE, Grubb TL, Greene SA, Meyer S, Valdez N, Bingman J, et al. Effects of intratesticular injection of bupivacaine and epidural administration of morphine in dogs undergoing castration. *J Am Vet Med Assoc* (2013) 242:631–42.
82. Steagall PV, Millette V, Mantovani FB, Gilbert P, Luna SP, Duke-Novakovski T. Antinociceptive effects of epidural buprenorphine or medetomidine, or the combination, in conscious cats. *J Vet Pharmacol Ther* (2009) 32:477–84.
83. Smith LJ, Yu JK. A comparison of epidural buprenorphine with epidural morphine for postoperative analgesia following stifle surgery in dogs. *Vet Anaesth Analg* (2001) 28:87–96.
84. Drenger B, Magora F. Urodynamic studies after intrathecal fentanyl and buprenorphine in the dog. *Anesth Analg* (1989) 69:348–53.
85. Wolff J, Carl P, Carwford ME. Epidural buprenorphine for postoperative analgesia. A controlled comparison with epidural morphine. *Anaesthesia* (1986) 41:76–9.
86. Popilskis S, Kohn D, Sanchez JA, Gorman P. Epidural vs. intramuscular oxymorphone analgesia after thoracotomy in dogs. *Vet Surg* (1991) 20:462–7.
87. Troncy E, Cuvelliez S, Blais D. Evaluation of analgesia and cardiorespiratory effects of epidurally administered butorphanol in isoflurane-anesthetized dogs. *Am J Vet Res* (1996) 57:1478–82.
88. Torske KE, Dyson DH, Conlon PD. Cardiovascular effects of epidurally administered oxymorphone and an oxymorphone-bupivacaine combination in halothane-anesthetized dogs. *Am J Vet Res* (1999) 60:194–200.
89. Yaksh TL, Provencher JC, Rathbun ML, Kohn FR. Pharmacokinetics and efficacy of epidurally delivered sustained-release encapsulated morphine in dogs. *Anesthesiology* (1999) 90:1402–12.

90. Yaksh TL, Provencher JC, Rathbun ML, Myers RR, Powell H, Richter P, et al. Safety assessment of encapsulated morphine delivered epidurally in a sustained-release multivesicular liposome preparation in dogs. *Drug Deliv* (2000) 7:27–36.
91. Leibetseder EN, Mosing M, Jones RS. A comparison of extradural and intravenous methadone on intraoperative isoflurane and postoperative analgesia requirements in dogs. *Vet Anaesth Analg* (2006) 33:128–36.
92. Hermeto LC, DeRossi R, Marques BC, Jardim PH. Potentiation of epidural lidocaine by co-administering tramadol by either intramuscular or epidural route in cats. *Can J Vet Res* (2015) 79:214–20.
93. Sarotti D, Rabozzi R, Franci P. Comparison of epidural versus intrathecal anaesthesia in dogs undergoing pelvic limb orthopaedic surgery. *Vet Anaesth Analg* (2015) 42:405–13.
94. Novello L, Corletto F. Combined spinal-epidural anesthesia in a dog. *Vet Surg* (2006) 35:191–7.
95. Novello L, Corletto F, Rabozzi R, Platt SR. Sparing effect of a low dose of intrathecal morphine on fentanyl requirements during spinal surgery: a preliminary clinical investigation in dogs. *Vet Surg* (2008) 37:153–60.
96. Wehrenberg A, Freeman L, Ko J, Payton M, Spivack R. Evaluation of topical epidural morphine for postoperative analgesia following hemilaminectomy in dogs. *Vet Ther* (2009) 10:E1–12.
97. Aprea F, Cherubini GB, Palus V, Vettorato E, Corletto F. Effect of extradurally administered morphine on postoperative analgesia in dogs undergoing surgery for thoracolumbar intervertebral disk extrusion. *J Am Vet Med Assoc* (2012) 241:754–9.
98. Barker JR, Clark-Price SC, Gordon-Evans WJ. Evaluation of topical epidural analgesia delivered in gelfoam for postoperative hemilaminectomy pain control. *Vet Surg* (2013) 42:79–84.
99. Bauquier SH. Hypotension and pruritus induced by neuraxial anaesthesia in a cat. *Aust Vet J* (2012) 90:402–3.
100. Iff I, Valeskini K, Mosing M. Severe pruritus and myoclonus following intrathecal morphine administration in a dog. *Can Vet J*(2012) 53:983–6.
101. Gent T, Iff I, Bettschart-Wolfensberger R, Mosing M. Neuraxial morphine induced pruritus in two cats and treatment with sub anaesthetic doses of propofol. *Vet Anaesth Analg* (2013) 40:517–20.
102. Borgeat A, Stirnemann HR. Ondansetron is effective to treat spinal or epidural morphine-induced pruritus. *Anesthesiology* (1999) 90:432–6.
103. da Cunha AF, Carter JE, Grafinger M, Montgomery H, Marks SL, Posner LP, et al. Intrathecal morphine overdose in a dog. *J Am Vet Med Assoc* (2007) 230:1665–8.
104. Peterson NW, Buote NJ, Bergman P. Effect of epidural analgesia with opioids on the prevalence of urinary retention in dogs undergoing surgery for cranial cruciate ligament rupture. *J Am Vet Med Assoc* (2014) 244:940–3.
105. Phillips LR, McAbee KP, Stephenson N, Stanke NJ, Booms ML, Degner DD. Evaluation of complications and feasibility of indwelling epidural catheter use for post-operative pain control in dogs in the home environment. *N Z Vet J* (2015) 63:86–91.
106. Branson KR, Ko JCH, Tranquilli WJ, Benson J, Thurmon JC. Duration of analgesia induced by epidurally administered morphine and medetomidine in dogs. *J Vet Pharmacol Ther* (1993) 16:369–72.
107. Soares JHN, Ascoli FO, Gremiao IDF, Gomez de Segura IA, Marsico Filho F. Isoflurane sparing action of epidurally administered xylazine hydrochloride in anesthetized dogs. *Am J Vet Res* (2004) 65:854–9.
108. Pohl VH, Carregaro AB, Lopes C, Gehrcke MI, Muller DC, Garlet CD. Epidural anesthesia and postoperatory analgesia with alpha-2 adrenergic agonists and lidocaine for ovariohysterectomy in bitches. *Can J Vet Res* (2012) 76:215–20.
109. Aithal H, Amarpal P, Suryawanshi S. Analgesic and cardiopulmonary effects of pethidine and medetomidine after epidural administration in dogs. *Indian J Anim Sci* (1999) 69:790–4.

110. Sabbe MB, Penning JP, Ozaki GT, Yaksh TL. Spinal and systemic action of the α_2 receptor agonist dexmedetomidine in dogs. *Anesthesiology* (1994) 80:1057–72.
111. Campagnol D, Teixeira Neto FJ, Giordano T, Ferreira TH, Monteiro ER. Effects of epidural administration of dexmedetomidine on the minimum alveolar concentration of isoflurane in dogs. *Am J Vet Res* (2007) 68:1308–18.
112. Martin-Bouyer V, Schauvliege S, Duchateau L, Bosmans T, Gasthuys F, Polis I. Cardiovascular effects following epidural injection of romifidine in isoflurane-anaesthetized dogs. *Vet Anaesth Analg* (2010) 37:87–96.
113. Konakci S, Adanir T, Yilmaz G, Rezanko T. The efficacy and neurotoxicity of dexmedetomidine administered via the epidural route. *Eur J Anaesth* (2008) 25:403–9.
114. Martin DD, Tranquilli WJ, Olson WA, Thurmon JC, Benson GJ. Hemodynamic effects of epidural ketamine in isoflurane-anesthetized dogs. *Vet Surg* (1997) 26:505–9.
115. Schmid RL, Sandler AN, Katz J. Use and efficacy of low-dose ketamine in the management of acute postoperative pain: a review of current techniques and outcomes. *Pain* (1999) 82:111–25.
116. Subramaniam B, Subramaniam K, Pawar DK, Sennaraj B. Preoperative epidural ketamine in combination with morphine does not have a clinically relevant intra- and postoperative opioid-sparing effect. *Anesth Analg* (2001) 93:1321–6.
117. Duque JC, Valadao CA, Farias A, De Almeida RM, Oleskovicz N. Pre-emptive epidural ketamine or S (+) ketamine in post-incisional pain in dogs: a comparative study. *Vet Surg* (2004) 33:361–7.
118. Hamilton SM, Johnston SA, Broadstone R. Evaluation of analgesia provided by the administration of epidural ketamine in dogs with a chemically induced synovitis. *Vet Anaesth Analg* (2005) 32:30–9.
119. Xie H, Wang X, Liu G, Wang G. Analgesic effects and pharmacokinetics of a low dose of ketamine preoperatively administered epidurally or intravenously. *Clin J Pain* (2003) 19:317–22.
120. Vranken JH, Troost D, Wegener JT, Kruis MR, van der Wegt MH. Neuropathological findings after continuous intrathecal administration of S(+)-ketamine for the management of neuropathic cancer pain. *Pain* (2005) 117:231–5.
121. Vranken JH, Troost D, de Haan P, Pennings FA, van der Wegt MH, Dijkgraaf MG, et al. Severe toxic damage to the rabbit spinal cord after intrathecal administration of preservative-free S(+)-ketamine. *Anesthesiology* (2006) 105:813–8.
122. Tejwani GA, Rattan AK, Mc Donald JS. Role of spinal opioid receptors in the antinociceptive interactions between intrathecal morphine and bupivacaine. *Anesth Analg* (1992) 74:726–34.
123. Bosmans T, Piron K, Oosterlinck M, Gasthuys F, Duchateau L, Waelbers T, et al. Comparison of analgesic efficacy of epidural methadone or ropivacaine/methadone with or without pre-operative oral tepoxalin in dogs undergoing tuberositas tibiae advancement surgery. *Vet Anaesth Analg* (2012) 39:618–27.
124. Tusell JM, Andaluz A, Prandi D, Costa C, García F. Effects of epidural anaesthesia-analgesia on intravenous anaesthesia with propofol. *Vet J* (2005) 169:108–12.
125. Campoy L, Martin-Flores M, Ludders JW, Erb HN, Gleed RD. Comparison of bupivacaine femoral and sciatic nerve block versus bupivacaine and morphine epidural for stifle surgery in dogs. *Vet Anaesth Analg* (2012) 39:91–8.
126. O O, Smith LJ. A comparison of epidural analgesia provided by bupivacaine alone, bupivacaine + morphine, or bupivacaine + dexmedetomidine for pelvic orthopedic surgery in dogs. *Vet Anaesth Analg* (2013) 40:527–36.
127. Lewis KA, Bednarski RM, Aarnes TK, Dyce J, Hubbell JA. Postoperative comparison of four perioperative analgesia protocols in dogs undergoing stifle joint surgery. *J Am Vet Med Assoc* (2014) 244:1041–6.
128. Arnholz M, Hungerbühler S, Weil C, Schütter AF, Rohn K, Tünsmeyer J, et al. Comparison of ultrasound guided femoral and sciatic nerve block versus epidural anaesthesia for orthopaedic surgery in dogs. *Tierarztl Prax Ausg K Kleintiere Heimtiere* (2016) 45:5–14.

129. Wendt-Hornickle E, Snyder LB. Comparison of anesthesia with a morphine-lidocaine-ketamine infusion or a morphine-lidocaine epidural on time to extubation in dogs. *Vet Anaesth Analg* (2016) 43:86–90.
130. Bartel AK, Campoy L, Martin-Flores M, Gleed RD, Walker KJ, Scanapico CE, et al. Comparison of bupivacaine and dexmedetomidine femoral and sciatic nerve blocks with bupivacaine and buprenorphine epidural injection for stifle arthroplasty in dogs. *Vet Anaesth Analg* (2016) 43:435–43.
131. Almeida TF, Fantoni DT, Mastrocinque S, Tatarunas AC, Imagawa VH. Epidural anesthesia with bupivacaine, bupivacaine and fentanyl, or bupivacaine and sufentanil during intravenous administration of propofol for ovariohysterectomy in dogs. *J Am Vet Med Assoc* (2007) 230:45–51.
132. Diniz MS, Kanashiro GP, Bernardi CA, Nicácio GM, Cassu RN. Extradural anesthesia with lidocaine combined with fentanyl or methadone to ovariohisterectomy in dogs. *Acta Cir Bras* (2013) 28:531–6.
133. Ossipov MH, Harris S, Lloyd P, Messineo E, Lin BS, Bagley J. Antinociceptive interaction between opioids and medetomidine: systemic additivity and spinal synergy. *Anesthesiology* (1990) 73:1227–35.
134. Pacharinsak C, Greene AS, Keegan RD, Kalivas PW. Postoperative analgesia in dogs receiving epidural morphine plus medetomidine. *J Vet Pharmacol Ther* (2003) 26:71–7.
135. Amarpal H, Aithal P, Kinjavdekar P, Singh GR. Interaction between epidurally administered ketamine and pethidine in dogs. *J Vet Med A Physiol Pathol Clin Med* (2003) 50:254–8.
136. Acosta AD, Gomar C, Correa-Natalini C, Bopp S, Polydoro A, Sala-Blanch X. Analgesic effects of epidurally administered levogyral ketamine alone or in combination with morphine on intraoperative and postoperative pain in dogs undergoing ovariohysterectomy. *Am J Vet Res* (2005) 66:54–61.
137. Nishiyama T, Yokoyama T, Hanaoka K. Midazolam improves postoperative epidural analgesia with continuous infusion of local anesthetics. *Can J Anaesth* (1998) 45:551–5.
138. Nishiyama T, Tamai H, Hanaoka K. Serum and cerebrospinal fluid concentrations of midazolam after epidural administration in dogs. *Anesth Analg* (2003) 96:159–62.
139. Gallivan ST, Johnston SJ, Broadstone R, Jortner BS, Reimer M. The clinical, cerebrospinal fluid, and histopathologic effects of epidural ketorolac in dogs. *Vet Surg* (1999) 29:436–41.
140. Marucio RL, Luna SPL, Teixeira Neto FJ, Minto BW, Hatschbach E. Postoperative analgesic effects of epidural administration of neostigmine alone or in combination with morphine in ovariohysterectomized dogs. *Am J Vet Res* (2008) 69:854–60.
141. Marucio RL, Monteiro ER, Moroz LR, Fantoni DT. Postoperative analgesic effects of epidural administration of neostigmine alone or in combination with morphine in dogs undergoing orthopedic surgery of the pelvic limbs. *Am J Vet Res* (2014) 75:956–63.

CHAPTER 4

Understanding the Osteosarcoma Pathobiology: A Comparative Oncology Approach

Jyotika Varshney [1,2], Milcah C. Scott [1,3,4], David A. Largaespada [1,5] and Subbaya Subramanian [1,2,*]

[1]*Masonic Cancer Center, University of Minnesota, Minneapolis, MN 55455, USA*

[2]*Department of Surgery, University of Minnesota Medical School, Moos Tower, 11-212420 Delaware Street, S.E.; MMC 195, Minneapolis, MN 55455, USA*

[3]*Animal Cancer Care and Research Program, University of Minnesota, St. Paul, MN 55455, USA*

[4]*Department of Veterinary Clinical Sciences, College of Veterinary Medicine, University of Minnesota, St. Paul, MN 55108, USA*

[5]*Department of Pediatrics, University of Minnesota, Minneapolis, MN 55455, USA*

ABSTRACT

Osteosarcoma is an aggressive primary bone tumor in humans and is among the most common cancer afflicting dogs. Despite surgical advancements and intensification of chemo- and targeted therapies, the survival outcome for osteosarcoma patients is, as of yet, suboptimal. The presence of metastatic disease at diagnosis or its recurrence after initial therapy is a major factor for the poor outcomes. It is thought that most human and canine patients have at least microscopic metastatic lesions at diagnosis. Osteosarcoma in dogs occurs naturally with greater frequency and shares many biological and clinical similarities with osteosarcoma in humans. From a genetic perspective, osteosarcoma in both humans and dogs is characterized by complex karyotypes with highly variable structural and numerical chromosomal aberrations. Similar molecular abnormalities have been observed in human and canine osteosarcoma. For instance, loss of *TP53* and *RB* regulated pathways are common. While there are several oncogenes that are commonly amplified in both humans and dogs, such as *MYC* and *RAS*, no commonly activated proto-oncogene has been identified that could form the basis for targeted therapies. It remains possible that recurrent aberrant gene expression changes due to gene amplification or epigenetic alterations could be uncovered and these could be used for developing new, targeted therapies. However, the remarkably high genomic complexity of osteosarcoma has precluded their definitive identification. Several advantageous murine models of osteosarcoma have been generated. These include spontaneous and genetically engineered mouse models, including a model based on forward genetics and transposon mutagenesis allowing new genes and genetic pathways to be implicated in osteosarcoma development. The proposition of this review is that careful comparative genomic studies between human, canine and mouse models of osteosarcoma may help identify commonly affected and targetable pathways for alternative therapies for osteosarcoma patients. Translational research may be found through a path that begins in mouse models, and then moves through canine patients, and then human patients.

Keywords: osteosarcoma; comparative oncology; microRNAs; prognosis; canine osteosarcoma

1. INTRODUCTION

Osteosarcoma is an aggressive primary bone tumor most prevalent in human patients. It mostly occurs during adolescence, with a second peak at middle age (older than 40) [1,2]. The tumor is characterized by increased production of osteoid (abnormal bone matrix) and by exceptionally complex karyotypes [3,4]. While the target cell for malignant transformation in the formation of osteosarcoma is not known with certainty, it is thought to be a mesenchymal stem cell (MSC) or a cell committed to the osteoblast lineage [5]. Evidence from mouse models indicates that any of several cell stages could serve as target cells for osteosarcoma development [6]. There are three common histologic types of osteosarcoma: osteoblastic, where tumor cells produce large amounts of tumor osteoid; chondroblastic, where tumor cells produce chondroid (cartilage) in addition to some amount of tumor osteoid; and fibroblastic, where tumor cells are predominantly fibroblasts and can produce both collagen and tumor osteoid [7]. The disease is highly metastatic, with distant spread mostly to lungs and other sites in bone, but osteosarcoma can also metastasize to lymph nodes and intra-abdominal organs [8,9]. The metastatic pattern (lungs, bones, lymph nodes) is similar for dogs and humans [10]. Osteosarcoma may present with macroscopic metastatic disease or metastatic disease can occur after therapy. In either case, prognosis is significantly worse once metastasis has been detected [11,12].

In canine patients, osteosarcoma is a disease primarily of adult dogs; the median age at diagnosis is approximately eight years, with a small peak of incidence in young animals (younger than 3 years) [10,13]. This differs from the situation in human patients where the peak age of incidence is in adolescence. Nevertheless, the natural history of the disease is similar in canine and human osteosarcoma patients [14]. Moreover, we have observed that genome wide expression profiles of canine osteosarcoma are indistinguishable from human pediatric osteosarcoma and more like human osteosarcoma than any other human cancer [14,15]. In dogs, there is generally a strong breed preference in the risk for cancer, such as osteosarcoma [16,17]. Many large breed dogs have an increased risk for osteosarcoma compared to other breeds [4,18,19]. The genetic determinants of osteosarcoma susceptibility in dogs are not fully elucidated, and it is possible that these factors will also be important in human osteosarcoma susceptibility. The development of treatment strategies in dogs and people has proven mutually beneficial for both species [20]. In fact, current treatment options are similar in both species.

Osteosarcoma is treated using surgery, radiation and chemotherapy. Surgical techniques have greatly improved with time, and in many cases limb salvage is feasible with low rates of local recurrence in humans [21]. Combination chemotherapies are vital for effective treatment of osteosarcoma [22]. However, despite decades of effort, nothing better than the most often used combination of three chemotherapeutic agents (*i.e.*, methotrexate, doxorubicin and cisplatin) has been found [22].The adjuvant chemotherapy is generally administered prior to surgery and, according to many studies, the extent of necrosis observed in the primary tumor after surgical removal is correlated with the outcome. [21]. The goal of the adjuvant chemotherapy is to eliminate the micrometastases that are thought to be present in >80% of all patients at the time of diagnosis, thus preventing relapse [1]. New research is underway to define specific molecular targets that could be utilized to treat this disease, but their elucidation has been proven difficult. It is possible that osteosarcoma may also be treated using recently developed immunotherapies, such as immune checkpoint blockade or chimeric antigen receptor-engineered T cells [21].

Osteosarcoma is characterized by aneuploidy and extensive genetic instability [23]. Due to this extreme genetic instability, common causes of osteosarcoma development have largely been limited to implicating loss of *RB* and *TP53*regulated cellular activities and upregulation of *MYC* transcriptional activity reviewed by Morrow *et al.* [24]. Beyond these, few alterations are common or well accepted as osteosarcoma drivers. Gene amplification and protein overexpression of the *RUNX2* transcription factor, which is a regulator of bone renewal, has been proposed to drive osteosarcoma [25]. Conflicting data on activation of beta-catenin dependent transcription in osteosarcoma has been reported [26]. In addition, TGF-β signaling has been proposed to promote the acquisition of metastatic disease [27]. Notable, expression of Ezrin at high levels has been found in metastatic human and canine osteosarcoma and functionally validated in model systems as a driver of osteosarcoma metastasis [28]. Recently, a large scale *Sleeping Beauty* (SB) transposon-based forward genetic screen was carried out for osteosarcoma in mice [29]. From this screen, a large number of new candidate osteosarcoma proto-oncogene and tumor suppressors were reported. These included the candidate oncogenes *SEMA4D* and *SEMA6D*, which were partially functionally validated as well as tumor suppressors like *NF1*, *NF2* and *PTEN* [29]. Beyond protein encoding genes, several micro RNAs (miRNAs) and miRNA clusters have been suggested as drivers or suppressors of osteosarcoma development and progression [30]. What remains unclear is which common oncogenic drivers for osteosarcoma should be targeted and how they would be targeted. Unlike the case for other malignancies, osteosarcoma is genetically heterogeneous and common molecular targets may not be present.

Osteosarcoma is much more common in dogs than in people (greater than 15 times) and can be experimentally induced in mice via transgenesis. Perceivably, the opportunity to study osteosarcoma in three different species may make it possible to identify core genes and pathways, whose alterations areto the osteosarcoma phenotype and that would make good targets for therapy. In the past, the main challenges were the development of improved surgical techniques for limb salvage, prevention of local recurrence and the identification of effective combination chemotherapies. Currently, a major pressing need is the development of new therapies to prevent the outgrowth of metastatic disease and to treat it once it has occurred. It is possible that germline or somatic mutations or tumor gene expression signatures identified through multi-species approaches could help predict which patients will respond well to current treatment regimes and which need to be referred for more intensive therapy or follow-up screening. However, much remains to be learned about what allows microscopic osteosarcoma lesions to resist chemotherapy and to suddenly break tumor dormancy and form a macroscopic mass. Until these questions are answered, we are not fully capable to help many human and canine osteosarcoma patients.

This review is an attempt to address many of the unanswered questions pertaining to osteosarcoma by focusing on recently acquired knowledge from a multi-species approach.

2. HUMAN OSTEOSARCOMA

Osteosarcoma in human patients is a rare tumor with a peak incidence in the second decade of life. The incidence is roughly 25 cases per 10 million per year in the United States [31]. A second wave of osteosarcoma is diagnosed in individuals between 70 and 80 years of age, often in association with Paget's disease. Paget's disease is an abnormality of bone homeostasis in which excessive bone reabsorption and reformation occur. Genetic and viral origins of Paget's disease have been suggested. While osteosarcoma is a rare manifestation of the disease, risk for osteosarcoma is increased. Osteosarcoma can occur in nearly any part of the skeleton, but commonly affects one of the long bones. As it occurs during a period of intense bone growth, there may be an association between this growth and sensitivity to osteoblast transformation [31]. Indeed, children with osteosarcoma are on average tall for their age [32]. Genetic studies also have suggested an association between the inheritance of certain forms of growth control genes and risk for osteosarcoma [33]. Other risk factors for human osteosarcoma include prior radiation therapy, male gender, and possibly African ancestry [34,35]. A genome-wide association study (GWAS) revealed possible risk alleles in the glutamate receptor, metaotropic 4 (*GRM4*) gene and a gene desert at chromosome 2p25.2 [36]. A recent GWAS also identified a potential high-risk allele for osteosarcoma metastasis at presentation within the transcriptional activator *NFIB* [37], a gene that was also identified as a potential tumor/metastasis suppressor in a transposon-based genetic screen in mice [29].

Some rare inherited cancer predisposition syndromes increase the risk for osteosarcoma. These rare cancers include Li-Fraumeni syndrome, hereditary retinoblastoma and Rothman-Thompson syndrome caused by germline mutations in *TP53, RB1*, and *RECQL4*, respectively [38]. Certain other cancer predisposition syndromes also seem to increase the risk of osteosarcoma development including Neurofibromatosis Type 1 syndrome, Werner's syndrome and Diamond-Blackfan anemia [24]. Patients with hereditary multiple osteochondromas, due to inheritance of mutations in the *EXT1* or *EXT2*genes, are predisposed to development of osteosarcoma also [24].

Human osteosarcoma is typically diagnosed due to swelling and pain often in the limbs. This can cause a limp if the involved limb is a leg or immobility of a joint. Pain may be constant or associated with exertion. In rare cases osteosarcoma will trigger a bone fracture. Diagnosis is made following a complete medical exam and imaging studies. Often an X-ray is done, followed by positron emission tomography or a bone scan. As osteosarcoma can often metastasize to the lungs, lung X-rays are utilized. A definitive diagnosis requires a biopsy. Staging of osteosarcoma is controversial. Most believe that all osteosarcoma should be considered high grade and treated as such although low grade osteosarcoma seems to be a distinct entity [38]. The only commonly accepted staging is localized *versus* metastatic disease.

Human osteosarcoma is commonly treated using surgery and chemotherapy [38]. Chemotherapy for roughly ten weeks that precedes surgery and the extent of necrosis in the surgically removed primary tumor has been suggested to be associated with long term risk for disease recurrence [38]. Chemotherapy is also given after surgery for a period of up to a year. Usually two or three chemotherapeutic agents are given for osteosarcoma treatment. Common combinations include high-dose methotrexate, doxorubicin and cisplatin (sometimes with ifosfamide); doxorubicin and cisplatin; ifosfamide and etoposide; ifosfamide, cisplatin (or carboplatin) and epirubicin [38]. Radiation therapy is used in some cases for non-resectable tumors [38].

The prognosis of human osteosarcoma is affected largely by the presence or absence of metastases. The 5-year survival rate for localized osteosarcoma is estimated to be between 50% and 60%. For metastatic disease the 5-year survival rate is 15% to 30% [39]. There is a better prognosis if metastatic osteosarcoma is present only in the lungs and is all metastatic lesions can be removed surgically [39]. That said, there is a dire need to develop therapies that can be used to treat patients who present with extensive metastatic disease or who recur with metastatic disease. At present, palliative care is often used in such cases or referral for clinical trials [38].

3. CANINE OSTEOSARCOMA

Canine osteosarcoma accounts for 80%–90% of canine primary bone tumors [40]. Dogs often present with a history of lameness or in some cases with a pathologic fracture of the affected bone. Diagnosis is based on clinical signs, imaging and biopsy. The metaphyseal region of long bones is the most common primary site with front limbs affected twice as often as rear limbs and the distal radius and proximal humerus being the two most common locations [41,42].

It is believed that external factors such as chemical carcinogens, ionization radiation, metallic implants to fix fractures and precedent skeletal disorders such as osteomyelitis and microscopic fractures can lead to canine osteosarcoma [10]. There are implications of genetic factors such as *TP53* and RB1 aberrations, as well as certain viruses and growth factor alterations [43]. Osteosarcoma can affect any breed of dog, but it is more commonly found in the larger breeds. Some breeds, such as the Scottish Deerhound, Great Dane, St. Bernard and Greyhound, are at high risk for developing primary bone tumors suggesting a genetic predilection [40]. In fact, the CDKN2A locus has been recently linked to osteosarcoma risk in dogs, and the risk allele is fixed in certain breeds like Rottweilers and Irish Wolfhounds [4].

Without treatment, the estimated survival for a dog with osteosarcoma is less than 3 months. On the other hand, survival times of approximately 1 year (or about 10% of a dog's lifetime) are achievable for 20%–50% of dogs with osteosarcoma treated using the current standard of care, and a small percentage of dogs can survive up to 5–6 years after diagnosis [44]. Standard of care for dogs is surgery (amputation of limb sparing surgery) with adjuvant chemotherapy [45]. The choice of chemotherapy drugs does not seem to influence survival, thus, toxicity, quality of life and cost tend to be the factors that guide treatment decisions [44]. Chemotherapy is only recommended when the primary tumor is removed and as of current, the drug of choice is carboplatin.

Many well-controlled studies show that a clinical response for osteosarcoma is only achievable with accepted standard of care. In very rare cases, dogs with osteosarcoma that receive palliative care may have prolonged (>1 year) survival, even in face of metastatic disease. Anecdotal benefits reported from herbal or "alternative" treatments, have not been reproducible, and no alternative therapies have been shown to have efficacy or provide consistent clinical benefit in controlled trials.

Metastatic bone cancer is the common cause of death or euthanasia, in 90% of dogs by 1 year [46]. Reducing the primary tumor's ability to metastasize and enhancing the antitumor activity of chemotherapy drugs and yet having minimal negative side effects is still a challenge for veterinary practitioners [10]. Treatment options for canine patients with metastases include pulmonary metastasectomy, but treatment for metastatic disease is only recommended if the primary tumor remains in complete remission [47]. The median survival after pulmonary metastasectomy can be up to 6 months; but without standard surgical procedure, survival outcome is usually less than 2 months [46].

4. MOUSE MODELS OF OSTEOSARCOMA

The number of genetically engineered or other mouse models of osteosarcoma are fairly restricted. However, these have been useful for studies on the cell of origin for osteosarcoma development and gaining insight into new genes and pathways that influence osteosarcoma initiation, progression and the process of metastasis. Most of the reported models utilize Cre/LoxP mediated deletion of *Trp53* and/or *Rb1* (reviewed in [48]). A few other models have been created by transgenic overexpression of various oncogenes including *Fos* and SV40 large T antigen [49], or co-deletion of one copy of *Twist* and mutation of *Apc* [50]. Osteosarcoma can be induced with high efficiency in mice by osteoblast lineage specific deletion of the *Trp53* gene, although the latency and penetrance is greatly enhanced by co-deletion of both copies of the *Rb1*gene [51].These data are consistent with the centrality of *TP53* and *RB1* loss of function alterations in human osteosarcoma [23]. Mouse osteosarcoma induced by deletion of *Trp53* is genomically unstable, as is seen in human osteosarcoma [52].

Others have created mouse cell lines from spontaneously occurring osteosarcoma. In cases in which these osteosarcoma were induced on a uniform strain background, the osteosarcoma cells can be grown in syngeneic hosts by intravenous, subcutaneous, or orthotopic injections into the tibia (reviewed in [53]). Examples include the K12 and K7M2 derivative that is highly metastatic, and the Dunn along with its metastatic derivative LM8. These cell lines have been useful for testing the role of various components of the immune system on primary osteosarcoma tumor growth and osteosarcoma metastasis [54]. Such cell lines have also been used to create matched pairs of poorly and highly metastatic cell lines for study [53]. These have been useful for developing candidate genes and pathways that may regulate the process of metastasis, such as high-level expression of ezrin [28].

Quist *et al.* reported that osteosarcoma could be induced in mice via homozygous Cre/LoxP-mediated deletion of *Trp53*and *Rb1* in undifferentiated mesenchyme, pre-osteoblasts or infrequently cycling mature osteoblasts [6]. These data suggest that a variety of cell types could be targets for osteosarcoma development. This plasticity may also be the basis for the ready acquisition of new osteosarcoma phenotypes in patients, including chemo/radiotherapy resistance and the ability to readily colonize the lung.

From osteosarcoma mouse models, we have learned that osteosarcoma cells have a predilection to metastatic colonization of the lung, but, interestedly, have observed a metastatic potential spectrum across various cell lines. Intriguingly, in mouse models of Li-Fraumeni, the rate of lung metastasis is influenced by the nature of the p53 mutant allele used and when in the lifetime of the mouse Cre is used to induce mutation of the p53 gene [51]. The rate of spontaneous development of osteosarcoma and the frequency of lung metastasis is higher in mice heterozygous for the $Trp53^{R272H}$ allele than in mice heterozygous for the carry $Trp53^{R270H}$ allele [55].

Recently, a new mouse model of osteosarcoma was developed which utilizes the *Sleeping Beauty* (SB) transposon mutagenesis system [29]. In this project, a SB transposon vector designed to activate proto-oncogenes or inactivate tumor suppressor genes by insertional mutagenesis, was mobilized specifically in *Osx1+* cells using a tissue-specific mutagenesis approach. Mutagenesis by SB was able to induce osteosarcoma in otherwise wild type mice or accelerate osteosarcoma in mice with tissue-specific induction of the $Trp53^{R270H}$ allele. Transposon insertion mutations in specific regions of the genome are found recurrently in SB induced tumors and these regions are called "common insertion sites" or CIS [29]. CIS result from selection for insertions that occur within or near a cancer gene in the right position and the right orientation so as to give the cell a selective advantage and drive the development of a tumor. Thus, the identification of genes at or near CIS defines new candidate drivers of cancer development.

5. CONSERVED DRIVERS OF OSTEOSARCOMA

5.1. Genomic Alterations

It has been reported that genomic alterations are necessary for development of osteosarcoma in mouse models (*TP53*) and are accepted as being commonly shared in humans and dogs [13,29,56].

Highly chaotic karyotypes encompass a key feature that characterizes some mouse models used for osteosarcoma studies, and naturally occurring osteosarcoma in dogs, as well as in human patients. The average human osteosarcoma harbors roughly 30 coding alterations [23,57] while in dog osteosarcoma and mouse models, this figure is not known with certainty. Many of the candidate genes implicated in the pathogenesis and progression of osteosarcoma in people have also been characterized in the canine disease. Notable examples of these genes are *PTEN* (phosphatase and tensin homolog), *RB1* (retinoblastoma), and *TP53* (tumor protein 53), and *MET* (mesenchymal-epithelial transition factor).

Genetic alterations of the retinoblastoma susceptibility (*RB1*) gene have been implicated in the development and progression of osteosarcoma. In people, almost 70% of osteosarcomas have at least one *RB1* gene alteration and the percentage is similar in dog osteosarcomas [13]. We recently reported that an aberrant RB-E2F1 regulatory pathway is predictive of biological behavior [58]. Since our work, another group has reported that RB1 alterations may serve as a prognostic marker for the management of osteosarcoma patients [59].

Approximately 50% of human osteosarcomas have been reported to have somatic *TP53* deletions or point mutations detected using exon sequencing strategies [23]. However, a recent study that used whole genome sequencing of human osteosarcomas suggests that nearly all osteosarcoma tumors have p53 pathways lesions, which in many cases are translocations that break in the first intron of *TP53* gene [57]. Thus, it would appear that p53 pathway loss might be a requirement for osteosarcoma development in people. Clearly, germline *TP53* mutations, which cause Li-Fraumeni syndrome, predispose to osteosarcoma. However, it is unclear whether *TP53* mutations act as initiating mutations or progression mutations in sporadic osteosarcoma. In any case, p53 pathway alterations are common in genetically engineered mouse models as well as spontaneous canine and human osteosarcoma. Therefore, mouse models could be used to develop therapies that exploit p53 pathway alterations in patient osteosarcoma.

In humans and dogs, several oncogenes have been identified as possibly playing a role in osteosarcoma including *MET*, *FOS*, *IGF1R* (Insulin-Like Growth Factor 1 Receptor), *PVT1/MYC*, *RUNX2*, and *HER2*. Some of these changes involve copy number alterations. Consistent genome and chromosome copy number changes have been reported in canine and human osteosarcoma [17]. For instance, a high copy number gain of the *RUNX2* locus has been reported in both human and dog osteosarcoma [56].

Some of these genes, or the pathways they regulate, including the *PVT1/MYC* locus, *MET* signaling, and *IGF1R* have been implicated in genetically engineered mouse models of osteosarcoma [29,52].

Recurrent point mutations have been observed in osteosarcoma, but there are few beyond *RB1* and *TP53* that reach statistical significance. This is in part because too few samples have been sequenced. However, point mutations or deletions of several tumor suppressor genes, likely to be drivers of osteosarcomagenesis, were observed in a recent report of human osteosarcoma including *NF1*, *NF2* and *PTEN*, all of which were also recovered in a transposon-based screen for osteosarcoma in mice [29]. Overlap exists between other reported candidate human osteosarcoma genes after genome-wide copy number and whole exome sequencing were done [23], and also after transposon based screening in a mouse model including activation of *PI3KCA*, and *AKT1*, and inactivation of *ARID1A* [29].

5.2. Deregulation of Micrornas in Canine and Human Osteosarcoma

microRNAs (miRNAs) are small non coding RNAs that regulate more than 60% of the genome post transcriptionally. Several studies have identified deregulated miRNAs in osteosarcoma [60,61,62,63,64,65,66]. One of the earliest studies reported four differentially expressed miRNAs in a handful of osteosarcoma samples compared to controls [67]. Since these initial studies, additional research has shown that miRNA deregulation is potentiallyto osteosarcoma development and progression [68]. Indeed, our group generated a Sarcoma MicroRNA Expression Database (S-MED) that represents 22 different sarcoma types, including osteosarcoma [69]. Through S-MED we found that osteosarcoma clustered separately from all other sarcoma types indicating the presence of potentially biologically important miRNAs. One of the unique miRNA signatures we found revealed the significant downregulation of around 50 miRNAs in the human 14q32 locus compared to the controls [65]. Most importantly, we have also shown a comparable decrease in expression of orthologous 14q32 miRNAs in canine osteosarcoma samples [65]. Interestingly we did not observe any copy number changes at the 14q32 locus, suggesting epigenetic changes at the locus. We confirmed that subset of these 50 miRNAs (miR-382, miR-369-5p, miR-544 and miR-134) could target the 3′ UTR of cMYC transcript. Additionally, overexpression of these miRNAs in osteosarcoma cells decreased the cMYC levels and induced apoptosis [65,70]. Also cMYC has been shown to transactivate a commonly known miRNA cluster, miR-17-92 [71]. We showed that restoration of the 14q32 miRNAs not only did decrease the cMYC levels but also significantly reduced the levels of mir-17-92 [70].

It was observed in other studies that miR-135b, -150, -370, -542-5p, -652, and -654 were highly expressed compared to osteoblasts [72,73,74]. Many of them play crucial roles in bone differentiation and key signaling pathways (miR-206 and -286) [75]. Additionally, miRNAs target *TP53*, which is an important tumor suppressor gene, was mutated in more than 60% osteosarcoma tumors [76]. One of the key miRNAs in the *TP53* pathway is miR-34, which is significantly downregulated in many osteosarcoma tumors that affects the cell cycle and proliferation [77,78,79].

The role of miRNAs in canine osteosarcoma remains to be fully elucidated. Since the discovery of miRNAs, there have been few studies that show a correlation between deregulation of miRNAs and canine osteosarcoma. A recent study investigated the role of miR-196a and its target Annexin V in human (143B, MG63) and canine (DAN) osteosarcoma cell lines to identify potential targets for new therapeutic agents in both the species [80]. They observed that miR-196a is downregulated in many of the canine osteosarcoma tumors compared to the normal bone, which is in contradiction with some earlier studies [66,81]. Further studies need to be performed on a larger cohort to establish that miR-196a is a potential therapeutic target and can impact multiple downstream targets in canine osteosarcoma.

A study led by Fenger *et al.*, sought to characterize miRNA expression in canine primary tumors among the major large breeds (greyhounds, rottweilers, golden retrievers and some mixed breeds) [43,82]. Using Nanominer software, they determined 189 miRNAs that were differentially expressed. Of these, miR-494 was highly expressed in all the breeds compared to normal canine osteoblasts.

As mentioned earlier, our group was among the first groups to observe a significant downregulation of around 50 miRNAs in the human 14q32 chromosomal region, which was highly conserved in the dog genome. Out of the 50 miRNAs, 2 miRNAs, miR-134 and mir-544, shared 100% conservation with the canine genome and mapped to the predicted synteny in canine osteosarcoma [14]. We have comprised a table of key miRNAs that play an important role in survival outcome and chemoresponse that can potentially lead to osteosarcoma diagnosis, subtyping and therapeutics in human and canine osteosarcoma (Table 1).

Table 1. miRNAs as potential markers for survival outcome and/ or chemosensitivity.

miRNAs Deregulated in OS	Expression Levels Compared to the Controls	Overall Function	References
miR-382	Down-regulated	Poor survival outcome and metastasis marker	[70,83]
miR-154	Down-regulated	Poor survival outcome	[70]
miR-33a	Up-regulated	Chemoresistance	[84]
miR-34c	Down-regulated	Chemoresistance	[85]

5.3. Epigenetic Changes in Osteosarcoma

Most forms of human cancer have changes in the epigenome compared to the normal cellular counterparts from which they are derived. Indeed, changes to the epigenome seem to be a ubiquitous feature of cancer. Epigenetic changes are revealed in alterations to the pattern of DNA methylation, histone modifications and nucleosome remodeling (reviewed in [24]). Osteosarcoma, like other forms of human cancer, seems to have many such epigenome changes compared to normal osteoblasts, the presumptive target cell for transformation. Several studies have analyzed the epigenome in a genome-wide manner, albeit with low numbers of cases [86,87]. These studies confirm that specific methylation events are heterogeneous among different cases of human osteosarcoma and that these differences may help explain differences in the clinical behavior of different osteosarcomas. These genome wide studies in human osteosarcoma have been restricted to the study of DNA methylation. Work on other chromatin modification remains to be done. The study of genome wide epigenetic changes has not been undertaken on a large scale in genetically engineered mouse models or canine osteosarcoma samples.

Many studies in osteosarcoma have focused on methylation of specific genes or gene regions. These studies suggest that certain genes and genetic pathways are subject to control by DNA methylation and attendant modifications to the histone code. Among these are alterations to the RB and p53 pathways. While these specific genes are not frequent targets of methylation based gene silencing, genes in these pathways have been reported as targets for pathogenic methylation. Specifically the *CDKN2A* locus, encoding the cyclin dependent kinase inhibitor $p16^{INK4a}$ and the Mdm2 inhibitor $p14^{Arf}$, has been reported [88,89]. Several genes that are targets of p53, or modulate p53 activity, have been shown to be methylated and silenced in osteosarcoma cell lines or xenograft tumors including *CDKN1A*, *HIC1* and *GADD45* [90,91,92]. Many other tumor suppressor genes are silenced by promoter hypermethylation in osteosarcoma cell lines including *RASSF1A*, *TIMP3*, DAPK1 and others (review in Morrow and Khanna, Critical Reviews in Oncogenesis, 2015). Methylation and silencing of tumor suppressive miRNA genes has also been observed as specific events in human osteosarcoma cell lines and primary tumors. Promoter hypomethyation, and probable overexpression, of oncogenes is less well studied in osteosarcoma, but several reports have been made suggesting such a mechanism, notably for the metastasis promoting gene *IRX1* and the growth factor gene *IGF2* [93,94]. Several candidate osteosarcoma oncogenes found in a transposon-based forward genetic screen in mice are also hypomethylated and overexpressed in human osteosarcoma compared to normal osteoblasts, including *SEMA4D*, *RAF1* and *PAK1* [29]. Some of these epigenetic changes, including aberrant repression and activation, are associated with loss of imprinting control at specific loci in osteosarcoma cells [95,96,97].

Genome wide studies of epigenetic alterations in human osteosarcoma need to incorporate many more samples before we learn to what extent they may explain molecular subtypes and clinical behavior of different cases. No such studies have been reported for canine osteosarcomas or mouse models either. When they are, they will allow cross-species comparisons that could help reveal the most meaningful alterations. Despite these limitations, it is apparent that osteosarcoma development is accompanied by changes to the epigenome that drive progression and confer specific cancer phenotypes. Published data does suggest that epigenome targeted therapies, including histone deacetylase inhibitors and DNA demethylating agents, can alter gene expression and inhibit osteosarcoma growth and/or metastasis [93,98]. Indeed, one study suggests that osteosarcoma cell lines can be reprogrammed using induced pluripotency genes and that the accompanying epigenetic changes reduce tumorigenic potential [99]. Taken together, these data suggest that modification of osteosarcoma epigenomes will be a useful therapeutic approach.

6. ADVANTAGES OF A MULTI-SPECIES, COMPARATIVE OSTEOSARCOMA STUDY APPROACH

Using comparative genetic studies in multiple organisms, it is possible to understand more about the inherited and somatically acquired genetic alterations that are informative of osteosarcoma risk and clinical behavior, especially metastasis. It is reasonable that genetic alterations that are present in human, dog, and mouse osteosarcomas represent the most likely true drivers of osteosarcoma progression, based on the concept of convergent evolution. Singly, mouse models offer opportunities to discover new genes and pathways important in disease initiation and progression. Dogs bring many other advantages to the table. Notably, since they show remarkable intra-breed homogeneity, which, together with noticeable interbreed heterogeneity, the dog offers distinctive opportunities to understand the genetics underlying osteosarcoma. Taken all together, an interspecies approach offers a possibly broader view and understanding of osteosarcoma since, for example, what is found to be altered at the genetic level in one species may be epigenetically silenced altered in another, but would be hard to pick out based on changes to gene expression alone.

Findings from mouse models and naturally occurring osteosarcoma in dogs need to be translated into therapeutic approaches in a clinically relevant model. Advantages of clinical trials in dogs are numerous and include that there are many more cases of canine osteosarcoma than human osteosarcoma [100,101]. In addition, disease progression in dogs, even with standard of care, is rapid so that assessments of improvement can be made in comparatively short periods of time [102].

It has been shown that histological response is not a good predictor (St. Jude, CTOS 20th Annual Meeting) in human osteosarcoma patients, and so there is an urgent need for informative biomarkers. The 14q32 locus is a new and promising marker, which was discovered by our group through a comparative species approach [65].

7. CONCLUSIONS

In more than 70% of osteosarcoma patients, current standard-of-care therapies ultimately fail to prevent relapse and metastasis. This dismal outcome has not improved over the past three decades, indicating a desperate need for novel drugs and treatment strategies. As osteosarcoma is genetically heterogeneous, successful therapies will need to target conserved pathobiology. In this review, we discussed the presence of multiple conserved signaling mechanisms as well as some of the approaches for understanding osteosarcoma pathobiology. Recent studies highlight the role of microRNAs in regulation of these conserved signaling pathways. In addition, we have just begun to understand the roles of the regulatory small RNAs and long-noncoding RNAs (lncRNAs). Further, intricacy in gene regulations does not end with identification of regulatory RNAs and its interacting partners. Molecular intricacy is also heavily dependent on our understanding of the implications of other layers of gene regulation such as role for competing endogenous RNAs along with pseudogenes, circular RNAs and lncRNAs. Acquiring such breadth and depth of knowledge is critical for developing therapies that will prevent bypass mechanisms of resistance to therapies. Moreover, cell non-autonomous functions of secreted factors such as exosomes are currently being investigated. In addition, emerging concepts of immune regulation by cancer and recent advancements in immunotherapy hold promise for treating and better outcomes in osteosarcoma.

ACKNOWLEDGMENTS

We thank Jaime Modiano and Aaron Sarver for helpful discussions and Anne Sarver for assisting in manuscript preparation. Due to the space restrictions we could not cite many other significant contributions made by numerous researchers and laboratories in this potentially important and rapidly progressing field. Subbaya Subramanian and David A. Largaespada are supported by research grants funded by American Cancer Society Grant (RSG 13-381-01-RMC), the Sobiech osteosarcoma fund, Children's cancer research funds, the Wyckoff Rein in Sarcoma Foundation.

AUTHOR CONTRIBUTIONS

Jyotika Varshney and Subbaya Subramanian developed the concept for the review. Milcah C. Scott and David A. Largaespada provided input in the canine osteosarcoma and epigenetics of osteosarcoma.

CONFLICTS OF INTEREST

The authors declare no conflict of interest.

REFERENCES

1. Moore, D.D.; Luu, H.H. Osteosarcoma. Cancer Treat Res. 2014, 162, 65–92.
2. Ottaviani, G.; Jaffe, N. The epidemiology of osteosarcoma. Cancer Treat Res. 2009, 152, 3–13.

3. Letson, G.D.; Muro-Cacho, C.A. Genetic and molecular abnormalities in tumors of the bone and soft tissues. Cancer Control 2001, 8, 239–251.
4. Karlsson, E.K.; Sigurdsson, S.; Ivansson, E.; Thomas, R.; Elvers, I.; Wright, J.; Howald, C.; Tonomura, N.; Perloski, M.; Swofford, R.; et al. Genome-wide analyses implicate 33 loci in heritable dog osteosarcoma, including regulatory variants near CDKN2A/B. Genome Biol. 2013, 14, R132.
5. Mutsaers, A.J.; Walkley, C.R. Cells of origin in osteosarcoma: Mesenchymal stem cells or osteoblast committed cells? Bone 2014, 62, 56–63.
6. Quist, T.; Jin, H.; Zhu, J.F.; Smith-Fry, K.; Capecchi, M.R.; Jones, K.B. The impact of osteoblastic differentiation on osteosarcomagenesis in the mouse. Oncogene 2015, 34, 4278–4284.
7. Yoshida, H.; Adachi, H.; Hamada, Y.; Aki, T.; Yumoto, T.; Morimoto, K.; Orido, T. Osteosarcoma. Ultrastructural and immunohistochemical studies on alkaline phosphatase-positive tumor cells constituting a variety of histologic types. Acta Pathol. Jpn. 1988, 38, 325–338.
8. Dirik, Y.; Cinar, A.; Yumrukcal, F.; Eralp, L. Popliteal lymph node metastasis of tibial osteoblastic osteosarcoma. Int. J. Surg. Case Rep. 2014, 5, 840–844.
9. Jeffree, G.M.; Price, C.H.; Sissons, H.A. The metastatic patterns of osteosarcoma. Brit. J. Cancer 1975, 32, 87–107.
10. Morello, E.; Martano, M.; Buracco, P. Biology, diagnosis and treatment of canine appendicular osteosarcoma: Similarities and differences with human osteosarcoma. Vet. J. 2011, 189, 268–277.
11. Gill, J.; Ahluwalia, M.K.; Geller, D.; Gorlick, R. New targets and approaches in osteosarcoma. Pharmacol. Ther. 2013, 137, 89–99.
12. Jaffe, N. Osteosarcoma: Review of the past, impact on the future. The American experience. Cancer Treat Res. 2009, 152, 239–262.
13. Modiano, J.F.; Breen, M.; Lana, S.E.; Ehrhart, N.; Fosmire, S.P.; Thomas, R.; Jubala, C.M.; Lamerato-Kozicki, A.R.; Ehrhart, E.J.; Schaack, J.; et al. Naturally occurring translational models for development of cancer therapy. Gene Ther. Mol. Biol. 2006, 10, 31–40.
14. Scott, M.C.; Sarver, A.L.; Gavin, K.J.; Thayanithy, V.; Getzy, D.M.; Newman, R.A.; Cutter, G.R.; Lindblad-Toh, K.; Kisseberth, W.C.; Hunter, L.E.; et al. Molecular subtypes of osteosarcoma identified by reducing tumor heterogeneity through an interspecies comparative approach. Bone 2011, 49, 356–367.
15. Paoloni, M.; Davis, S.; Lana, S.; Withrow, S.; Sangiorgi, L.; Picci, P.; Hewitt, S.; Triche, T.; Meltzer, P.; Khanna, C. Canine tumor cross species genomics uncovers targets linked to osteosarcoma progression. BMC Genom. 2009, 10, 625.
16. Schiffman, J.D.; Breen, M. Comparative oncology: What dogs and other species can teach us about humans with cancer. Philos. Trans. R. Soc. Lond. B Biol. Sci. 2015, 370.
17. Thomas, R.; Wang, H.J.; Tsai, P.C.; Langford, C.F.; Fosmire, S.P.; Jubala, C.M.; Getzy, D.M.; Cutter, G.R.; Modiano, J.F.; Breen, M. Influence of genetic background on tumor karyotypes: Evidence for breed-associated cytogenetic aberrations in canine appendicular osteosarcoma. Chromosome Res. 2009, 17, 365–377.
18. McNeill, C.J.; Overley, B.; Shofer, F.S.; Kent, M.S.; Clifford, C.A.; Samluk, M.; Haney, S.; van Winkle, T.J.; Sorenmo, K.U. Characterization of the biological behaviour of appendicular osteosarcoma in Rottweilers and a comparison with other breeds: A review of 258 dogs. Vet. Comp. Oncol. 2007, 5, 90–98.
19. Dobson, J.M. Breed-predispositions to cancer in pedigree dogs. ISRN Vet. Sci. 2013, 2013, 941275.
20. Paoloni, M.; Khanna, C. Translation of new cancer treatments from pet dogs to humans. Nat. Rev. Cancer 2008, 8, 147–156.
21. Isakoff, M.S.; Bielack, S.S.; Meltzer, P.; Gorlick, R. Osteosarcoma: Current treatment and a collaborative pathway to success. J. Clin. Oncol. 2015, 33, 3029–3035.
22. Ferrari, S.; Serra, M. An update on chemotherapy for osteosarcoma. Expert Opin. Pharmacother. 2015, 16, 2727–2736.

23. Perry, J.A.; Kiezun, A.; Tonzi, P.; van Allen, E.M.; Carter, S.L.; Baca, S.C.; Cowley, G.S.; Bhatt, A.S.; Rheinbay, E.; Pedamallu, C.S.; et al. Complementary genomic approaches highlight the PI3K/mTOR pathway as a common vulnerability in osteosarcoma. Proc. Natl. Acad. Sci. USA 2014, 111, 5564–5573.
24. Morrow, J.J.; Khanna, C. Osteosarcoma genetics and epigenetics: Emerging biology and candidate therapies. Crit. Rev. Oncog. 2015, 20, 173–197.
25. Li, N.; Luo, D.; Hu, X.; Luo, W.; Lei, G.; Wang, Q.; Zhu, T.; Gu, J.; Lu, Y.; Zheng, Q. RUNX2 and osteosarcoma. Anticancer Agents Med. Chem. 2015, 15, 881–887.
26. Tian, J.; He, H.; Lei, G. Wnt/beta-catenin pathway in bone cancers. Tumour Biol. 2014, 35, 9439–9445.
27. Lamora, A.; Talbot, J.; Bougras, G.; Amiaud, J.; Leduc, M.; Chesneau, J.; Taurelle, J.; Stresing, V.; Le Deley, M.C.; Heymann, M.F.; et al. Overexpression of smad7 blocks primary tumor growth and lung metastasis development in osteosarcoma. Clin. Cancer Res. 2014, 20, 5097–5112.
28. Khanna, C.; Wan, X.; Bose, S.; Cassaday, R.; Olomu, O.; Mendoza, A.; Yeung, C.; Gorlick, R.; Hewitt, S.M.; Helman, L.J. The membrane-cytoskeleton linker ezrin is necessary for osteosarcoma metastasis. Nat. Med. 2004, 10, 182–186.
29. Moriarity, B.S.; Otto, G.M.; Rahrmann, E.P.; Rathe, S.K.; Wolf, N.K.; Weg, M.T.; Manlove, L.A.; LaRue, R.S.; Temiz, N.A. A Sleeping Beauty forward genetic screen identifies new genes and pathways driving osteosarcoma development and metastasis. 2015, 47, 615–624.
30. Chang, L.; Shrestha, S.; LaChaud, G.; Scott, M.A.; James, A.W. Review of microRNA in osteosarcoma and chondrosarcoma. Med. Oncol. 2015, 32, 613.
31. Geller, D.S.; Gorlick, R. Osteosarcoma: A review of diagnosis, management, and treatment strategies. Clin. Adv. Hematol. Oncol. 2010, 8, 705–718.
32. Mirabello, L.; Pfeiffer, R.; Murphy, G.; Daw, N.C.; Patino-Garcia, A.; Troisi, R.J.; Hoover, R.N.; Douglass, C.; Schuz, J.; Craft, A.W.; et al. Height at diagnosis and birth-weight as risk factors for osteosarcoma. Cancer Causes Control 2011, 22, 899–908.
33. Savage, S.A.; Woodson, K.; Walk, E.; Modi, W.; Liao, J.; Douglass, C.; Hoover, R.N.; Chanock, S.J. Analysis of genes critical for growth regulation identifies Insulin-like Growth Factor 2 Receptor variations with possible functional significance as risk factors for osteosarcoma. Cancer Epidemiol. Biomark. Prev. 2007, 16, 1667–1674.
34. Miller, R.W. Contrasting epidemiology of childhood osteosarcoma, Ewing's tumor, and rhabdomyosarcoma. Natl. Cancer Inst. Monogr. 1981, 56, 9–15.
35. Jawad, M.U.; Cheung, M.C.; Min, E.S.; Schneiderbauer, M.M.; Koniaris, L.G.; Scully, S.P. Ewing sarcoma demonstrates racial disparities in incidence-related and sex-related differences in outcome: An analysis of 1631 cases from the SEER database, 1973–2005. Cancer 2009, 115, 3526–3536.
36. Savage, S.A.; Mirabello, L.; Wang, Z.; Gastier-Foster, J.M.; Gorlick, R.; Khanna, C.; Flanagan, A.M.; Tirabosco, R.; Andrulis, I.L.; Wunder, J.S.; et al. Genome-wide association study identifies two susceptibility loci for osteosarcoma. Nat. Genet. 2013, 45, 799–803.
37. Mirabello, L.; Koster, R.; Moriarity, B.S.; Spector, L.G.; Meltzer, P.S.; Gary, J.; Machiela, M.J.; Pankratz, N.; Panagiotou, O.A.; Largaespada, D.; et al. A Genome-wide scan identifies variants in NFIB associated with metastasis in patients with osteosarcoma. Cancer Discov. 2015, 5, 920–931.
38. Gorlick, R.; Khanna, C. Osteosarcoma. J. Bone Miner. Res. 2010, 25, 683–691.
39. Hameed, M.; Dorfman, H. Primary malignant bone tumors—Recent developments. Semin. Diagn. Pathol. 2011, 28, 86–101.
40. Anfinsen, K.P.; Grotmol, T.; Bruland, O.S.; Jonasdottir, T.J. Breed-specific incidence rates of canine primary bone tumors—A population based survey of dogs in Norway. Can. J. Vet. Res. 2011, 75, 209–215.
41. Liptak, J.M.; Dernell, W.S.; Straw, R.C.; Rizzo, S.A.; Lafferty, M.H.; Withrow, S.J. Proximal radial and distal humeral osteosarcoma in 12 dogs. J. Am. Anim. Hosp. Assoc. 2004, 40, 461–467.

42. Boerman, I.; Selvarajah, G.T.; Nielen, M.; Kirpensteijn, J. Prognostic factors in canine appendicular osteosarcoma—A meta-analysis. BMC Vet. Res. 2012, 8, 56–56.
43. Fenger, J.M.; London, C.A.; Kisseberth, W.C. Canine osteosarcoma: A naturally occurring disease to inform pediatric oncology. ILAR J. 2014, 55, 69–85.
44. Szewczyk, M.; Lechowski, R.; Zabielska, K. What do we know about canine osteosarcoma treatment? Review. Vet. Res. Commun. 2015, 39, 61–67.
45. Berg, J. Canine osteosarcoma: Amputation and chemotherapy. Vet. Clin. N. Am. Small Anim. Pract. 1996, 26, 111–121.
46. Straw, R.C.; Withrow, S.J.; Powers, B.E. Management of canine appendicular osteosarcoma. Vet. Clin. North Am. Small Anim. Pract. 1990, 20, 1141–1161.
47. Gilson, S.D. Principles of surgery for cancer palliation and treatment of metastases. Clin. Tech. Small Anim. Pract.1998, 13, 65–69.
48. Guijarro, M.V.; Ghivizzani, S.C.; Gibbs, C.P. Animal models in osteosarcoma. Front. Oncol. 2014, 4, 189.
49. Ruther, U.; Komitowski, D.; Schubert, F.R.; Wagner, E.F. c-fos expression induces bone tumors in transgenic mice. Oncogene 1989, 4, 861–865.
50. Entz-Werle, N.; Choquet, P.; Neuville, A.; Kuchler-Bopp, S.; Clauss, F.; Danse, J.M.; Simo-Noumbissie, P.; Guerin, E.; Gaub, M.P.; Freund, J.N.; et al. Targeted apc;twist double-mutant mice: A new model of spontaneous osteosarcoma that mimics the human disease. Transl. Oncol. 2010, 3, 344–353.
51. Walkley, C.R.; Qudsi, R.; Sankaran, V.G.; Perry, J.A.; Gostissa, M.; Roth, S.I.; Rodda, S.J.; Snay, E.; Dunning, P.; Fahey, F.H.; et al. Conditional mouse osteosarcoma, dependent on p53 loss and potentiated by loss of Rb, mimics the human disease. Genes Dev 2008, 22, 1662–1676.
52. Rao, P.H.; Zhao, S.; Zhao, Y.J.; Yu, A.; Rainusso, N.; Trucco, M.; Allen-Rhoades, W.; Satterfield, L.; Fuja, D.; Borra, V.J.; et al. Coamplification of Myc/Pvt1 and homozygous deletion of Nlrp1 locus are frequent genetics changes in mouse osteosarcoma. Genes Chromosomes Cancer 2015, 54, 796–808.
53. Mohseny, A.B.; Hogendoorn, P.C.; Cleton-Jansen, A.M. Osteosarcoma models: From cell lines to zebrafish. Sarcoma2012, 2012, 417271.
54. Merchant, M.S.; Melchionda, F.; Sinha, M.; Khanna, C.; Helman, L.; Mackall, C.L. Immune reconstitution prevents metastatic recurrence of murine osteosarcoma. Cancer Immunol. Immunother. 2007, 56, 1037–1046.
55. Olive, K.P.; Tuveson, D.A.; Ruhe, Z.C.; Yin, B.; Willis, N.A.; Bronson, R.T.; Crowley, D.; Jacks, T. Mutant p53 gain of function in two mouse models of Li-Fraumeni syndrome. Cell 2004, 119, 847–860.
56. Angstadt, A.Y.; Thayanithy, V.; Subramanian, S.; Modiano, J.F.; Breen, M. A genome-wide approach to comparative oncology: High-resolution oligonucleotide aCGH of canine and human osteosarcoma pinpoints shared microaberrations. Cancer Genet. 2012, 205, 572–587.
57. Chen, X.; Bahrami, A.; Pappo, A.; Easton, J.; Dalton, J.; Hedlund, E.; Ellison, D.; Shurtleff, S.; Wu, G.; Wei, L.; et al. Recurrent somatic structural variations contribute to tumorigenesis in pediatric osteosarcoma. Cell Rep. 2014, 7, 104–112.
58. Scott, M.C.; Sarver, A.L.; Tomiyasu, H.; Cornax, I.; van Etten, J.; Varshney, J.; O'Sullivan, M.G.; Subramanian, S.; Modiano, J.F. Aberrant RB-E2F transcriptional regulation defines molecular phenotypes of osteosarcoma. J. Biol. Chem. 2015.
59. Ren, W.; Gu, G. Prognostic implications of RB1 tumour suppressor gene alterations in the clinical outcome of human osteosarcoma: A meta-analysis. Eur. J. Cancer Care 2015.
60. Sampson, V.B.; Yoo, S.; Kumar, A.; Vetter, N.S.; Kolb, E.A. MicroRNAs and potential targets in osteosarcoma: Review. Front. Pediatr. 2015, 3, 69.
61. Zhang, J.; Yan, Y.G.; Wang, C.; Zhang, S.J.; Yu, X.H.; Wang, W.J. MicroRNAs in osteosarcoma. Clin. Chim. Acta 2015, 444, 9–17.
62. Miao, J.; Wu, S.; Peng, Z.; Tania, M.; and Zhang, C. MicroRNAs in osteosarcoma: Diagnostic and therapeutic aspects. Tumour Biol. 2013, 34, 2093–2098.

63. Zhou, G.; Shi, X.; Zhang, J.; Wu, S.; Zhao, J. MicroRNAs in osteosarcoma: From biological players to clinical contributors, a review. J. Int. Med. Res. 2013, 41, 1–12.
64. Li, Y.; Zhang, J.; Zhang, L.; Si, M.; Yin, H.; Li, J. Diallyl trisulfide inhibits proliferation, invasion and angiogenesis of osteosarcoma cells by switching on suppressor microRNAs and inactivating of Notch-1 signaling. Carcinogenesis2013, 34, 1601–1610.
65. Sarver, A.L.; Thayanithy, V.; Scott, M.C.; Cleton-Jansen, A.M.; Hogendoorn, P.C.; Modiano, J.F.; Subramanian, S. MicroRNAs at the human 14q32 locus have prognostic significance in osteosarcoma. Orphanet J. Rare Dis. 2013, 8, 7.
66. Namlos, H.M.; Meza-Zepeda, L.A.; Baroy, T.; Ostensen, I.H.; Kresse, S.H.; Kuijjer, M.L.; Serra, M.; Burger, H.; Cleton-Jansen, A.M.; Myklebost, O. Modulation of the osteosarcoma expression phenotype by microRNAs. PloS ONE 2012, 7, e48086.
67. Lulla, R.R.; Costa, F.F.; Bischof, J.M.; Chou, P.M.; de, F.B.M.; Vanin, E.F.; Soares, M.B. Identification of differentially expressed MicroRNAs in osteosarcoma. Sarcoma 2011, 2011, 732690.
68. Varshney, J.; Subramanian, S. MicroRNAs as potential target in human bone and soft tissue sarcoma therapeutics. Front. Mol. Biosci. 2015, 2, 31.
69. Sarver, A.L.; Phalak, R.; Thayanithy, V.; Subramanian, S. S-MED: Sarcoma microRNA expression database. Lab. Investig. J. Tech. Methods Pathol. 2010, 90, 753–761.
70. Thayanithy, V.; Sarver, A.L.; Kartha, R.V.; Li, L.; Angstadt, A.Y.; Breen, M.; Steer, C.J.; Modiano, J.F.; Subramanian, S. Perturbation of 14q32 miRNAs-cMYC gene network in osteosarcoma. Bone 2012, 50, 171–181.
71. Li, Y.; Choi, P.S.; Casey, S.C.; Dill, D.L.; Felsher, D.W. MYC through miR-17–92 suppresses specific target genes to maintain survival, autonomous proliferation, and a neoplastic state. Cancer Cell 2014, 26, 262–272.
72. Schaap-Oziemlak, A.M.; Raymakers, R.A.; Bergevoet, S.M.; Gilissen, C.; Jansen, B.J.; Adema, G.J.; Kogler, G.; le Sage, C.; Agami, R.; van der Reijden, B.A.; et al. MicroRNA hsa-miR-135b regulates mineralization in osteogenic differentiation of human unrestricted somatic stem cells. Stem. Cells Dev. 2010, 19, 877–885.
73. Wei, J.Q.; Chen, H.; Zheng, X.F.; Zhang, B.X.; Wang, Y.; Tang, P.F.; She, F.; Song, Q.; Li, T.S. Hsa-miR-654–5p regulates osteogenic differentiation of human bone marrow mesenchymal stem cells by repressing bone morphogenetic protein 2. J. Southern Med. Univ. 2012, 32, 291–295.
74. Fang, S.; Deng, Y.; Gu, P.; Fan, X. MicroRNAs regulate bone development and regeneration. Int. J. Mol. Sci. 2015, 16, 8227–8253.
75. Inose, H.; Ochi, H.; Kimura, A.; Fujita, K.; Xu, R.; Sato, S.; Iwasaki, M.; Sunamura, S.; Takeuchi, Y.; Fukumoto, S.; et al. A microRNA regulatory mechanism of osteoblast differentiation. Proc. Natl. Acad. Sci. USA 2009, 106, 20794–20799.
76. Feng, Z.; Zhang, C.; Wu, R.; Hu, W. Tumor suppressor p53 meets microRNAs. J. Mol. Cell Biol. 2011, 3, 44–50.
77. Wang, Y.; Jia, L.S.; Yuan, W.; Wu, Z.; Wang, H.B.; Xu, T.; Sun, J.C.; Cheng, K.F.; Shi, J.G. Low miR-34a and miR-192 are associated with unfavorable prognosis in patients suffering from osteosarcoma. Am. J. Transl. Res. 2015, 7, 111–119.
78. Zhao, H.; Ma, B.; Wang, Y.; Han, T.; Zheng, L.; Sun, C.; Liu, T.; Zhang, Y.; Qiu, X.; Fan, Q. miR-34a inhibits the metastasis of osteosarcoma cells by repressing the expression of CD44. Oncol. Rep. 2013, 29, 1027–1036.
79. He, C.; Xiong, J.; Xu, X.; Lu, W.; Liu, L.; Xiao, D.; Wang, D. Functional elucidation of MiR-34 in osteosarcoma cells and primary tumor samples. Biochem. Biophys. Res. Commun. 2009, 388, 35–40.
80. Pazzaglia, L.; Leonardi, L.; Conti, A.; Novello, C.; Quattrini, I.; Montanini, L.; Roperto, F.; del Piero, F.; di Guardo, G.; Piro, F.; et al. miR-196a expression in human and canine osteosarcomas: A comparative study. Res. Vet. Sci. 2015, 99, 112–119.
81. Zhang, W.; Zhang, C.; Chen, H.; Li, L.; Tu, Y.; Liu, C.; Shi, S.; Zen, K.; Liu, Z. Evaluation of microRNAs miR-196a, miR-30a-5P, and miR-490 as biomarkers of disease activity among patients with FSGS. Clin. J. Am. Soc. Nephrol. 2014, 9, 1545–1552.

82. Gardner, H.L.; Fenger, J.M.; London, C.A. Dogs as a model for cancer. Annu. Rev. Anim. Biosci. 2015. in print.
83. Xu, M.; Jin, H.; Xu, C.X.; Sun, B.; Song, Z.G.; Bi, W.Z.; Wang, Y. miR-382 inhibits osteosarcoma metastasis and relapse by targeting Y box-binding protein 1. Mol. Ther. 2015, 23, 89–98.
84. Zhou, Y.; Huang, Z.; Wu, S.; Zang, X.; Liu, M.; Shi, J. miR-33a is up-regulated in chemoresistant osteosarcoma and promotes osteosarcoma cell resistance to cisplatin by down-regulating TWIST. J. Exp. Clin. Cancer Res. 2014, 33, 12.
85. Xu, M.; Jin, H.; Xu, C.X.; Bi, W.Z.; Wang, Y. MiR-34c inhibits osteosarcoma metastasis and chemoresistance. Med. Oncol. 2014, 31, 972.
86. Rosenblum, J.M.; Wijetunga, N.A.; Fazzari, M.J.; Krailo, M.; Barkauskas, D.A.; Gorlick, R.; Greally, J.M. Predictive properties of DNA methylation patterns in primary tumor samples for osteosarcoma relapse status. Epigenetics 2015, 10, 31–39.
87. Kresse, S.H.; Rydbeck, H.; Skarn, M.; Namlos, H.M.; Barragan-Polania, A.H.; Cleton-Jansen, A.M.; Serra, M.; Liestol, K.; Hogendoorn, P.C.; Hovig, E.; et al. Integrative analysis reveals relationships of genetic and epigenetic alterations in osteosarcoma. PLoS ONE 2012, 7, e48262.
88. Oh, J.H.; Kim, H.S.; Kim, H.H.; Kim, W.H.; Lee, S.H. Aberrant methylation of p14ARF gene correlates with poor survival in osteosarcoma. Clin. Orthop. Relat. Res. 2006, 442, 216–222.
89. Park, Y.B.; Park, M.J.; Kimura, K.; Shimizu, K.; Lee, S.H.; Yokota, J. Alterations in the INK4a/ARF locus and their effects on the growth of human osteosarcoma cell lines. Cancer Genet. Cytogenet. 2002, 133, 105–111.
90. Al-Romaih, K.; Sadikovic, B.; Yoshimoto, M.; Wang, Y.; Zielenska, M.; Squire, J.A. Decitabine-induced demethylation of 5′ CpG island in GADD45A leads to apoptosis in osteosarcoma cells. Neoplasia 2008, 10, 471–480.
91. Badal, V.; Menendez, S.; Coomber, D.; Lane, D.P. Regulation of the p14ARF promoter by DNA methylation. Cell Cycle2008, 7, 112–119.
92. Chen, W.; Cooper, T.K.; Zahnow, C.A.; Overholtzer, M.; Zhao, Z.; Ladanyi, M.; Karp, J.E.; Gokgoz, N.; Wunder, J.S.; Andrulis, I.L.; et al. Epigenetic and genetic loss of Hic1 function accentuates the role of p53 in tumorigenesis. Cancer Cell 2004, 6, 387–398.
93. Lu, J.; Song, G.; Tang, Q.; Zou, C.; Han, F.; Zhao, Z.; Yong, B.; Yin, J.; Xu, H.; Xie, X.; et al. IRX1 hypomethylation promotes osteosarcoma metastasis via induction of CXCL14/NF-kappaB signaling. J. Clin. Investig. 2015, 125, 1839–1856.
94. Li, Y.; Meng, G.; Huang, L.; Guo, Q.N. Hypomethylation of the P3 promoter is associated with up-regulation of IGF2 expression in human osteosarcoma. Hum. Pathol. 2009, 40, 1441–1447.
95. Ulaner, G.A.; Vu, T.H.; Li, T.; Hu, J.F.; Yao, X.M.; Yang, Y.; Gorlick, R.; Meyers, P.; Healey, J.; Ladanyi, M.; et al. Loss of imprinting of IGF2 and H19 in osteosarcoma is accompanied by reciprocal methylation changes of a CTCF-binding site. Hum. Mol. Genet. 2003, 12, 535–549.
96. Li, Y.; Meng, G.; Guo, Q.N. Changes in genomic imprinting and gene expression associated with transformation in a model of human osteosarcoma. Exp. Mol. Pathol. 2008, 84, 234–239.
97. Lee, D.F.; Su, J.; Kim, H.S.; Chang, B.; Papatsenko, D.; Zhao, R.; Yuan, Y.; Gingold, J.; Xia, W.; Darr, H.; et al. Modeling familial cancer with induced pluripotent stem cells. Cell 2015, 161, 240–254.
98. Thayanithy, V.; Park, C.; Sarver, A.L.; Kartha, R.V.; Korpela, D.M.; Graef, A.J.; Steer, C.J.; Modiano, J.F.; Subramanian, S. Combinatorial treatment of DNA and chromatin-modifying drugs cause cell death in human and canine osteosarcoma cell lines. PLoS ONE 2012, 7, e43720.
99. Zhang, X.; Cruz, F.D.; Terry, M.; Remotti, F.; Matushansky, I. Terminal differentiation and loss of tumorigenicity of human cancers via pluripotency-based reprogramming. Oncogene 2013, 32, 2249–2260.
100. Maniscalco, L. Canine osteosarcoma: Understanding its variability to improve treatment. Vet. J. 2015, 203, 135–136.

101. Rodriguez, C.O., Jr. Using canine osteosarcoma as a model to assess efficacy of novel therapies: Can old dogs teach us new tricks? Adv. Exp. Med. Biol. 2014, 804, 237–256.

102. Modiano, J.F.; Bellgrau, D.; Cutter, G.R.; Lana, S.E.; Ehrhart, N.P.; Ehrhart, E.; Wilke, V.L.; Charles, J.B.; Munson, S.; Scott, M.C.; et al. Inflammation, apoptosis, and necrosis induced by neoadjuvant fas ligand gene therapy improves survival of dogs with spontaneous bone cancer. Mol. Ther. 2012, 20, 2234–2243.

CHAPTER 5

Using Bronson Equation to Accurately Predict the Dog Brain Weight Based on Body Weight Parameter

L. Miguel Carreira [1,2,3,4]

[1]*Faculty of Veterinary Medicine, University of Lisbon (FMV/ULisboa), Av. da Universidade Técnica de Lisboa, Polo Universitário Alto da Ajuda, Lisbon 1300-477, Portugal*

[2]*Centre for Interdisciplinary Research in Animal Health (CIISA), FMV/ULisboa, Av. da Universidade Técnica de Lisboa, Polo Universitário Alto da Ajuda, Lisbon 1300-477, Portugal*

[3]*Anjos of Assis Veterinary Medicine Centre (CMVAA), Rua Dª. Francisca da Azambuja Nº 9–9A, Barreiro 2830-077, Portugal*

[4]*Private Practice in Human Dentistry, Oral Medicine and Surgery Clinic, Lisbon 1300-477, Portugal*

ABSTRACT

The study used 69 brains (n = 69) from adult dog cadavers, divided by their skull type into three groups, brachi (B), dolicho (D) and mesaticephalic (M) (n = 23 each), and aimed: (1) to determine whether the Bronson equation may be applied, without reservation, to estimate brain weight (BW) in brachy (B), dolicho (D), and mesaticephalic (M) dog breeds; and (2) to evaluate which breeds are more closely related to each other in an evolutionary scenario. All subjects were identified by sex, age, breed, and body weight (bw). An oscillating saw was used for a circumferential craniotomy to open the skulls; the brains were removed and weighed using a digital scale. For statistical analysis, p-values < 0.05 were considered significant. The work demonstrated a strong relationship between the observed and predicted BW by using the Bronson equation. It was possible to hypothesize that groups B and D present a greater encephalization level than M breeds, that B and D dog breeds are more closely related to each other than to M, and from the three groups, the D individuals presented the highest brain mass mean.

Keywords: dog; brain; brachycephalic; dolichocephalic; mesaticephalic; body weight

1. INTRODUCTION

Body weight influences almost every aspect of organism biology, including brain mass [1]. Variations in brain mass occur in proportion to the increase of body mass of the animal (allometric scaling); by increasing body size and weight, an increase of peripheral receptor conduction will promote a rise in the corresponding representation fields in the cerebral cortex (both motor and somatosensory cortices) [2,3,4]. Several studies demonstrated a strong positive correlation between brain weight (BW) and body weight (bw) in general and specifically in dogs [5,6,7,8,9,10,11,12]. The brain is at adult size when body growth is only 40 percent complete [13,14,15,16,17]. Bronson developed a mathematical equation to explain the encephalic allometry, demonstrating that the mean BW of dogs is related to mean bw and therefore BW could be calculated without previous models' inconsistencies The evolution of diversity in brain size [3,18,19,20,21,22,23,24]. According to Bronson's dog equation, average BW (y) can be expressed by the following allometric function: $Y = 0.39X^{0.27}$, where 0.39 is the integration constant, X is the bw mean, and 0.27 is the allometric exponent or the potency associated with the intra-species ratio of BW and bw means [3,18,25,26,27]. This function may serve as a baseline for measuring increases or decreases in encephalization during species evolution [3]. Encephalization level (El) as a measure of relative brain size—defined as the ratio between the actual and the predicted brain mass for an individual El—is also used as a rough estimate of the intelligence of the animal and is useful for comparison within species or between fairly closely related species [28]. Since the allometric coefficient often exhibits intraspecific variation, which can be heritable, some theories assume that greater BW differences are a result of selection for differences in bw, registered among more distantly than closely related species [5,15]. The present study was developed in dog specimens and aimed: (1) to determine whether the Bronson equation may be applied, without reservation, to estimate BW in all the brachy (B), dolicho (D), and mesaticephalic (M) dog breeds; and (2) to evaluate which breeds are more closely related to each other in an evolutionary scenario.

2. MATERIALS AND METHODS

The study used 69 brains (n = 69) from adult dog cadavers, divided by their skull type into three groups, B, D, and M (n = 23 each), obtained at the teaching hospital of Faculty of Veterinary Medicine, University of Lisbon (FMV-ULisboa), Portugal, and Anjos of Assis Veterinary Medicine Centre (CMVAA), Barreiro, Portugal, after verification of death by a veterinary surgeon, authorization by the dog owners with signed consent forms, and approval by the ethical committee. All subjects were identified by sex, age, breed, and bw. An oscillating saw was used for a circumferential craniotomy to open the skulls; the brains were removed and weighed using a digital scale. The measurements were recorded onto recording forms, and the data were put into an SPSS® database (IBM, Armonk, North Castle, NY, USA). The Kolgomorov-Smirnov test (KS) was used to test for normality and ANOVA was used to test if BW differed among B, D, and M breeds. To test for pair-wise differences between groups, we used a post hoc Bonferroni correction. A p-value < 0.05 was considered statistically significant.

3. RESULTS

Summary statistics for all considered variables (age, bw, and BW) are listed in Table 1. From all the evaluated parameters, only the BW of the brachycephalic group did not show data normality. The Bronson equation was used to calculate the expected BW from each specimen. Table 2 presents the one-way ANOVA results that tested if BW differed among B, D, and M breeds, and the post hoc Bonferroni correction to test for significant pair-wise differences between groups. Differences were registered for the bw and BW parameters. Statistically significant differences were registered for bw between the pair groups B-D ($p < 0.00$) and D-M ($p < 0.00$), and for BW only between the pair groups D-M ($p < 0.00$).

Table 1. Descriptive statistics with mean, standard deviation, minimum and maximum values, sigma and *t*, regarding the parameters of age, living weight, brain weight, and breeds in brachy (B), dolicho (D) and mesaticephalic (M) dogs.

Group	Parameter	*n*		$\bar{x} \pm SD$	95%CI		σ for $p > 0.05$	*t*
					Min (mm)	Max (mm)		
B	age	23		8.00 ± 1.65 ¥	4.00	10.00	$\sigma > 0.10$	20.13
	bw	23		15.79 ± 6.46 §	4.70	26.10	$\sigma > 0.10$	11.71
	BW	23		84.91 ± 31.29 *	26.62	147.27	$\sigma = 0.08$	13.02
	Breed	23	8	French Bulldog				
			7	Boxer				
			4	Pekingese				
			2	Pug Carlin				
			2	Shitzu				
D	age	23		8.70 ± 2.40 ¥	5.00	14.00	$\sigma > 0.10$	17.46
	bw	23		23.63 ± 2.59 §	18.70	29.39	$\sigma > 0.10$	37.79
	BW	23		92.50 ± 8.60 *	71.60	102.10	$\sigma = 0.00$	50.60
	Breed	23	11	Doberman Pinsher				
			5	Rough Collie				
			4	Whippet				
			3	Miniature Bull Terrier				
M	age	23		9.50 ± 2.50 ¥	5.00	15.00	$\sigma = 0.02$	17.70
	bw	23		13.80 ± 7.20 §	7.30	31.50	$\sigma < 0.01$	8.74
	BW	23		69.90 ± 28.70 *	35.70	143.0	$\sigma = 0.04$	11.68
	Breed	23	7	Beagle				
			6	Golden Retriever				
			5	Yorkshire Terrier				
			3	Border Collie				
			2	Dalmatian				

Body weight (bw); brain weight (BW), minimum (min); maximum (max); confidence interval (CI). Measurements are in millimetres. ¥ Years, § kilograms, * grams.

4. DISCUSSION

According to the results, BW mean differed little, with no statistical significant differences, between the three groups considered; however, from all the groups, the M group presented a greater mean age (9.5 years), followed by group D (8.7 years), and then B (8.0 years). This is a very important parameter, since aging is associated with decreased brain volume (about 0.29% per year) and weight (about 5%–10% over the lifetime), due to atrophy of the gray and white matters, enhanced by multiple factors [17,29,30,31,32]. Aged dog brains show increased cortical atrophy, ventricular dilation, decreased total brain volume, and

Table 2. One-way ANOVA and post hoc Bonferroni corrected tests for differences in age, living weight, brain weight, maximum width, length and height between the brachy (B), dolicho (D) and mesaticephalic (M) dogs. The comparison between groups was made with an n = 46 (23 specimens within each group). Also, the comparison between observed and expected brain weight values obtained by using the Bronson equation in brachy (B), dolicho (D) and mesaticephalic (M) dogs. The correlation coefficient (Pearson's product moment correlation) between both values in the three considered groups. The difference was significant at p-values < 0.05.

Sample Characteristics

Parameter	Type of Test	Differences Between Groups		n	DM (I–J)	EP	SS	MS	F	Sig.	CI 95%	
											Min	Max
Age	ANOVA	-		69	-	-	28.17	14.08	2.77	0.06	-	-
	Bonferroni	B	D	46	−0.78	1.17	-	-	-	1.00	−2.41	0.85
		B	M	46	−1.56	2.35	-	-	-	0.09	−3.19	0.06
		D	M	46	−0.78	1.17	-	-	-	0.39	−2.41	0.85
bw	ANOVA	-		69	-	-	3303	447.7	12.27	0.00	-	-
	Bonferroni	B	D	46	−6.62 *	3.72	-	-	-	0.00	−11.00	2.25
		B	M	46	1.73	0.97	-	-	-	1.00	−2.64	6.10
		D	M	46	8.36 *	4.69	-	-	-	0.00	−3.98	12.73
BW	ANOVA	-		69	-	-	4739	3026	4.83	0.01	-	-
	Bonferroni	B	D	46	−7.62	1.03	-	-	-	1.00	−25.75	10.50
		B	M	46	14.92	2.02	-	-	-	>0.05	−3.20	33.06
		D	M	46	22.55 *	3.05	-	-	-	0.00	4.41	40.68

Brain Weight

Group	n	Observed BW (g)			Expected BW (g)			p	r	χ^2 for $\chi^{2c} > 5.99$
		$\overline{x}$	min	max	$\overline{x}$	min	max			
B	23	84.91	53.60	116.20	82.20	71.30	90.10	0.68	0.99	2.47
D	23	92.50	83.90	101.10	91.30	88.50	94.00	0.35	0.81	1.28
M	23	69.90	41.20	98.60	79.20	64.60	88.80	0.17	0.96	5.85

Body weight (bw); brain weight (BW); Mean Difference (I–J); confidence interval (CI); minimum (min); maximum (max); Pearson correlation coefficient (r); Chi square (χ^c); Chi square critical (χ^{2c}); * statistically significant.

decreased frontal lobe volume, similar to humans [29,32]. In previous studies, age-related reduction in gray matter volume was observed bilaterally in the frontal gyrus, orbitalis gyrus, ectosylvius hyrus, olfactory bulb, superior olivaris nucleus in brainstem, and unilaterally in the posreal gyrus, sylvium gyrus, suprasylvian gyrus, cerebellum, and brainstem nuclei. Variations related to white matter loss were largely bilateral and included the internal capsula, tracts of anterior cingulate, and the alveus of the hypocampus [17,29,32]. Additionally, high body mass index, biochemical changes (in particular, those related to dopamine loss), and the loss of neurons and myelinated axons in different brain parenchyma regions decrease the brain mass with age promoting an enlargement of the ventricular system and the grooves on the brain surface [17,30,31,32,33,34]. In accordance with Schmidt et al. [35], the Bronson equation may serve as a baseline for measuring the El of species during their evolution. According to Kruska [5] and Pagel and Harvey [15], greater BW differences are registered among distantly related species as a result of selection for differences in bw. The study results showed that the biggest differences were registered between the pair-wise groups M-B and M-D; thus we can conclude that B and D dog breeds are more closely related to each other than to M. This is supported by the fact that the B and D breeds were the result of natural and artificial selection over the M breeds, which are most closely related to the gray wolf—*Canis lupus*—and from which they were domesticated [36]. This selection acted directly on the brain itself, and also depended on the selective forces involved, promoting the decreasing size of one part of the brain in order to increase the size of another [35,37]. This is the result of simultaneous isocortical neurogenesis onset and terminal neurogenesis that varies substantially between the rostral and caudal poles in some species, allowing for disproportionate expansion of some encephalic cortical regions relative to others [38,39,40,41,42,43]. With the Bronson equation, it was possible to verify that D and B specimens presented the best ratio between the registered and the predicted BW, and according the study results, we found that D has the highest mean brain mass, heavier than the B and the M specimens.

5. CONCLUSIONS

In conclusion, the work demonstrated a strong relationship between the observed and predicted BW by using the Bronson equation. It was possible to hypothesize that groups B and D present a greater El than M breeds, and from the three groups, the D individuals presented the highest brain mass mean.

ACKNOWLEDGMENTS

The author thank Centre for Interdisciplinary Research in Animal Health, CIISA, of the Faculty of Veterinary Medicine, University of Lisbon (FMV-ULisboa), Portugal; and Anjos of Assis Veterinary Medicine Centre (CMVAA), Barreiro, Portugal.

CONFLICTS OF INTEREST

The author declare that he has no conflict of interest. This research received no specific grant from any funding agency in the public, commercial, or not-for-profit sectors.

REFERENCES

1. Lindstedt, S.L.; Schaeffer, P.J. Use of allometry in predicting anatomical and physiological parameters of mammals. Lab. Anim. 2002, 36, 1–19.
2. Jerison, H.J. Evolution of the Brain and Intelligence; Academic: New York, NY, USA; London, UK, 1973; pp. 406–435.
3. Bronson, R.T. Brain weight-body weight scaling in breeds of dogs and cats. Brain Behav. Evol. 1979, 16, 227–236.
4. Schoenemann, P.T. Brain size scaling and body composition in mammals. Brain Behav. Evol. 2004, 63, 47–60.
5. Kruska, D. Effects of domestication on brain structure and behavior in mammals. Hum. Evol. 1988, 3, 473–485.
6. Radinsky, L. Evolution of brain size in carnivores and ungulates. Am. Nat. 1978, 112, 815–831.
7. Röhrs, M. Cephalisation bei Caniden. Z. Zool. Syst. Evolutionsforsch. 1986, 24, 300–307.
8. Carrera, I.; Dennis, R.; Mellor, D.J.; Penderis, J.; Sullivan, M. Use of magnetic resonance imaging for morphometric analysis of the caudal cranial fossa in Cavalier King Charles Spaniels. Am. J. Vet. Res. 2009, 70, 340–345.
9. Carreira, L.M. Anatomotopography Cranioencephalic Relationships for Intracranial Neurosurgery Procedures in Dog. The Ultrasound as Neuronavigation Technique for Brain Surgery in Real Time. Ph.D. Thesis, Technical University of Lisbon (FMV/UTL), Lisbon, Portugal, 2011.
10. Carreira, L.M.; Ferreira, A. Reference values for dog sagittal and transverse cephalic indices in different skull types and their importance. J. Anim. Vet. Adv. 2015, 14, 91–94.
11. Hart, B.L.; Hart, L.A.; McCoy, M.; Sarath, C.R. Cognitive behaviour in Asian elephants: Use and modification of branches for fly switching. Anim. Behav. 2011, 62, 839–847.
12. Seid, M.A.; Castillo, A.; Wcislo, W.T. The allometry of brain miniaturization in ants. Brain Behav. Evol. 2011, 77, 5–13.
13. Armstrong, E. Relative brain size and metabolism in mammals. Science 1983, 4603, 1302–1304.
14. Martin, R.D.; Harvey, P. Brain size allometry: Ontogeny and phylogeny. In Size and Scaling in Primate Biology; Jungers, W., Ed.; Plenum Press: New York, NY, USA, 1985.
15. Pagel, M.D.; Harvey, P.H. Taxonomic differences in the scaling of brain on body weight among mammals. Science1989, 30, 1589–1593.
16. Wikswo, J.P.; Curtis, E.L.; Eagleton, Z.E.; Evans, B.C.; Kole, A.; Hofmeister, L.H.; Matloff, W.J. Scaling and systems biology for integrating multiple organs-on-a-chip. Lab Chip 2013, 13, 3496–3511.
17. Conn, P.M. Handbook of Models for Human Aging; Academic Press: Cambridge, MA, USA, 2006.
18. Gould, S.J. Ontogeny and Phylogeny; Harvard University Press: Estados Unidos, MA, USA, 1977.

19. Gould, S.J. An allometric interpretation of species-area curves: The meaning of the coefficient. Am. Nat. 1979, 114, 335–343.
20. Jerison, H.J. Allometry, brain size, cortical surface, and convolutedness. In Primate Brain Evolution; Armstrong, E., Falk, D., Eds.; Plenum Press: New York, NY, USA, 1982.
21. Jerison, H.J. The evolutionary biology of intelligence: Afterthoughts. In Intelligence and Evolutionary Biology; Jerison, H., Jerison, I., Eds.; Springer: Berlin, Germany, 1988.
22. Jerison, H.J. The theory of Encephalization. Ann. N. Y. Acad. Sci. 1977, 299, 146–160.
23. Gayon, J. History of the concept of allometry. Integr. Comp. Biol. (ICB) 2000, 40, 748–758.
24. Williams, M.F. Primate Encephalization and intelligence. Med. Hypotheses 2002, 58, 284–290.
25. Tan, U.; Caliskan, S. Allometry and asymmetry in the dog brain: The right hemisphere is heavier regardless of paw preference. Int. J. Neurosci. 1987, 35, 189–194.
26. Rushton, J.P.; Ankney, C.A. Whole brain size and general mental ability: A review. Int. J. Neurosci. 2009, 119, 692–732.
27. Bayat, P.D.; Ghanbari, A.; Sohouli, P.; Amiri, S.; Sari-aslani, P. Correlation of skull size and brain volume, with age, weight, height and body mass index of Arak Medical Sciences students. Int. J. Morphol. 2012, 30, 157–161.
28. Marino, L.; Sol, D.; Toren, K.; Lefebvre, L. Does diving limit brain size in cetaceans? Mar. Mammal Sci. 2006, 22, 413–425.
29. Su, M.Y.; Head, E.; Brooks, W.M.; Wang, Z.; Muggenburg, B.A.; Adam, G.E. Magnetic resonance imaging of anatomic and vascular characteristics in a canine model of human aging. Neurobiol. Aging 1988, 19, 479–485.
30. Anderton, B.H. Ageing of the brain. Mech. Ageing Dev. 2002, 123, 811–817.
31. Enzinger, C.; Fazekas, F.; Matthews, P.M.; Ropele, S.; Schmidt, H.; Smith, S.; Schmidt, R. Risk factors for progression of brain atrophy in aging: Six-year follow-up of normal subjects. Neurology 2005, 64, 1704–1711.
32. Tapp, P.D.; Siwak, C.T.; Gao, F.Q.; Chiou, J.Y.; Black, S.E.; Head, E.; Muggenburg, B.A.; Cotman, C.W.; Milgram, N.W.; Su, M.Y. Frontal lobe volume, function, and beta-amyloid pathology in a canine model of aging. J. Neurosci. 2004, 24, 8205–8213.
33. Conde, J.R.; Streit, W.J. Microglia in the aging brain. J. Neuropathol. Exp. Neurol. 2006, 65, 199–203.
34. Raz, N.; Rodrigue, K.M.; Head, D.; Kennedy, K.M.; Acker, J.D. Differential aging of the medial temporal lobe: A study of a five-year change. Neurology 2004, 62, 433–438.
35. Schmidt, M.J.; Amort, K.H.; Failing, K.; Klingler, M.; Kramer, M.; Ondreka, N. Comparison of the endocranial- and brain volumes in brachycephalic dogs, mesaticephalic dogs and Cavalier King Charles spaniels in relation to their body weight. Acta Vet. Scand. 2014, 56, 30.
36. Hemmer, H. Domestication: The Decline of Environmental Appreciation, 2nd edition; Cambridge University Press: Cambridge, UK, 1990.
37. Healy, S.D.; Harvey, P.H. Comparative studies of the brain and its components. Neth. J. Zool. 1990, 40, 203–214.
38. Rakic, P. Neurons in rhesus monkey visual cortex: Systematic relation between time of origin and eventual disposition. Science 2002, 183, 425–427.
39. Bush, E.C.; Allman, J.M. The scaling of frontal cortex in primates and carnivores. Proc. Natl. Acad. Sci. USA 2004, 101, 3962–3966.
40. Hutsler, J.J.; Lee, D.G.; Porter, K.K. Comparative analysis of cortical layering and supragranular layer enlargement in rodent carnivore and primate species. Brain Res. 2005, 1052, 71–81.
41. Carreira, L.M.; Ferreira, A.; Burilo, F.L. The dorsal sagittal venous sinus anatomical variations in brachycephalic, dolichocephalic, and mesocephalic dogs and their significance for brain surgery. Anat. Rec. Adv. Integr. Anat. Evol. Biol. (Hoboken) 2011, 294, 1920–1929.

42. Carreira, L.M.; Ferreira, A. The importance of the longitudinal cerebral fissure anatomy variations in brachy-, dolicho- and mesaticephalic dogs and their importance to brain surgery. Anat. Rec. Adv. Integr. Anat. Evol. Biol. (Hoboken) 2015, 298, 1612–1621.
43. Workman, A.D.; Charvet, C.J.; Clancy, B.; Darlington, R.B.; Finlay, B.L. Modeling transformations of neurodevelopmental sequences across mammalian species. J. Neurosci. 2013, 33, 7368–7383.

CHAPTER 6

An Ambulatory Electroencephalography System for Freely Moving Horses: An Innovating Approach

Hugo Cousillas[1]*, Martial Oger[2], Céline Rochais[3], Claire Pettoello[3], Mathilde Ménoret[1], Séverine Henry[3] and Martine Hausberger[1]

[1]*Université de Rennes 1, CNRS UMR 6552 – Ethologie Animale et Humaine EthoS, Rennes Cedex, France*

[2]*IETR, Université de Rennes 1, UMR CNRS 6164, Rennes Cedex, France*

[3]*Université de Rennes 1, CNRS UMR 6552 – Ethologie Animale et Humaine EthoS, Station Biologique, Paimpont, France*

ABSTRACT

Electroencephalography (EEG) that has been extensively studied in humans presents also a large interest for studies on animal brain processes. However, since the quality of the recordings is altered by muscular activity, most EEG recordings on animals are obtained using invasive methods with deeply implanted electrodes. This requires anesthesia and can thus only be used in laboratory or clinical settings. As EEG is a very useful tool both for detecting brain alterations due to diseases or accidents and to evaluate the arousal and attentional state of the animal, it seemed crucial to develop a tool that would make such recordings possible in the horse's home environment, with a freely moving horse. Such a tool should neither be invasive nor cause discomforts to the horse as the usual other practice which consists, after shaving the zone, in gluing the electrodes to the skin.

To fulfill these requirements, we developed a novel EEG headset adapted to the horse's head that allows an easy and fast positioning of the electrodes and that can be used in the home environment on a freely moving horse. In this study, we show that this EEG headset allows to obtain reliable recordings, and we propose an original evaluation of an animal's "EEG profile" that allows comparisons between individuals and situations. This EEG headset opens new possibilities of investigation on horse cognition, and it can also become a useful tool for veterinarians to evaluate cerebral disorders or check the anesthesia level during a surgery.

Keywords: electroencephalography, horses, freely moving, EEG headset, Telemetry

INTRODUCTION

Electroencephalography (EEG) has been extensively studied in humans over the last decades (1). It has been especially used in human medicine as a diagnostic tool to assess cerebral dysfunctions like epilepsy (2, 3). EEG is especially useful to characterize different levels of vigilance from the different stages of sleep to wakefulness (4). In humans, EEG recordings are mostly based on nontraumatic external electrodes placed on the head's skin. The quality of these human EEG recordings depends on the subject's quietness, and hence, the level of muscular activity may interfere: the subject must be quiet and avoid any movement. Adult humans can be instructed to do so.

Of course, more difficulties are, thus, encountered with animals. EEG presents also a large interest for studies of animal brain processes from basic research on sleep; attention; and awareness to applied issues, such as the impact of anesthesia, brain damages, induced or spontaneous brain diseases, and epilepsy (5–7). Although EEG recording techniques have been developed for animals, there are important constraints that may prevent to exploit the full range of possibilities offered by EEG.

Thus, in awake animals, it is almost impossible to avoid movements, and therefore, most EEG recordings in animals have been done invasively using deeply implanted electrodes (8, 9), in which case, the electrodes record directly the cortical activity (electrocorticogram), and hence, produce high-quality data. Otherwise, subdermal needle could be used, which may also induce discomfort or even pain, leaving potentially negative memories (10, 11). Such techniques have been used in equids and have allowed to characterize the EEG waves of the different stages of vigilance (4). Of course, this requires a surgery under anesthesia to implant the electrodes that can be quite traumatic. This, of course, is not adapted for "routine" use, especially in large domestic species.

Other authors have developed a noninvasive approach on awake animals: the animals are kept immobile (generally in metal stocks), the electrodes are glued on the animal's head and any movement has negative consequences (12).

Often the horse's forehead has to be shaved, which in the case of horses may be a problem for both esthetical reasons (owners may refuse if the EEG is for non-pathological reasons) and be a source of fear for untrained (to the procedure) horses, which limit the population that can be tested. Moreover, the affixed electrodes may, at the time of removal, create a "bad memory" of the procedure that makes the horse reluctant to accept them again on further occasions. This anticipation may also impact the quality of the EEG recordings (see further).

However, EEG is a precious tool not only for evaluating brain alterations or dysfunction due to diseases or trauma (7), but also to study the vigilance state of the animal, for testing the impact of drugs, welfare reasons (13, 14), or studies on cognition. It is then crucial to be able to use it in a field situation that is in the home environment of the horse and preferably on freely moving animals. Thus, it is absolutely necessary that the recording apparatus (1) can be used in the home environment, (2) is easy and rapid to adapt to each horse, (3) can be used on a freely moving animal (that can orient toward a stimulus), and (4) does not put more constraint on the horses than the usual gear such as halter of snaffle (i.e., rapid habituation).

In this study, we present a novel EEG headset adapted to the horse's head that allows an easy and fast (less than 5 min) positioning of the electrodes that solves all the above-mentioned issues (15). The aim of this article is to describe this novel system and display the EEG recordings obtained. We also describe here an original approach for analyzing the EEG data through an evaluation of the relative proportion of the different types of waves. This approach allows to examine intra- and interindividual variations in the "EEG profile" within a given situation and intraindividual variations between situations. Here, we show EEG recordings obtained from five relaxed standing adult sport horses that were freely moving in their home stall over two sessions performed on two successive days.

MATERIALS AND METHODS

Material and Technological Developments

The main innovation was the EEG headset (Figure 1A; patent # R23701WO) that allowed us to place easily five electrodes on the horse's forehead over the parietal and frontal bones. The design of this headset, a necklace made of large rubber bands, allowed a good and stable electrode contact with the horse skin and was adjustable for different head sizes (Figure 1B).

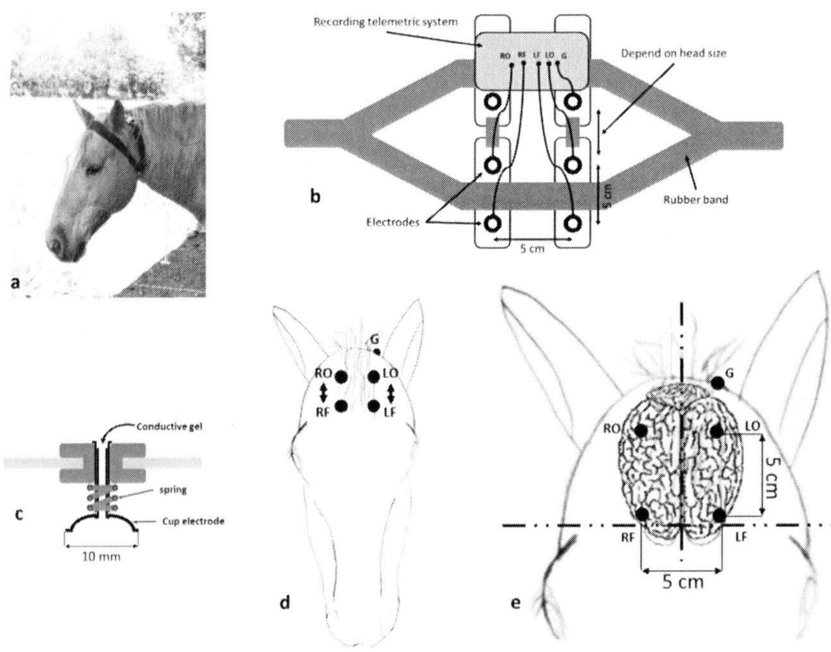

FIGURE 1. (A) Electroencephalography (EEG) headset, a necklace made of large rubber bands, placed on the horse's head. **(B)** Schematic representation of the EEG headset. **(C)** Cup gold electrodes with a 10-mm diameter (DSGSAS102600 Spes Medica) mounted with a spring on a tube that can be filled with a conductive gel (NEURGEL250F Spes Medica) in order to have a good electrical contact. **(D)**Position of the electrodes, RO, right occipital; RF, right frontal; LO, left occipital; LF, left frontal; G, body ground. **(E)** The electrodes locations with the anatomical landmarks and the measurements are shown on a schematic representation of the horse brain that is located just underneath the frontal bone.

Electrode Positioning

The electrodes (Figure 1C) were cup gold electrodes with a 10-mm diameter similar to those used on humans (DSGSAS102600 Spes Medica). They were filled with a conductive gel (NEURGEL250F Spes Medica) in order to have a good electrical contact. They were positioned on each side of the horse forehead to allow separate hemispheric recordings (Figure 1D). On each side, the electrodes were placed on occipital and frontal positions. This electrode positioning, similar to the one used by Mysinger et al. (16), allowed us to record the differential activity between the most occipital part of the brain and the most frontal one. On an adult horse, the most rostral electrodes were positioned on a reference line located between the zygomatic processes of the frontal bones (Figure 1E). The occipital electrodes were located 5 cm backward from the frontal ones. Left and right electrodes, positioned on either side of the sagittal plane, were separated by 5 cm. The ground electrode was placed on the back of the left ear that allowed us to avoid most of the muscular artifacts. Using these anatomical landmarks allowed a precise repositioning of the electrodes on repetition with less than 2 mm differences.

Telemetric Recording Setup

The electrophysiological recordings were performed, thanks to a homemade telemetric recording setup (L, 110 mm; l, 90 mm; h, 30 mm; weighting, 110 g) developed by Martial Oger (Figure 2). The homemade EEG amplifier was based on the two-way recordings “ModularEEG” from the openEEG project (http://openeeg.sourceforge.net/doc/index.html). This amplifier was connected to the telemetric radio transmitter. The whole telemetric recording setup was fixed on the headset.

Any existing telemetric system could fit with the headset. In the example described here, we used a homemade system. After being amplified the EEG signal was chopped in order to modulate two oscillators working with two nearby frequencies (around 433 MHz). The output signals of the oscillators were added and delivered to a quarter of wave whip antenna. The oscillators were both stabilized and modulated in frequency. Thefrequencies of the oscillators were stabilized using a frequency synthesizer in the 50 Hz bandwidth. In order to modulate the frequency of the oscillators, the EEG signal frequencies that are too low to modulate these oscillators were shifted toward higher frequencies. Thus, the EEG signals were chopped with a period of 1 msec (1 kHz). A microcontroller was used to drive the two frequency synthesizers and to command the choppers.

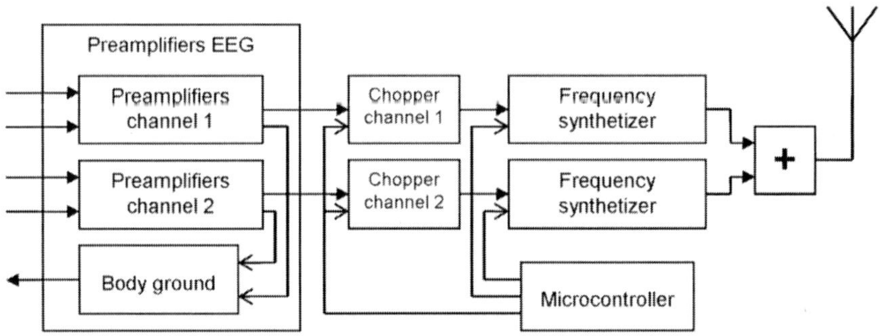

FIGURE 2. SCHEMATIC DIAGRAM OF THE HOMEMADE RECORDING TELEMETRIC SYSTEM USED IN THIS STUDY. The electroencephalography (EEG) amplifier (left) was based on the two-way recordings "ModularEEG" from the openEEG project (http://openeeg.sourceforge.net /doc/ index.html).

To avoid wiring around the head and along the neck of the horse and to allow the animal to move freely, a homemade EEG amplifier associated with a radioemitter was also developed. With this telemetric EEG amplifier that was fixed directly on the recording headset on the head, the animal was completely detached from any monitor.

To obtain the EEG signals, we used a classical FM receiver associated with an envelope demodulator. The demodulation of the signal received gives an audio signal chopped at 1 kHz, and to obtain the EEG signals, we made an envelope demodulation of this audio signal.

The EEG signals were then recorded in a computer using a digital oscilloscope "Picoscope 2200 Series" (pico^R Technology) and the Picoscope Oscilloscope Software. The sampling rate was 1 kHz.

The Animals and the Experimental Procedure Used in the Test Study

The recordings were performed at the French national equitation school of Saumur on five horses (three French saddlebred, two Anglo-Arab) aged 7–9 years (four females and one gelding), trained for competition [show jumping (*N* = 3) and eventing (*N* = 2)].

In the test study, the horses were quietly standing. Each horse was tested during two sessions of 15 min, one on the first day (D1) and the other on the following day (D2). All experiments were video-recorded using the media-recorder software that allowed to record and synchronize several videos. Behaviors were then encoded and synchronized with EEG recordings using "The Observer" software from Noldus.

Data Analysis

We processed all data using FieldTrip (17), a free toolbox of Matlab 7.0 (MathWorks, Natick, MA). For each horse, the videos were screened, and the periods when the animal was standing relaxed were identified. EEG signals during these time periods were visually inspected, and large artifacts, due to movements (body, head, ears), were removed. The data were then segmented into 500 ms epochs, and a second visual artifact rejection procedure was performed to exclude potential remaining artifacts. There were an average of 86.45 ± 9.05 epochs/horse/session. Then, a complex fast-Fourier transform was applied on these 500 ms epochs in which, for each frequency, outlying values ±3SD were identified, and the corresponding epoch was removed from the analysis. The FFT values for each horse were averaged separately for each session. All usual waves were present (delta, theta, alpha, beta, and gamma), but their relative weight varies according to the animal state. In the present case, we concentrated on the pertinent waves and extracted the power values of the alpha (8–12 Hz), beta (12–30 Hz), and gamma (>30 Hz) frequency bands, characteristic of awake animals. We calculated the proportion of the mean power values of these three wave types (alpha, beta, and gamma) to express these values as a percentage. Then, to evaluate the stability of the "EEG profiles" of the different individuals and over the time, the coefficients of variation (Cv = SD/mean) of these proportions were calculated. Coefficients of variations have long been shown to be a reliable way of broaching the question of the levels of variation (18).

RESULTS

The EEG recordings obtained were similar to those produced by other methods (Figure 3). This EEG, similar for all horses, was characteristic of a quiet awake state and in accordance with those obtained in previous studies (16) using external or implanted electrodes, with a clear desynchronization, low to medium voltages, and slow and more rapid waves superimposed (Figure 3A). There was no difference between hemispheres, and thus, the data were pooled for further analysis (see last part).

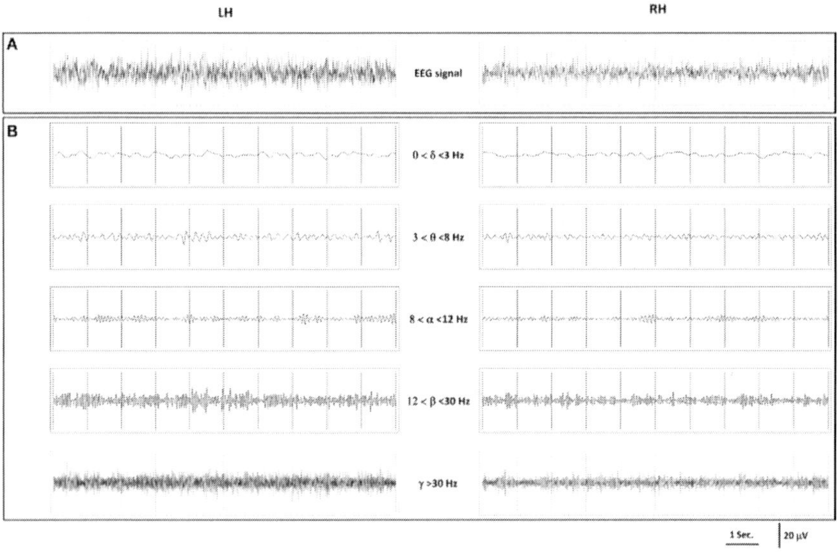

FIGURE 3. (A) Example of 10 s electroencephalography (EEG) recordings obtained in a relaxed standing horse; **(B)** representation of each wave range of these recordings.

To disentangle the relative part of the different waves, the EEG recordings were further filtered and revealed that low frequencies (delta and theta) that are characteristic of slow wave sleep were less represented and of low amplitude (<10 µV), while three frequency ranges (among the five) dominated in all horses: the alpha waves (8–12 Hz) that are characteristic of a relaxed state, the beta (12–30 Hz), and the gamma waves (>30 Hz) that are characteristic of higher awareness state with amplitudes of 10–30 µV (Figure 3B).

We then tried to evaluate the potential interindividual variations in the EEG pattern by calculating the relative proportion of the three dominating wave ranges (Figure 4). On day 1, the proportion of alpha, beta, and gamma waves was, respectively, of 37.63% (±3.5), 34.38% (±2.19), and 27.99% (±1.97) and on day 2, 36.62% (±5.6), 34.57 (±1.66), and 28.81 (±4.92).

The interindividual variation was very low with coefficients of variations of 12.29%, 5.59%, and 12.06%. Obviously, in a same context, the EEG pattern of different horses was very similar.

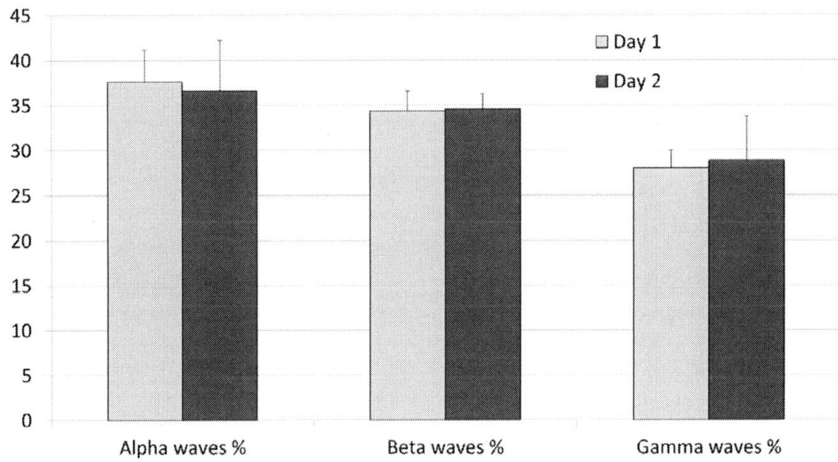

FIGURE 4. PROPORTIONS OF THE POWER SPECTRUM OF ALPHA, BETA, AND GAMMA WAVES RECORDED ON THE FIVE RELAXED STANDING HORSES ON TWO DIFFERENT RECORDING DAYS. Error bars indicate mean ± SD.

Interestingly, the intraindividual variation (same individual's day 1/day 2) was still lower with mean coefficients of variation of 6.39%, 4.69%, and 9.22% for alpha, beta, and gamma waves, respectively. It is worth noting that the proportion of beta waves appeared the most stable feature both at the inter- and intraindividual levels.

DISCUSSION

The novel EEG headset for EEG recordings proved to be quite adapted for recordings of freely moving animals in their home environment (15). The use of a telemetric device enabled us to avoid the usual constraints related to cabling. Typical EEG recordings were obtained for the five relaxed standing horses: the major part in the EEG recordings was composed of frequencies higher than 8 Hz. Since the recordings have been performed on awake horses, low frequencies (delta and theta waves) that are characteristic of slow wave sleep were weakly represented here. We recorded small amplitude alpha waves, which is in accordance with Williams et al.'s (12) study. All horses showed alpha waves, which is interesting as according to Mysinger et al. (16) only half of the horses presented alpha waves in their study. The authors noted that the horses that were lacking these waves were apprehensive during the recordings, which is logical because cortical arousal associated with anxiety would increase beta and decrease alpha frequencies. One reason why we observed them in all horses may be because of the conditions of recordings that differ from the abovementioned study: here, the horses were totally freely moving in their usual stall and the device was very light, whereas in the other study, they were restrained and had a heavier device to wear. For example, in equids, several authors have developed nontraumatic EEG recordings (16, 19), with the aim of studying EEG waves during anesthesia, sleep, or even epilepsy (7, 12, 20–22). The electrodes were positioned and glued on the head, a conductive paste was added. In general, the horse's forehead was shaved to have a good contact. Long wires were used to connect the electrodes to the amplifiers, and the recording system was placed on the back of the horse. These EEG recording systems are better and easier to use than those using implanted electrodes, but they still need a relatively long preparation. Moreover, the horses were placed between bars preventing them to move their body. However, the temporal region of the horse's head is heavily muscled and movements of the ears, the eyes, and the jaws impacted strongly the recordings, and a good part had to be left aside for these reasons. Being able to record EEG in freely moving animals in their familiar settings may, therefore, considerably change the knowledge we have of brain processing in horses.

The lower variability of intraindividual recordings shows that this headset can be placed precisely on the horse head and that it allows a precise repositioning of the electrodes.

Among the three types of waves, the beta waves were more stable. The horses were awake, quiet, and looking around, which means that attentional processes may vary as the animal is taking care or not of what is going on around it. The beta waves are characteristic of awake states, whereas the gamma waves are involved on higher awareness states and alpha waves on quiet awake states. The higher variability found on alpha and gamma waves may reflect changes on attentional processes due to environment changes.

The aim of the system we developed was as follows: (1) the system can be used in the home environment, (2) easy and rapid to adapt to each horse, (3) can be used on a freely moving animal, and (4) does not put more constraint on the horses than the usual gear. We answer all these questions: the system can be used on familiar environment, on a freely moving horse, without more constraint than the usual gear, easy and rapid to adapt on each horse and that allows reliable recordings.

This apparatus will be a precious tool for future studies aiming at evaluating, in the "field" situation, individual differences in cognitive processes (Rochais et al., in preparation) or vigilance states in different contexts such as sleep or anesthesia. For the latter, further studies will be needed to examine the reliability of data over prolonged periods of time, as this study, as a first step, was based on short-time periods.

ETHICS STATEMENT

All procedures were in accordance with the European Communities Council Directive of 22 September 2010 (2010/63/UE) and the French law relative to the protection of the animal used in scientific experiment (Décret no. 2013-118 13 février 2013; Article R. 21488). According to these European and French laws, our experiment did not require an authorization request. Indeed, our manipulations did not cause any physical or mental pain, they consisted only of a common practice, which is the positioning of a necklace on a horse head that was freely moving in its home environment.

AUTHOR CONTRIBUTIONS

HC, CP, and MH conceived the EEG helmet; MO conceived the telemetric system; HC, CR, SH, and MH conceived the study and designed the experiments. HC and CR conducted experiments. MM, HC, CR, and MH analyzed the data; HC and MH wrote the manuscript, and all the authors participated in the revision of the manuscript.

CONFLICT OF INTEREST STATEMENT

The authors declare that the research was conducted in the absence of any commercial or financial relationships that could be construed as a potential conflict of interest.

ACKNOWLEDGMENTS

The study was funded by the IFCE (Institut français du cheval et de l'équitation), the University of Rennes 1, the CNRS (Centre National de la Recherche Scientifique), and the French Ministry of Research. The authors are grateful to the director of the National School of Equitation and especially Mrs. S. Biau, as well as the school staff who made this study possible. The authors also thank Melissa Sebilleau for her technical help.

REFERENCES

1. Dietrich A, Kanso R. A review of EEG, ERP, and neuroimaging studies of creativity and insight. *Psychol Bull* (2010) 136:822–48.
2. Lascano AM, Vulliemoz S, Lantz G, Spinelli L, Michel C, Seeck M. A review on non-invasive localization of focal epileptic activity using EEG source imaging. *Epileptology* (2012) 29:80–9.
3. Smith SJM. EEG in the diagnosis, classification, and management of patients with epilepsy. *J Neurol Neurosurg Psychiatry*(2005) 76(Suppl II):ii2–7.
4. Rodenbeck A, Binder R, Geisler P, Danker-Hopfe H, Lund R, Raschke F, et al. A review of sleep EEG patterns. Part I: a compilation of amended rules for their visual recognition according to Rechtschaffen and Kales. *Somnologie* (2006) 10:159–75.
5. Lacombe VA, Mayes M, Mosseri S, Reed SM, Fenner WR, OU HT. Epilepsy in horses: aetiological classification and predictive factors. *Equine Vet J* (1999) 44(6):646–51.
6. Grint NJ, Johnson CB, Clutton RE, Whay HR, Murrell JC. Spontaneous electroencephalographic changes in a castration model as an indicator of nociception: a comparison between donkeys and ponies. *Equine Vet J* (2014) 47(1):36–42.
7. Wijnberg ID, van der Ree M, van Someren P. The applicability of ambulatory electroencephalography (AEEG) in healthy horses and horses with abnormal behaviour or clinical signs of epilepsy. *Vet Q* (2014) 33:121–31.
8. Ruckebusch Y. Etude EEG et comportementale des alternances veille-sommeil chez l'âne. *CR Soc Biol* (1963) 157:840–4.
9. Dallaire A, Ruckebusch Y. Sleep and wakefulness in the housed pony under different dietary conditions. *Can J Comp Med* (1974) 38:65–71.
10. Murrell JC, Johnsony CB, White KL, Taylor PM, Haberham ZL, Waterman-Pearson AE. Changes in the EEG during castration in horses and ponies anaesthetized with halothane. *Vet Anaesth Analg* (2003) 30:138–46.
11. Murrell JC, White KL, Johnson CB, Taylor PM, Doherty TJ, Waterman-Pearson AE. Investigation of the EEG effects of intravenous lidocaine during halothane anaesthesia in ponies. *Vet Anaesth Analg* (2005) 32:212–21.
12. Williams D, Aleman M, Holliday T, Fletcher D, Tharp B, Kass P, et al. Qualitative and quantitative characteristics of the electroencephalogram in normal horses during spontaneous drowsiness and sleep. *J Vet Intern Med* (2008) 22:630–8.
13. Rochais C, Henry S, Fureix C, Hausberger M. Investigating attentional processes in depressive-like domestic horses (Equus caballus). *Behav Processes* (2016) 124:93–6.
14. Rochais C, Fureix C, Lesimple C, Hausberger M. Lower attention to daily environment: a novel cue for detecting chronic horses' back pain? *Sci Rep* (2016) 6:20117.
15. Cousillas H, Oger M, Rochais C, Pettoello C, Henry S, Hausberger M. An ambulatory electroencephalography system for free moving horses: an innovating approach. *12th ISES Conference*. Saumur (2016). p. 143–4. Demonstration 8.
16. Mysinger PW, Redding RW, Vaughan JT, Purohit RC, Holladay JA. Electroencephalographic patterns of clinically normal, sedated and tranquilized newborn foals and adult horses. *Am J Vet Res* (1985) 46:36–41.
17. Oostenveld R, Fries P, Maris E, Schoffelen J-M. FieldTrip: open source software for advanced analysis of MEG, EEG, and invasive electrophysiological data. *Comput Intell Neurosci* (2011) 2011:156869.
18. Barlow GW. Ethological units of behavior. In: Ingle D, editor. *TheNervous System and Fish Behavior*. Chicago: University of Chicago Press (1968). p. 217–32.
19. Hale LA, Huggins SE. The electroencephalogram of the normal "grade" pony in sleep and wakefulness. *Comp Biochem Physiol*(1980) 66A:251–7.

20. Williams DC, Aleman M, Tharp B, Fletcher DJ, Kass PH, Steffey EP, et al. Qualitative and quantitative characteristics of the electroencephalogram in normal horses after sedation. *J Vet Intern Med* (2012) 26(3):645–53.
21. Williams DC, Brosnan RJ, Fletcher DJ, Aleman M, Holliday TA, Tharp B, et al. Qualitative and quantitative characteristics of the electroencephalogram in normal horses during administration of inhaled anesthesia. *J Vet Intern Med* (2016) 30:289–303.
22. Williams DC, Aleman MR, Brosnan RJ, Fletcher DJ, Holliday TA, Tharp B, et al. Electroencephalogram of healthy horses during inhaled anesthesia. *J Vet Intern Med* (2016) 30:304–8.

CHAPTER 7

Utility of MRI for Characterizing Articular Cartilage Pathology in Dogs with Medial Coronoid Process Disease

Samuel P. Franklin[1,2]*, Emily E. Burke[3] and Shannon P. Holmes[3]

[1]*Department of Small Animal Medicine and Surgery, Veterinary Teaching Hospital, University of Georgia, Athens, GA, USA*

[2]*Regenerative Bioscience Center, University of Georgia, Athens, GA, USA*

[3]*Veterinary Biosciences and Diagnostic Imaging, University of Georgia, Athens, GA, USA*

ABSTRACT

Objective: To determine whether assessment of morphological MRI sequences or delayed gadolinium-enhanced MRI of cartilage (dGEMRIC) would have strong correlations with arthroscopic assessment of cartilage pathology in dogs with naturally occurring medial compartment pathology of the elbow.

Methods: Dogs tentatively diagnosed with medial coronoid disease had evaluation of their affected elbows using radiography, morphological MRI sequences, and dGEMRIC MRI evaluation prior to arthroscopy. Elbow radiographs were graded 0–6 for severity of changes. Cartilage of the medial coronoid process (MCP) and humeral trochlea (HT) were scored on a 0–3 scale using anatomical MRI sequences. The T1 relaxation times for the MCP and trochlea were quantified using dGEMRIC. Cartilage pathology was graded arthroscopically using a modified Outerbridge score (MOS) by a surgeon blinded to MRI assessment. Correlations between radiography and MOS, and between MRI and MOS, were quantified.

Results: Twenty-six elbows in 14 dogs were evaluated. There were statistically significant ($p < 0.05$) moderate correlations between radiographic scores and MOS for the MCP ($r = 0.71$) and HT (0.57). There was a statistically significant moderate correlation between morphological MRI scoring and MOS for the HT ($r = 0.54$; $p < 0.05$), but not for the MCP ($p > 0.05$). There was a weak, but significant correlation, between the dGEMRIC value and MOS of the MCP ($r = 0.41$; $p < 0.05$), but no correlation between the dGEMRIC values and MOS for the HT ($p > 0.05$).

Clinical relevance: Statistically significant correlations to MOS were identified for both radiography and MRI but neither diagnostic modality provided sufficiently strong correlations to serve as a substitute for arthroscopic evaluation of the articular cartilage.

Keywords: MRI, medial coronoid process, arthroscopy, dGEMRIC, elbow dysplasia

INTRODUCTION

There is great variability in the character and severity of pathology affecting dogs with medial compartment disease of the elbow. Such variability includes varying degrees and distribution of cartilage damage on the medial coronoid process (MCP) or humeral trochlea (HT) with some dogs having arthroscopically normal or near normal articular cartilage in the medial compartment while other dogs have complete loss of articular cartilage throughout the medial compartment (1, 2). There is a need to accurately characterize pathology in the canine elbow non-invasively so that treatment planning can be performed prior to performing any surgery, including arthroscopic assessment. This is clinically relevant because approximately 50% of dogs with medial coronoid disease have some pathology of cartilage on the HT and there are a growing number of surgical options available for treating dogs when pathology is not limited solely to the MCP (3–7). An ability to accurately characterize pathology of the medial compartment non-invasively would facilitate selection of such treatments before diagnostic arthroscopy is performed. In addition, accurate non-invasive characterization of pathology would enable performance of clinical trials in which dogs managed non-surgically have pathology that is well defined. In turn, their response to non-surgical treatment could be compared to dogs managed surgically while being confident that the pathology affecting both groups were similar or equivalent.

Currently, radiography, computed tomography (CT), and arthroscopic surgery are the primary diagnostics used to make a diagnosis of elbow dysplasia (8). Of these, only arthroscopy provides an ability to assess the articular cartilage and is generally considered the gold standard for assessment of pathology in the canine elbow in dogs with MCP disease (2, 9, 10). Unfortunately, arthroscopy, although a relatively non-invasive treatment, is a relatively invasive diagnostic imaging assessment and staging tool. Conversely, MRI could potentially be used to non-invasively determine the pathology in the canine elbow. A few studies have evaluated the use of MRI for such purpose (11–14). Initial work demonstrated good sensitivity and specificity of MRI for diagnosing pathology of the MCP and HT in comparison to radiography while using surgical findings as the gold standard (11). However, there were limitations of the technique and so use of MRI arthrography was subsequently evaluated but failed to identify suspected clinically relevant information in some cases (12).

A more recent investigation evaluated standard morphological MRI sequences (using a 1.5-T scanner) and their correlation to arthroscopic and histopathologic assessments in dogs with subtle radiographic changes and only those having arthroscopically normal HT. There was a statistically significant but only moderate correlation between the MRI score and the modified Outerbridge score (MOS) (14). The authors concluded that the results were inadequate to validate use of MRI as a stand-alone staging tool. Another study evaluated the repeatability of values obtained using quantitative MRI techniques, delayed gadolinium-enhanced MRI for cartilage [delayed gadolinium enhanced MRI of cartilage (dGEMRIC)], and T2 mapping specifically, in six research dogs with normal elbows (15). Both intra- and interobserver precisions were found to be good. These results are encouraging and raise the possibility that MRI could potentially be used to identify pathology of the articular cartilage early in the course of disease and could also potentially monitor response to treatments. However, no study has yet to demonstrate a sufficiently strong correlation between MRI scoring and either arthroscopy or biochemical analysis of the articular cartilage in dogs with naturally occurring pathology of the medial compartment to justify routine clinical application.

The purpose of this study was to determine whether morphological or quantitative MRI assessment of articular cartilage pathology correlates with arthroscopic assessments in dogs with naturally occurring pathology of the medial compartment. Our second objective was to determine whether radiographic assessment of pathology correlated with arthroscopic assessed severity of joint pathology as has been previously demonstrated (1). We hypothesized that correlations between MRI and arthroscopic scoring and between radiographic and arthroscopic scoring would both be statistically significant. We hypothesized that the strength of correlation would be greater between MRI and arthroscopy than between radiography and arthroscopy.

ANIMALS AND METHODS

This study was approved by the Clinical Research Committee at the University of Georgia.

Animals

Dogs that were tentatively diagnosed with naturally occurring medial coronoid disease based upon history, physical examination, and radiography and that were candidates for general anesthesia and arthroscopy were recruited for enrollment. Owners were provided with a written description of the study and owners provided written consent for inclusion of their dog in the study. These dogs would not have had MRI at our institution for suspected medial coronoid disease if they were not enrolled in this research study. The owner did not pay for the MRI as funding for the MRI was provided by the study sponsor.

Radiography

If dogs had recent (within 6 weeks), well-positioned, adequate quality medial–lateral and cranial–caudal radiographs of the affected elbows prior to presentation to the UGA VMC, radiography was not repeated. If appropriate radiographs were not available at the time of presentation, dogs were sedated with 0.005 mg/kg of Dexmedetomidine (Pfizer, NY, USA) and 0.5 mg/kg Nalbuphine (Pfizer, NY, USA) given intravenously and orthogonal radiographs of the affected elbows were made.

MRI Protocol

Dogs were anesthetized for MRI evaluation with routine anesthetic protocols that typically included premedication with an opioid and a sedative followed by induction with Propofol (PropoFlo 28, Zoetis, Kalamazoo, MI, USA) or Ketamine (Ketaset, Zoetis, Kalamazoo, MI, USA) and maintained with Isoflurane or Sevoflurane dependent upon anesthesiologist choice. The MRI examinations were performed using a 3.0-T MRI unit (Skyra, Siemens). Since bilateral studies were performed, the dogs were positioned in sternal recumbency with their thoracic limbs extended forward. Care was taken to ensure straight alignment of the joints along the long axis of the thoracic limbs. Parallel imaging was employed with a combination of a phased-array spine coil and a flexible body matrix coil, respectively positioned ventral and dorsal to the elbows. The imaging protocol started with the morphologic sequences, including intermediate-weighted (IW) fat-suppressed fast spine echo (FSE) (sagittal plane), IW fat-suppressed FSE (dorsal plane), proton density (PD) FSE (sagittal plane), and PD SPACE (sagittal plane) imaging. The dGEMRIC series were collected following the morphologic series and T1 MapIT sequence in the sagittal plane. A pre-contrast series was collected of both elbows after which the dogs were administered 0.01 mmol/kg of gadopentetate dimeglumine (Magnevista®, Bayer, Pittsburgh, PA, USA) intravenously. The elbows were cycled through a range of motion (flexion, extension, and rotational motion) for 10 min while the dog was still anesthetized, similar to a previously described protocol (15). If both elbows were to be evaluated, they were cycled through range of motion simultaneously. Following range of motion, the dogs were repositioned in sternal recumbency with the limbs pulled forward and the post-contrast dGEMRIC images were collected.

Arthroscopic Assessment and Treatment

Immediately following completion of MRI and under the same general anesthesia, dogs were aseptically prepared for surgery. Arthroscopy was performed by a surgeon blinded to MRI findings using a standard caudomedial instrument portal and a craniomedial instrument portal and using 1.9 mm 30° foreoblique arthroscope (Arthrex Vet Systems, Naples, FL, USA). The specific structures evaluated included the synovium and articular surface of the radial head, humeral capitulum, HT, MCP, lateral coronoid process, and the semilunar notch. The arthroscopic procedure was recorded. Treatment of the affected MCP included a combination of fragment removal and/or subtotal coronoid ostectomy depending upon the extent of pathology. All dogs were recovered from general anesthesia and postoperative pain control provided with intermittent opioid administration for 24 h and approximately 14 days of non-steroidal anti-inflammatory medication.

Arthroscopic Scoring

The recorded videos were reviewed by the surgeon who was blinded to MRI findings for characterization of pathology. Pathology of the MCP and the HT was scored using a MOS (0–5) that has been described and used in prior studies for characterization of elbow pathology in dogs with medial compartment pathology (Table 1) (1, 3).

TABLE 1. Modified outerbridge scoring system used for arthroscopic evaluation of cartilage pathology (**1**, **3**).

Modified Outerbridge score	Description of gross cartilage quality
0	Normal
1	Chondromalacia, determined in part by probing with an arthroscopic probe
2	Partial thickness fibrillation
3	Deep fibrillation
4	Full thickness cartilage loss
5	Subchondral bone eburnation

Radiographic Scoring

All radiographs had all identifying information removed, were randomized, and assigned a novel identification number to create blinded assessment. Radiographs were assessed by the surgeon a minimum of 6 months following the last arthroscopy. The medial–lateral radiographs were scored on a 4-point (0–3) scale using a previously described system (1). Cranial–caudal radiographs were also scored on a 4-point scale in which osteophyte formation was considered with a score of 0 representing a normal elbow and 1, 2, and 3 were consistent with mild, moderate, and severe radiographic changes, respectively. Scores for both images were summed to provide a maximum score of 6, representing the most substantial radiographic changes.

MRI Scoring

The morphologic MRI images were graded by a radiologist blinded to arthroscopic and radiographic findings. Cartilage of the MCP and HT were each scored on a 0–3 scale (0, no change in signal from the cartilage; 1, signal change, but no change in thickness of the associated articular cartilage; 2, signal change and partial thinning of the cartilage; 3, full thickness cartilage loss).

Evaluation of the dGEMRIC images was performed by the same radiologist blinded to arthroscopic and radiographic findings. The radiologist evaluated the dGEMRIC images a minimum of 3 months following evaluation of the morphologic MRI images and did not review the morphologic images while evaluating the dGEMRIC images so as to obviate any influence on the dGEMRIC evaluations by prior assessment of the morphologic images. Three consecutive sagittal plane slices were assessed that included the MCP and HT. On each of these slices, three different methods of creating regions of interest (ROIs) were performed. The first method included creating one small ROI on the MCP and two small ROIs of the same dimension on the weight-bearing articular surface of the HT (Figures 1A,B). Each of these ROIs was created and placed to include the articular cartilage without including underlying subchondral bone or overlying joint fluid. Accordingly, three and six quantitative measurements of the T1 relaxation time were obtained for the MCP and HT, respectively. The means of these were calculated and used in subsequent statistical analyses. Small circular ROIs were used to ensure that the area encircled with each anatomic region was constant and to ensure that none of these ROIs spanned the joint space to include the opposing articular surface. In addition, use of small ROIs enabled placement to include the articular cartilage while excluding measurement from the adjacent synovial fluid or subchondral bone.

In an effort to more closely replicate the previous study documenting consistency in ROI measurements within and between observers using dGMERIC in the canine elbow, measurements were also taken by placing a single larger square ROI that spanned the joint space and incorporated the cartilage surface of the HT, any intervening joint fluid, and the articular surface of the MCPs (Figure 1C) (15). Such measurements were made on the same three sagittal images and the mean calculated for subsequent statistical analysis. For the third method of creating ROIs, free-form traces of the MCP and the HT were performed (Figure 1D). Measurements were made on the same three sagittal images and the means calculated.

Statistical Analysis

All correlations were assessed using Pearson product-moment correlations. Specifically, the correlation between the MOS for the MCP and the radiographic score and between the MOS for the HT and radiographic score were quantified. Similarly, correlations between the arthroscopic MOS for the MCP and the MRI score for the MCP using the morphologic images was quantified. Likewise, the correlation between the arthroscopic MOS for the HT and the MRI score for the HT using the morphologic images was determined. Correlations between the MOS and the dGEMRIC scores were similarly calculated using the dGEMRIC values obtained from each of the three different techniques of drawing ROIs (Table 2).

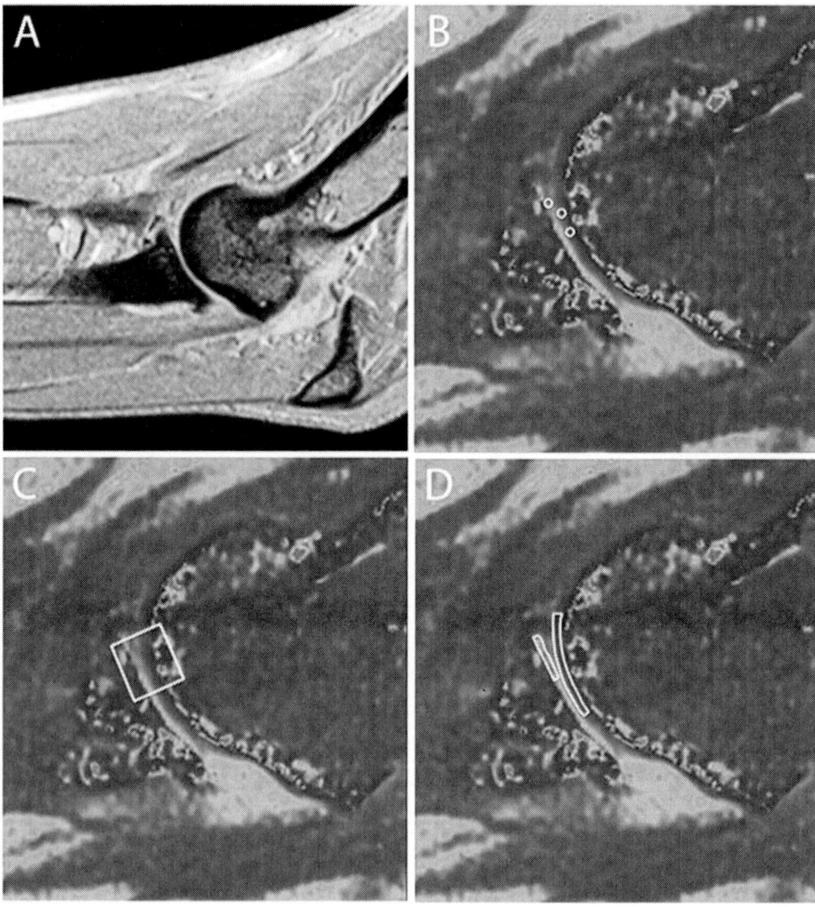

FIGURE 1. REPRESENTATIVE SAGITTAL PLANE MRI IMAGES. **(A)** T1-weighted anatomical reference image and **(B–D)** magnified delayed gadolinium-enhanced MRI of cartilage images depicting three methods for drawing regions of interest (ROIs) for cartilage scoring. **(B)** One ROI is on the medial coronoid process (MCP) and two on the humeral trochlea (HT). Note that the ROIs include the articular cartilage layer without substantial inclusion of adjacent subchondral bone, joint fluid, or opposing articular surface. **(C)** A single square ROI is used that spans the joint space and includes the articular cartilage of the medial HT and MCP. **(D)** Two free-form ROIs are used that trace the MCP and HT.

TABLE 2. Correlations of diagnostic evaluations to arthroscopic assessment of cartilage pathology.

Diagnostic imaging modality	Anatomic location	MOS location	*r*	*p*-Value
Radiography	Global elbow score	MCP	0.71	<0.0001
Radiography	Global elbow score	Trochlea	0.57	<0.01
MRI	MCP (morphologic images)	MCP	0.22	0.28
MRI	Trochlea (morphologic images)	Trochlea	0.54	<0.01
Delayed gadolinium enhanced MRI of cartilage (dGEMRIC)	MCP circular region of interest (ROI)	MCP	0.25	0.23
dGEMRIC	Trochlea circular ROIs	Trochlea	−0.14	0.50
dGEMRIC	Square ROI spanning MCP/human trochlea (HT)	MCP	0.41	0.04
dGEMRIC	Square ROI spanning MCP/HT	Trochlea	0.10	0.65
dGEMRIC	MCP free-form trace	MCP	0.35	0.09
dGEMRIC	Trochlea free-form trace	Trochlea	−0.06	0.76

MOS, modified Outerbridge score; MCP, medial coronoid process.
Trochlea refers to the HT.

RESULTS

Demographics

Radiography, MRI, and arthroscopic evaluations were performed in 26 elbows in 14 dogs; 2 dogs had only one elbow evaluated. There were 2 females, 6 spayed females, 4 castrated males, and 2 intact males included. Breeds included Labrador Retriever ($n = 5$), Golden Retriever ($n = 2$), mixed breed (3), and one each of Anatolian Shepherd, Newfoundland, Pit Bull, and Rottweiler. The mean age was 3.1 years (±SD 4.0 years). The mean body weight was 35.0 kg (±10.1). T1 relaxation times using dGEMRIC were assessed from 25 elbows in 14 dogs because the associated sequence was lost prior to analysis for one elbow of 1 dog.

Diagnostic Imaging Assessment

Radiographic and MRI Assessment

The mean radiographic score for each elbow was 3.0 (±1.8) out of 6. The mean grade for articular cartilage on the MCP using MRI anatomical sequences was 2.2 (±0.6). The mean MRI score for the humeral trochlear cartilage using MRI anatomical sequences was 1.68 (±0.8).

Delayed Gadolinium-Enhanced MRI of Cartilage

The mean time between gadolinium administration and associated imaging was 37 min (±13). If only those elbows imaged first following contrast administration were considered the mean time between contrast administration and the beginning of image acquisition was 29 min (±8) and the mean delay between contrast administration and beginning of image acquisition for the second elbow was 48 min (±12). The mean T1 relaxation times for the assessed areas are presented in Table 3.

TABLE 3. Mean delayed gadolinium enhanced mri of cartilage t1 relaxation times.

Anatomic location	Region of interest shape	Mean (±SD) T1 relaxation time
MCP	Small circular	728.1 (±228.5)
Trochlea	Small circular (#1)	723.2 (±274.8)
Trochlea	Small circular (#2)	734.0 (±267.5)
MCP and trochlea	Square	647.5 (±159.7)
MCP	Free-form trace	808.5 (±337.4)
Trochlea	Free-form trace	788.2 (±258.5)

MCP, medial coronoid process.
Trochlea (humeral trochlea).

Arthroscopic Assessment

The mean MOS for cartilage pathology on the MCP was 1.8 (±1.4). The mean MOS for cartilage pathology on the HT was 1.5 (±1.6).

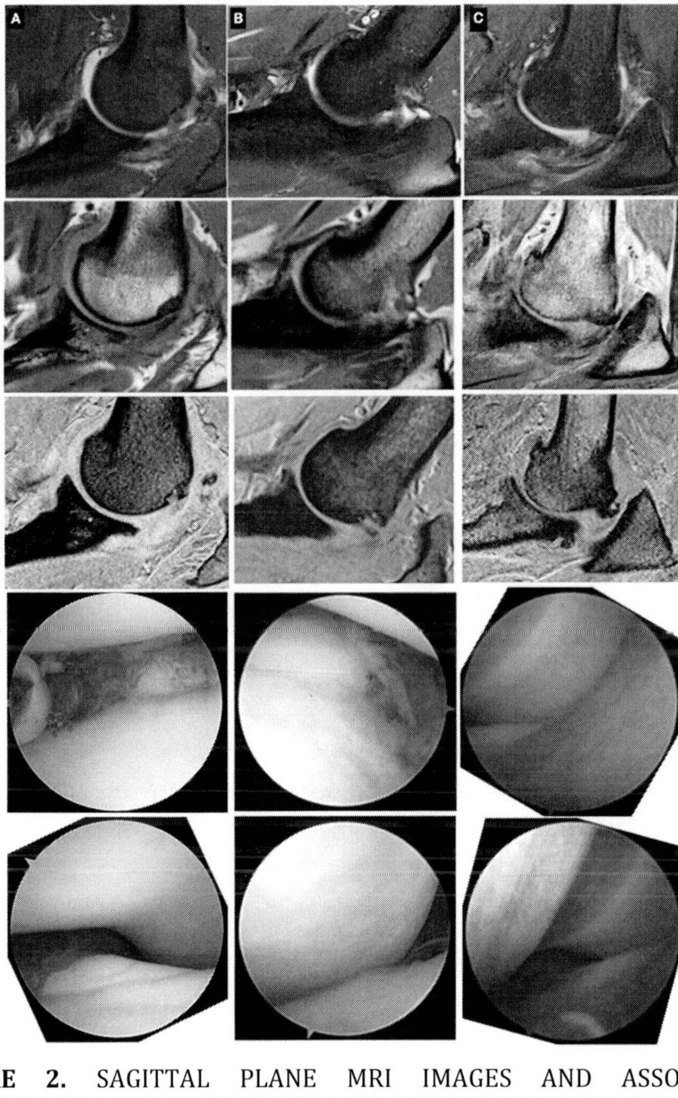

FIGURE 2. SAGITTAL PLANE MRI IMAGES AND ASSOCIATED ARTHROSCOPIC IMAGES FROM THREE PATIENTS (PATIENTS A, B, AND C) WITH CARTILAGE PATHOLOGY OF VARYING SEVERITY. From top to bottom within a column, the images are fat-suppressed proton density (PD)-weighted (2.5 mm thick), PD-weighted (1.5 mm thick), and T1-weigthed (1.0 mm thick) images followed by a representative arthroscopic image of the medial coronoid process and the humeral trochlea. Note that the spectrum of cartilage pathology is readily apparent with the arthroscopic images, but less apparent with the MRI images.

Correlations

There were numerous statistically significant correlations between the arthroscopic MOS and diagnostic imaging (Table 2). The strength of correlations between radiographic scoring and arthroscopic MOS were moderate. There was moderately strong ($r = 0.54$) statistically significant ($p < 0.05$) correlation between MRI cartilage assessment of the HT and MOS using morphological sequences; there was not a significant correlation between MRI cartilage assessment of the MCP and MOS of the MCP. Three out of five MCPs that had full thickness cartilage loss (MOS Grade IV or V) were assessed as having only partial thickness cartilage loss using the morphological MRI sequences (see Figure 2). Two out of three HTs that had MOS Grade IV or V change were assessed as only having partial thickness cartilage loss using the morphological MRI sequences (Figure 2). There was a weak to moderate ($r = 0.41$) statistically significant ($p < 0.05$) correlation between the T1 relaxation time using the dGEMRIC sequences and MOS for the MCP when using a single square ROI that spanned the joint space and included the articular surface of the MCP and HT. There were no other statistically significant correlations between the MOS and T1 relaxation times using dGEMRIC sequences (Table 2).

DISCUSSION

The demographic data demonstrate inclusion of dogs of typical breeds (large breeds) affected with elbow pathology but with a wide range in age and degree of arthroscopic assessed cartilage pathology. The mean MOS score for the MCP and HT were relatively low at 1.8 and 1.5, respectively. Although these values are somewhat low and might be interpreted as hampering our ability to detect significant correlations between the MOS and MRI scores that SDs of these values were relatively high and the MOS values ranged from 0 to 5 for both the MCP and HT. The inclusion of dogs with notable variability in severity of cartilage pathology was specifically desired to maximize the likelihood of detecting positive correlations between non-invasive diagnostic imaging modalities and the arthroscopic assessed cartilage pathology. Indeed, these data do demonstrate statistically significant but only moderate correlations between radiographic assessment and arthroscopic assessments of cartilage pathology in the medial compartment of the canine elbow. These results warrant accepting our first hypothesis and are not surprising given results from a previous study demonstrating a relationship between the arthroscopic global cartilage pathology score and radiographic assessment of disease severity in a large group of Labrador Retrievers (1). The results from that study enables one to make probability statements regarding severity of cartilage pathology based upon radiographs but do not enable accurate prediction of cartilage pathology in many dogs with mild or moderate (radiographic score of 1 or 2) radiographic changes (1). The data from this study are based upon far fewer dogs and do not provide evidence for greater clinical utility of radiographic examination than that provided in the previous study (1). As a result, we conclude that there is a moderate correlation between radiographic findings and arthroscopically assessed cartilage pathology but there remains a notable possibility that dogs with minimal or moderate radiographic change have severe cartilage pathology as assessed arthroscopically (16). As a result, radiographs remain insufficiently sensitive for identification of all dogs with end stage medial compartment disease. In turn, radiographs in isolation are inadequate for staging or outcome assessment in clinical trials comparing non-surgical management to surgical treatment for dogs with pathology of the medial compartment.

There was also a statistically significant correlation between MOS for the HT and assessment of cartilage pathology using the MRI morphological images. However, this correlation was weak to moderate and judged to be clinically irrelevant and there was no correlation between MRI assessment using the morphological images and the MOS for the MCP. As a result, we do not accept our hypothesis that correlations between MRI assessment (using morphological images) and arthroscopy would be stronger than the correlations between radiographic scoring and arthroscopic assessments of cartilage pathology. These results, although disappointing, are not unprecedented as a recent study with similar objectives also failed to find highly clinically relevant correlations between MRI assessment of canine elbows using morphological images and arthroscopic assessment of the elbows (14). The data from these two studies demonstrate that the morphpological MRI protocols used in these studies cannot be reliably used to fully stage elbow pathology. This conclusion is made with consideration to the fact that the signal to noise ratio, high bandwidth, and image resolution were maximized in this 3.0-T MRI study using longer scan times than typical for routine clinical scans. Specifically, image resolution and bandwidth were used to have multiple pixels within the 0.4–0.6 mm thick cartilage layer to facilitate identification of partial and full thickness cartilage loss (17). Thinner slices were also obtained but partial volume averaging effects were noted, particularly on the HT, which potentially could mask lesions with the mild curvature of the structure in the sagittal plane. We do not believe that poor quality images can be an explanation for failing to more accurately identify cartilage pathology with MRI. Possible explanations for inaccuracy, or rather opportunities for re-evaluation and improvement of MRI for cartilage assessment in the canine elbow may include using different sequences (i.e., 3D submillimeter isotropic FSE images) and imaging in different planes than what was used in this study. The sagittal plane was selected to obtain high in-plane resolution of cartilage contours perpendicular to the orientation of the majority of the cartilage line the MCP and HT. Based on this study, the MRI reviewer's impression was that at most one morphologic slice was truly perpendicular to the articular surface and would not suffer partial volume effects due to the small anatomy and the bidirectional curvature of the MCP and HT.

There was also a moderate and statistically significant correlation between the dGEMRIC T1 relaxation time and MOS for the MCP using a single ROI that spanned the articular surface. This result is interesting in that this was the only method of drawing ROIs, of those assessed in this study, that demonstrated a significant correlation and most closely resembles the ROI methodology used in previous study demonstrating repeatability of measurements in dogs with normal elbows (15). For those reasons, this significant finding is encouraging that this methodology could ultimately become clinically useful with further validation. However, the correlation identified in this study is considered too weak to be clinically applicable. As a result, we are unaware of any studies demonstrating a clinically relevant correlation between dGEMRIC values and arthroscopic assessment in canine elbows with naturally occurring disease.

There are several possible explanations as to why the dGEMRIC values did not correlate more strongly with the arthroscopic assessed images in these elbows. First, it is possible that the arthroscopic assessment does not correlate well with the biochemical composition and histology of the cartilage and that the MRI values may have correlated more closely with biochemical asssessments of the cartilage.

A recent study demonstrated only weak correlations between histology and MRI interpretations from anatomical sequences (14). Second, it is possible that the protocol for gadolinium administration, passive range of motion, and relatively short delay until image acquisition did not result in sufficient gadolinium penetration into the joint and affected cartilage. When considering that the time frame between gadolinium administration and image acquisition is shorter in this study than in a previous study evaluating dGEMRIC for normal canine elbows, this seems a possible explanation (15). However, comparison of the dGEMRIC values for the HT and humeral capitulum from this study (data not shown) demonstrated significantly higher T1 relaxation rates in the lateral compartment, consistent with at least some gadolinium entering the joint and decreasing the T1 relaxation times in the medial compartment in comparison to the lateral compartment. Last and with relevance to use of morphological images as well as quantitative MRI, the cartilage in the canine elbow is quite thin with mean cartilage thickness of just 0.51 mm in one study (17). Assessment of cartilage status in this anatomic location using either morphological images or quantitative MRI may always be hampered by the thinness of the cartilage.

Conclusions that can be drawn from negative results are tempered by associated study limitations. Further limitations of the study beyond those mentioned above include that intra- and interobserver repeated measures were not quantified for the radiographic, arthroscopic, or MRI assessments. However, demonstration of precision is more pertinent when positive results are obtained, and it is necessary to show that such results can be repeated. The greatest limitation is that biochemistry and histologic assessment of tissue samples was not possible in this study using client-owned dogs and it remains feasible that MRI may have correlated more closely with histology or biochemistry than with the arthroscopic evaluations. No tissue was removed from the HT and the subtotal coronoidecotmies performed in this study were typically performed with an arthroscopic burr to mitigate the risk of creating intra-articular free bodies as has been documented following arthroscopic treatment of medial coronoid disease (18). Large subtotal coronoidectomies were not performed based upon some concern that such approach may not provide an optimal outcome (19). As a result, larger portions of bone and cartilage were not available for biochemical or histological evaluation.

Despite these limitations, we conclude that the correlations between arthroscopic findings and interpretations of MRI images were absent to moderate. Importantly, the morphological MRI images evaluated in this study did not consistently enable identification of elbows with full thickness cartilage loss on the HT (Figure 2). As a result, MRI cannot be substituted for arthroscopic assessment at this point in time, a conclusion also obtained by other investigators using morphological MRI assessments as well as using CT evaluations of canine elbows (10, 14).

AUTHOR CONTRIBUTIONS

SF devised the study, performed surgery, evaluated data, and prepared the manuscript. EB performed MRI evaluations. SH developed the MRI protocols, evaluated the MRI images, and contributed to manuscript preparation.

CONFLICT OF INTEREST STATEMENT

The authors declare that the research was conducted in the absence of any commercial or financial relationships that could be construed as a potential conflict of interest.

FUNDING

The authors acknowledge the generous funding support of the AKC Canine Health Foundation (Grant 2142).

REFERENCES

1. Farrell M, Heller J, Solano M, Fitzpatrick N, Sparrow T, Kowaleski M. Does radiographic arthrosis correlate with cartilage pathology in Labrador Retrievers affected by medial coronoid process disease? *Vet Surg* (2014) 43:155–65.
2. Farrell M, Fitzpatrick N. Conservative versus arthroscopic management for medial coronoid process disease in dogs: a prospective gait evaluation. *Vet Surg* (2012) 41:541–2; author reply 542–3.
3. Fitzpatrick N, Smith TJ, Evans RB, Yeadon R. Radiographic and arthroscopic findings in the elbow joints of 263 dogs with medial coronoid disease. *Vet Surg* (2009) 38:213–23.
4. Franklin SP, Schulz KS, Karnes J, Cook JL. Theory and development of a unicompartmental resurfacing system for treatment of medial compartment disease of the canine elbow. *Vet Surg* (2014) 43(7):765–73.
5. Fitzpatrick N, Yeadon R, Smith T, Schulz K. Techniques of application and initial clinical experience with sliding humeral osteotomy for treatment of medial compartment disease of the canine elbow. *Vet Surg* (2009) 38:261–78.
6. Dejardin L, Guillou R. Total elbow replacement in dogs. In: Tobias K, Johnston S, editors. *Small Animal Veterinary Surgery*. St. Louis, MO: Elsevier Saunders (2012). p. 752–59.
7. Fitzpatrick N, Yeadon R. Working algorithm for treatment decision making for developmental disease of the medial compartment of the elbow in dogs. *Vet Surg* (2009) 38:285–300.
8. Cook CR, Cook JL. Diagnostic imaging of canine elbow dysplasia: a review. *Vet Surg* (2009) 38:144–53.
9. Moores AP, Benigni L, Lamb CR. Computed tomography versus arthroscopy for detection of canine elbow dysplasia lesions. *Vet Surg* (2008) 37:390–8.
10. Coppieters E, Van Ryssen B, van Bree H, Verhoeven G, Broeckx B, de Bakker E, et al. Computed tomographic findings in canine elbows arthroscopically diagnosed with erosion of the medial compartment: an analytical method comparison study. *Vet Radiol Ultrasound* (2016) 57:572–581.
11. Snaps FR, Balligand MH, Saunders JH, Park RD, Dondelinger RF. Comparison of radiography, magnetic resonance imaging, and surgical findings in dogs with elbow dysplasia. *Am J Vet Res* (1997) 58:1367–70.

12. Snaps FR, Park RD, Saunders JH, Balligand MH, Dondelinger RF. Magnetic resonance arthrography of the cubital joint in dogs affected with fragmented medial coronoid processes. *Am J Vet Res* (1999) 60:190–3.
13. Snaps FR, Saunders JH, Park RD, Daenen B, Balligand MH, Dondelinger RF. Comparison of spin echo, gradient echo and fat saturation magnetic resonance imaging sequences for imaging the canine elbow. *Vet Radiol Ultrasound* (1998) 39:518–23.
14. Wavreille V, Fitzpatrick N, Drost WT, Russell D, Allen MJ. Correlation between histopathologic, arthroscopic, and magnetic resonance imaging findings in dogs with medial coronoid disease. *Vet Surg* (2015) 44:501–10.
15. Wucherer KL, Ober CP, Conzemius MG. The use of delayed gadolinium enhanced magnetic resonance imaging of cartilage and T2 mapping to evaluate articular cartilage in the normal canine elbow. *Vet Radiol Ultrasound* (2012) 53:57–63.
16. Goldhammer MA, Smith SH, Fitzpatrick N, Clements DN. A comparison of radiographic, arthroscopic and histological measures of articular pathology in the canine elbow joint. *Vet J* (2010) 186:96–103.
17. Zeissler M, Maierl J, Grevel V, Oechtering G, Bottcher P. Cartilage thickness and split-line pattern at the canine humeral trochlea. *Vet Comp Orthop Traumatol* (2010) 23:343–7.
18. Coppieters E, Seghers H, Verhoeven G, Gielen I, Samoy Y, de Bakker E, et al. Arthroscopic, computed tomography, and radiographic findings in 25 dogs with lameness after arthroscopic treatment of medial coronoid disease. *Vet Surg* (2016) 45:246–53.
19. Brauer S, Bottcher P. Focal humero-ulnar impingement following subtotal coronoid ostectomy in six dogs with fragmented medial coronoid process. *Tierarztl Prax Ausg K Kleintiere Heimtiere* (2015) 43:156–60.

CHAPTER 8

Forced-Air Warming Provides Better Control of Body Temperature in Porcine Surgical Patients

Brian T. Dent [1], Karla A. Stevens [2] and Jeffrey W. Clymer [2,*]

[1]*College of Veterinary Medicine, The Ohio State University, Columbus, OH 43210, USA*

[2]*Ethicon, Inc., Cincinnati, OH 45242, USA*

ABSTRACT

Background: Maintaining normothermia during porcine surgery is critical in ensuring subject welfare and recovery, reducing the risk of immune system compromise and surgical-site infection that can result from hypothermia. In humans, various methods of patient heating have been demonstrated to be useful, but less evaluation has been performed in techniques to prevent hypothermia perioperatively in pigs. *Methods:* We compared body temperature regulation during surgery before and after modification of the ambient temperature of the operating laboratories. Three different methods of heating were then compared; a standard circulating water mattress, a resistive fabric blanket, and a forced hot air system. The primary measure was percentage of temperature readings outside a specification range of

36.7–40.0 °C. *Results:* Tighter control of the ambient temperature while using a circulating water mattress reduced the occurrence of out-of-specification body temperature readings from 20.8% to 5.0%, with most of these the result of hypothermia. Use of a resistive fabric blanket further reduced out-of-specification readings to 1.5%, with a slight increase in the occurrence of hyperthermia. Use of a forced air system reduced out-of-specification readings to less 0.1%. *Conclusions:* Maintenance of normothermia perioperatively in pig can be improved by tightly controlling ambient temperatures. Use of a resistive blanket or a forced air system can lead to better control than a circulating water mattress, with the forced air system providing a faster response to temperature variations and less chance of hyperthermia.

Keywords: porcine; surgery; body temperature; hypothermia; resistive blanket; forced air heating

1. INTRODUCTION

Perioperative hypothermia is a recurrent challenge which is critically important to patient welfare, recovery, and thus, reliable data collection. As such, it must be proactively and thoroughly addressed. Hypothermia occurs in nearly all surgical patients who do not receive supplemental heat, primarily as a result of the cooler operating room temperatures required for surgeon and staff comfort along with the thermoregulatory impairment experienced by patients under anesthesia [1,2,3,4]. Anesthetic-induced tonic vasoconstriction inhibition leads to peripheral vasodilation, allowing the rapid redistribution and loss of core body heat within the first hour of surgery [2]. Approximately three hours into surgery, the linear decline in core temperature reaches a plateau in response to the development of a thermal equilibrium or a sufficiently hypothermic state which reactivates peripheral vasoconstriction [5,6,7]. Under such conditions, tissue oxygenation is significantly reduced along with leukocyte migration, neutrophil phagocytosis, as well as antibody and cytokine production [8,9]. Given such immunocompromised conditions at the tissue level, it is unsurprising that hypothermic patients experience higher wound infection rates [10,11,12]. One retrospective review of over 500 laparotomies identified a 221 percent increase in surgical site infections in its hypothermic patients [13]. In addition, hypothermia-induced catecholamine release and systemic vasoconstriction can dramatically increase blood pressure [14], cardiac demand, and subsequently lead to increased incidence of myocardial ischemia [10]. Uncorrected perioperative hypothermia has also been demonstrated to prolong the duration of pharmaceutical action [15] and induce coagulopathy [16], as in a recent review of intraoperative warming trials which identified a more than 60 percent increase in the mean surgical blood loss in patients not receiving supplemental heat [17].

At Ethicon, the concern of hypothermia is paramount in porcine surgical patients which are relied upon to produce dependable, consistent results regarding the safety and efficacy of experimental medical devices. Though perioperative hypothermia has not been extensively studied in pigs, it has been demonstrated to reduce systolic and diastolic cardiac function as well as myocardial compliance in porcine and canine models [18,19,20]. In order to maintain normothermia in pigs, between 38 and 39.5 °C [21], the heat loss associated with surgery must be counterbalanced by externally supplied heat. Such supplemental heat can come from a variety of patient warming systems; including forced-air, resistive fabrics, and circulating hot water mattresses. The most widely utilized perioperative warming systems provide forced-air heating, the effectiveness and safety of which have been well reported [22,23]. However, disposable forced-air covers can generate prohibitive long-term cost, indicating a potential benefit of non-disposable warming technologies. Conductive blankets and circulating water mattresses are reusable heat sources, but various human studies have demonstrated a lack of consensus regarding the efficacies of available products [17]. Product evaluation for the purpose of maintaining porcine perioperative normothermia has not, to the authors' knowledge, been previously described.

This study investigated the incidence of hypothermia in anesthetized porcine surgical patients under a standard warming protocol and evaluated the efficacy of various warming methods, including ambient temperature adjustment as well as the use of circulating hot water mattresses, conductive heating blankets, and forced-air warming systems.

2. MATERIALS AND METHODS

Animals. Domestic pigs (*Sus scofa domesticus*), cross-breeds of Yorkshire and Hampshire, weighing 35–55 kg were utilized in the development and evaluation of surgical devices. Though the pigs included in this experiment were purchased from different suppliers, all animals shared a vaccination history of *Erysipelothrix rhusiopathiae*, several serovars of *Leptospira*, and *Parvovirus.* All animals are brought onsite and housed according to the Association for Assessment and Accreditation of

Laboratory Animal Care International guidelines for a minimum of a seven day quarantine period with other pigs from the same vendor prior to their use. Pigs were housed in pairs in slotted-floor cages and maintained on a 12:12-h light: dark cycle at 22 ± 5 °C and 30%–70% relative humidity, with room air changes occurring at least ten times every hour. During this time, they were fed two scoops of Purina Mills Lab Porcine Grower Diet once daily. Water was provided *ad libitum* and enrichment devices were made available. All pigs were examined by the attending veterinarian prior to their procedures and any pig determined to be unfit for surgery was removed from study.

Surgical procedures. All procedures were covered by protocols reviewed and approved by the Ethicon Institutional Animal Care and Use Committee. In the spirit of our 3R's program (reduction, refinement, replacement) to improve animal testing, we sought to make best use of data already acquired, hence our control was historical data. All included surgeries were part of an educational course which repeats periodically, thus maintaining a relatively standardized set of surgery parameters throughout the year. Surgeries covered in this program consist of various abdominal and thoracic surgeries approximately split between open and laparoscopic access. Duration of surgeries were at least 3 hours and up to 6 hours and all animals were euthanized post-operatively. Anesthesia was induced with an intramuscularly administered mixture of 5 mg/kg telazol and xylazine, (Telazol, Rompun, Lloyd Inc., Shenandoah, IA, USA) and 0.01 mg/kg glycopyrrolate (Robinul, Baxter Labs, Deerfield, IL, USA); anesthesia was maintained with 1.5%–3.0% isoflurane gas with a target minimum alveolar concentration of 1.5%.

Ambient temperature. Four different surgical suites were utilized. Historical ambient temperature measurements were used from each suite six times a day for nine days for a baseline evaluation. Temperature ranges and averages were determined for each suite and compared to a standard operating room temperature of 20–22.8 °C, as recommended by the Association of periOperative Registered Nurses and the American Institute of Architects [24]. As needed, surgical suite temperatures were adjusted to within the acceptable range and verified for one month prior to the collection of patient temperature data.

Body temperatures prior to ambient adjustment. A historical review of porcine surgeries within the Ethicon facility in 2010 ($n = 159$) was utilized to determine baseline hypothermia rate. For the purposes of this study, porcine hypothermia was defined as a core body temperature below 36.7 °C (98.0 °F). Historically, we have observed serious physiological changes at body temperatures below 96 °F (35.6 °C), so we set our action level for hypothermia 2.0 °F (1.1 °C) higher. According to the standard operating procedures, any animals which would have experienced hypothermia during these surgical laboratories would have been supplemented with heat from a circulating water mattress placed below the patient pre-operatively. However the recovery from hypothermia was typically slow or incomplete, leading to our search for improved methods of body temperature control. Intraoperative temperature data, taken with a calibrated nasal probe (Cardiocap 5 Monitor withTemperature Probe 165612, Datex-Ohmeda, Madison, WI, USA) inserted until contact is made with the nasal turbinate at the time of anesthetic induction, was collected from the records of 160 pigs at a consistent time point three hours into surgery.

Body temperatures post ambient adjustment. The effect of the tighter control of ambient temperature on body temperature was determined through the intraoperative sampling of 32 pigs three hours into surgery. Timing of sampling was based on prior experience that hypothermia was infrequently observed during the first three hours of surgery. All incidents of hypothermia (i.e., <36.7 °C) were addressed using a circulating hot water mattress (Temp Pump Professional TP700 with Temperature Therapy Pad TP22G, Gaymar Industries, Inc., Orchard Park, NY, USA) placed under the animal per the established warming protocol in effect for the surgical labs included in the in-house historical review.

Body temperatures with supplemental heating. Once a new frequency of hypothermia was determined after the ambient temperature changes, intraoperative temperature measurements were taken for surgeries being supplied supplemental heat from a conductive fabric warming blanket (n = 19, HotDog Patient Warming System, Augustine Temperature Management, Eden Prairie, MN, USA) or a forced-air warming system (n = 25, Bair Hugger Animal Health Warming Unit 50577, 3M, Maplewood, MN, USA). The Bair Hugger was used without a perforated mattress by inserting the output tube under a standard surgical drape that cocooned the animal. When in use, the Bair Hugger was set to Medium (100 °F /38 °C) or High (110 °F /43 °C). Both devices were used in a similar distribution of open and laparoscopic abdominal and thoracic surgeries.

Statistical Analyses

Primary measures were parametric estimates of the percentage of temperature readings outside of control limits. For ambient temperature, the control range was 20.0–22.8 °C. For porcine body temperature, the control range was 36.7–40.0 °C. For both the ambient temperature and body temperature measurements, we determined the percent occurrence of out-of-specification temperatures using the mean and standard deviation.

Based on the near normal distribution of data, comparisons of means before and after adjustments were performed using Student's t-test, and comparison of standard deviation using the F-test, with $\alpha = 0.05$. For ambient temperature measurements, we compared the historical baseline mean and standard deviation to adjusted ambient temperature measurements during use of the circulating hot water mattress, the HotDog Patient Warming System and the Bair Hugger Animal Health Warming Unit. For body temperature measurements, we compared means of the historical baseline temperatures to after ambient temperature adjustment both with the use of the hot water mattress, and then compared both HotDog and Bair Hugger to the hot water mattress values.

3. RESULTS

Ambient temperature. Baseline ambient temperature across the four surgical suites utilized in this study had an average value of 21.0 ± 0.62 °C, with a range of 19.7–22.0 °C. Most of the lowest data was collected from one particular suite (Figure 1). Following adjustment of the ambient temperatures, the average was 21.1 ± 0.11 °C, with a range of 20.8–21.4 °C. The average was slightly, but significantly increased ($p = 0.005$), while the overall standard deviation was greatly reduced ($p < 0.001$). Using the recommended AORN temperatures (20.0–22.78 °C) as limits, the percentage of out-of-specification readings decreased from a baseline value of 6.4% to 0.0% after room temperature adjustments.

Body temperatures prior to ambient adjustment. Initial in-house historical review of temperature data from 159 porcine surgeries supplemented with heat from a circulating hot water mattress revealed an estimated 20.4% hypothermia incidence with a mean intraoperative temperature of 37.4 °C and a low incidence (0.4%) of hyperthermia (Table 1). The total percentage of out-of-specification body temperatures (hypothermia and hyperthermia) was 20.8%.

Body temperatures post ambient adjustment. Among pigs that underwent surgery in the newly temperature-regulated suites, still receiving traditional circulating water heat as needed, there was an increase in mean body temperature of +0.6 °C ($p = 0.002$) with hypothermia incidence decreased to 4.6%, a five-fold reduction (Table 1). Hyperthermia occurrence remained at 0.4%, for a total out-of-specification rate of 5.0%. This improvement from merely adjusting the room temperature met our initial target for hypothermia, however, direct control of the individual animal's environment offered the prospect of complete compliance with the control limits. Hence the conductive fabric (HotDog) and forced-air (Bair Hugger) warming systems were compared to the standard circulating water mattress.

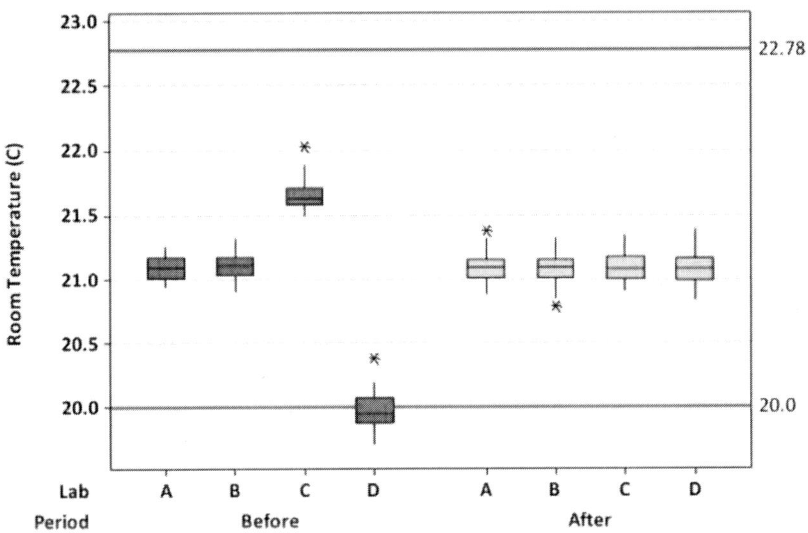

Figure 1. Ambient room temperature of the four rooms before and after temperature control adjustments. The control limits are those recommended by the Association of periOperative Registered Nurses (AORN). Asterisks (*) in the graph represent outlier points, i.e., at least 1.5 times the interquartile range; these points were not excluded from the analysis.

Table 1. Summary of temperature measurements.

Measure	Before RT Adj. (Baseline)	After RT Adj.	HotDog	Bair Hugger
Room Temperature				
Mean ± St. Dev.	21.0 ± 0.6 °C	21.1 ± 0.1 °C	21.0 ± 0.1 °C	21.2 ± 0.1 °C
Range	19.7–22.0 °C	20.8–21.4 °C	20.7–21.3 °C	20.8–21.6 °C
% Out-of-Spec (20.0–22.8 °C)	6.4%	0.0%	0.0%	0.0%
t-test, Mean vs. Baseline	-	$p = 0.005$	$p = 0.007$	$p = 0.004$
F-test, St. Dev. vs. Baseline	-	$p < 0.001$	$p < 0.001$	$p < 0.001$
Body Temperature n	159	32	19	25
Mean ± St. Dev.	37.4 ± 0.9 °C	38.0 ± 0.8 °C	38.6 ± 0.6 °C	38.3 ± 0.4 °C
% Hypothermic (<36.7 °C)		4.6%	0.2%	0.0%
% Hyperthermic (>40.0 °C)	20.4%	0.4%	1.3%	0.0%
t-test, Mean	0.4%	$p = 0.002$	$p = 0.005$	$p = 0.025$
	-	vs. Baseline	vs. After Adj.	vs. After Adj.

Body temperatures with supplemental heating. For pigs that received intraoperative heat supplementation (Table 1, Figure 2) from the HotDog conductive fabric system, mean body temperature increased by +0.6 °C compared to the standard circulating water mattress ($p = 0.005$) with 0.16% hypothermia and 1.3% hyperthermia, with hyperthermia being considered body temperatures above 40.0 °C. Hence the total out-of-specification results were less than 1.5%. For pigs supplemented with the Bair Hugger forced-air system the mean body temperature increased +0.3 °C compared to the circulating water mattress ($p = 0.025$) with 0.0% incidence of hypothermia and hyperthermia.

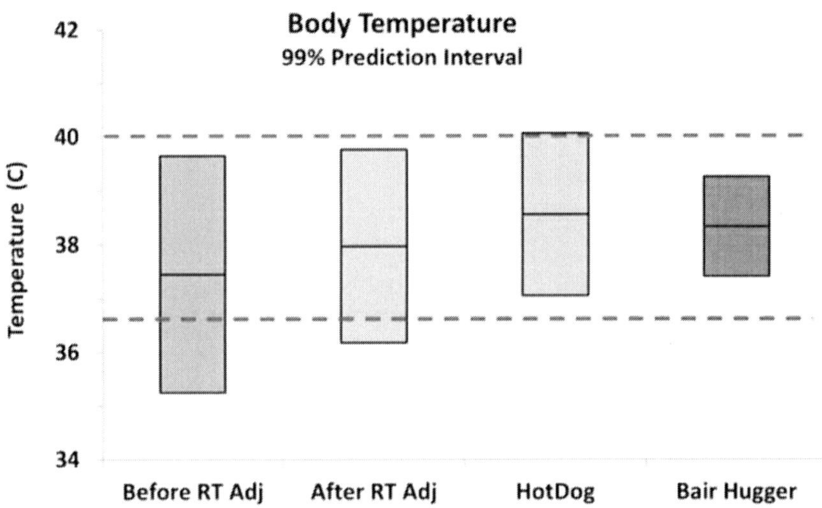

Figure 2. Intraoperative porcine body temperatures during surgery before and after room temperature adjustments, and after room temperature adjustments with addition of a supplemental heat source. For the first two measurements a circulating hot water mattress was used. Without use of a supplemental heat source there were occurrences of hypothermia. With the HotDog warmer there were a few cases of hyperthermia. There was no hypothermia or hyperthermia for the Bair Hugger. The middle line of each box is the mean value, and the range represents the 99% prediction interval from a parametric estimation.

4. DISCUSSION

Though there has been much inconsistency in the results of numerous studies comparing various warming systems, these studies have not previously been conducted in porcine surgical models, hence for our review we used studies primarily performed in humans. While our results seems to correlate with a number of publications supporting the use of forced-air warming systems [25,26], the differences between our results and other apparently contrary preceding work [27,28] are difficult to compare directly, given the dissimilar circumstances of these experiments such as skin conductivity, body positioning of the warming apparatus and total contact surface area, as well as the difference between preventative and corrective use of the devices.

In the development of surgical equipment, the importance of having a reliable porcine surgical model is paramount, ensuring the consistency and validity of data which is generated and used to endorse the development of human surgical instruments and techniques. Anesthesia and operating conditions tremendously predispose subjects to intraoperative hypothermia, which can present a very significant dilemma, both to the process of data collection and the maintenance of animal welfare throughout experimentation. Of paramount importance in complying with animal welfare protocols is the commitment to reducing and eliminating any unnecessary animal use in the laboratory. As a result, many of the pigs which undergo surgery at our facility are under anesthesia for prolonged periods of time, over five hours on average, in order to maximize the usefulness of every specimen. This fact only increases the likelihood of intraoperative hypothermia, which has been demonstrated to occur more frequently in human subjects under anesthesia for longer than three hours [29].

Our initial retrospective review determined that there was, in fact, a substantial incidence of hypothermia among our porcine subjects, as was anticipated based on the widespread incidence in human surgical patients reported in the literature. This lends credence to the application of conclusions regarding human anesthesia complications to our porcine model as well as indicates the ineffectiveness of our traditional patient warming protocol, the circulating water mattress.

Additionally, we found that in our institute a simple and effective means for reducing, though not eliminating, intraoperative hypothermia was by adjusting the ambient temperature of the operating suite to the recommended range. The cost of this temperature change is insignificant, as the additional spending required for heat in the winter is partially compensated by savings on cooling in the summer.

This considerable reduction in hypothermia incidence indicates a cost-effective means of better ensuring the welfare of most surgical patients; however, the inherent predisposition to hypothermia makes it improbable that the condition will be avoided completely with such simple corrective measures. Furthermore, it does not circumvent the need for reliable means of resolving hypothermia in those remaining patients that do experience it.

With regard to the most efficacious warming system, our results point to forced-air warming as the method of choice, as it completely eliminated hypothermia from our surgical patients. Although the conductive fabric system reduced the occurrence of hypothermia, an unexpected result was the development of hyperthermia in some cases, which was unprecedented with the circulating water mattress. While it was likely that the relatively superior warming abilities of the other two warming systems predisposed them somewhat to the possibility of inducing hyperthermia, this effect was not observed with the forced-air system despite producing a higher mean body temperature than the circulating water mattress. One apparent explanation for this is the more immediate and precise temperature control provided by forced-air systems, as opposed to the more gradual temperature change in the resistive blankets. Often, technician overestimation of patients' supplemental heat requirements leads to the blankets being turned to higher than necessary temperatures from which the heat dissipates too slowly to be able to quickly adjust patient temperature back down. This, in turn, leads to the larger range of temperature fluctuations seen with this device. In contrast, forced-air systems are better equipped to prevent this problem through the option of blowing cool air over the patient to more rapidly dissipate any excess heat.

While one might consider the potential for turbulent air movement while using forced-air warming to cause increased surgical site contamination, it has been demonstrated in multiple studies to reduce post-operative infection rates [8,12], probably due to the physiologic effects of its superior body temperature maintenance. This potential obstacle is somewhat beyond the scope of our investigation however, as all of the surgeries included were acute and neither patient survival nor post-operative complications were considered.

A limitation of this study is that it was performed retrospectively without the benefit of a controlled design. Hence we cannot be certain whether other factors varied between the treatments, and these factors may have influenced the outcomes in an unknown fashion. However, since the time of these initial observations, we have continued to use a forced air system with success, so we suspect that the actual treatments used in the study were the most critical. Additionally, we have not observed body temperature in recovering animals, since all subjects were euthanized immediately after surgery. Again, forced airs systems appear to be useful during recovery in survival studies. Furthermore, there were a variety of procedures performed and it is generally observed that open procedures, depending upon the size of the incision, would require more attention to body temperature than laparoscopic ones. The distribution of procedures was similar during the different phases of the study, so we feel the comparisons between conditions/devices were, in general, fair.

5. CONCLUSIONS

Overall our findings indicate that careful monitoring of ambient conditions and use of forced-air warming can reduce the incidence of hypothermia during porcine surgery, and this is expected to result in less patient morbidity and faster recovery. Since the completion of this study, our laboratory has continued to use forced-air warming for intraoperative heat supplementation of its porcine patients. Hypothermia incidence was monitored for several months following the study conclusion with no reported morbidity.

ACKNOWLEDGMENTS

This study was funded by Ethicon, Inc., Cincinnati, OH, USA.

AUTHOR CONTRIBUTIONS

Karla A. Stevens conceived and designed the experiments; Brian T. Dent and Jeffrey W. Clymer analyzed the data; Brian T. Dent and Jeffrey W. Clymer wrote the paper. All authors read and approved the paper.

CONFLICTS OF INTEREST

The authors declare no conflict of interest.

REFERENCES

1. Buggy, D.J.; Crossley, A.W. Thermoregulation, mild perioperative hypothermia and postanaesthetic shivering. Br. J. Anaesth. 2000, 84, 615–628.
2. Kurz, A.; Sessler, D.I.; Narzt, E.; Lenhardt, R.; Lackner, F. Morphometric influences on intraoperative core temperature changes. Anesth. Analg. 1995, 80, 562–567.

3. Arkilic, C.F.; Akca, O.; Taguchi, A.; Sessler, D.I.; Kurz, A. Temperature monitoring and management during neuraxial anesthesia: An observational study. Anesth. Analg. 2000, 91, 662–666.
4. Sessler, D.I. Complications and treatment of mild hypothermia. Anesthesiology 2001, 95, 531–543.
5. Sessler, D.I.; Rubinstein, E.H.; Eger, E.I., 2nd. Core temperature changes during N2O fentanyl and halothane/O2anesthesia. Anesthesiology 1987, 67, 137–139.
6. Kurz, A.; Sessler, D.I.; Christensen, R.; Dechert, M. Heat balance and distribution during the core-temperature plateau in anesthetized humans. Anesthesiology 1995, 83, 491–499.
7. Sessler, D.I. Mild perioperative hypothermia. N. Engl. J. Med. 1997, 336, 1730–1737.
8. Kurz, A.; Sessler, D.I.; Lenhardt, R. Perioperative normothermia to reduce the incidence of surgical-wound infection and shorten hospitalization. Study of wound infection and temperature group. N. Engl. J. Med. 1996, 334, 1209–1215.
9. Kumar, S.; Wong, P.F.; Melling, A.C.; Leaper, D.J. Effects of perioperative hypothermia and warming in surgical practice. Int. Wound J. 2005, 2, 193–204.
10. Frank, S.M.; Fleisher, L.A.; Breslow, M.J.; Higgins, M.S.; Olson, K.F.; Kelly, S.; Beattie, C. Perioperative maintenance of normothermia reduces the incidence of morbid cardiac events. A randomized clinical trial. JAMA 1997, 277, 1127–1134.
11. Vaughan, M.S.; Vaughan, R.W.; Cork, R.C. Postoperative hypothermia in adults: Relationship of age, anesthesia, and shivering to rewarming. Anesth. Analg. 1981, 60, 746–751.
12. Melling, A.C.; Ali, B.; Scott, E.M.; Leaper, D.J. Effects of preoperative warming on the incidence of wound infection after clean surgery: A randomised controlled trial. Lancet 2001, 358, 876–880.
13. Seamon, M.J.; Wobb, J.; Gaughan, J.P.; Kulp, H.; Kamel, I.; Dempsey, D.T. The effects of intraoperative hypothermia on surgical site infection: An analysis of 524 trauma laparotomies. Ann. Surg. 2012, 255, 789–795.
14. Frank, S.M.; Higgins, M.S.; Breslow, M.J.; Fleisher, L.A.; Gorman, R.B.; Sitzmann, J.V.; Raff, H.; Beattie, C. The catecholamine, cortisol, and hemodynamic responses to mild perioperative hypothermia. A randomized clinical trial. Anesthesiology 1995, 82, 83–93.
15. Heier, T.; Caldwell, J.; Sessler, D.; Miller, R. Mild intraoperative hypothermia increases duration of action and spontaneous recovery of vecuronium blockade during nitrous oxide-isoflurane anesthesia in humans. Anesthesiology1991, 74, 815–819.
16. Schmied, H.; Reiter, A.; Kurz, A.; Sessler, D.; Kozek, S. Mild hypothermia increases blood loss and transfusion requirements during total hip arthroplasty. Lancet 1996, 347, 289–292.
17. Sajid, M.S.; Shakir, A.J.; Khatri, K.; Baig, M.K. The role of perioperative warming in surgery: A systematic review. São Paulo Med. J. 2009, 127, 231–237.
18. Espinoza, A.; Kerans, V.; Opdahl, A.; Halvorsen, P.; Skulstad, H.; Hoff, L.; Bugge, J.; Edvardsen, T. Moderate hypothermia induces severe diastolic changes in a porcine model. Eur. Heart J. 2009, 30, 21.
19. Fischer, U.M.; Cox, C.S., Jr.; Laine, G.A.; Mehlhorn, U.; Allen, S.J. Mild hypothermia impairs left ventricular diastolic but not systolic function. Investig. Surg. 2005, 18, 291–296.
20. Berne, R.M. Myocardial function in severe hypothermia. Circ. Res. 1954, 2, 90–95.
21. Straw, B.E.; Zimmerman, J.; D'Allaire, S.; Taylor, D. Diseases of Swine, 9th ed.; Blackwell Publishing Professional: Ames, IA, USA, 2006.
22. Andrzejowski, J.; Hoyle, J.; Eapen, G.; Turnbull, D. Effect of prewarming on post-induction core temperature and the incidence of inadvertent perioperative hypothermia in patients undergoing general anaesthesia. Br. J. Anaesth. 2008, 101, 627–631.
23. Perez-Protto, S.; Sessler, D.; Reynolds, L.; Bakri, M.; Mascha, E.; Cywinski, J.; Parker, B.; Argalious, M. Circulating-water garment or the combination of a circulating-

water mattress and forced-air cover to maintain core temperature during major upper-abdominal surgery. Br. J. Anaesth. 2010, 105, 466–470.

24. Ninomura, P.; Bartley, J. New ventilation guidelines for health-care facilities. ASHRAE J. 2001, 43, 29.
25. Röder, G.; Sessler, D.; Roth, G.; Schopper, C.; Mascha, E.; Plattner, O. Intra-operative rewarming with hot dog®resistive heating and forced-air heating: A trial of lower-body warming. Anaesthesia 2011, 66, 667–674.
26. Giesbrecht, G.G.; Ducharme, M.B.; McGuire, J.P. Comparison of forced-air patient warming systems for perioperative use. Anesthesiology 1994, 80, 671–679.
27. Brandt, S.; Oguz, R.; Hüttner, H.; Waglechner, G.; Chiari, A.; Greif, R.; Kurz, A.; Kimberger, O. Resistive-polymer vs. forced-air warming: Comparable efficacy in orthopedic patients. Anesth. Analg. 2010, 110, 834–838.
28. Kimberger, O.; Held, C.; Stadelmann, K.; Mayer, N.; Hunkeler, C.; Sessler, D.I.; Kurz, A. Resistive polymer vs. forced-air warming: Comparable heat transfer and core rewarming rates in volunteers. Anesth. Analg. 2008, 107, 1621–1626.
29. Abelha, F.J.; Castro, M.A.; Neves, A.M.; Landeiro, N.M.; Santos, C.C. Hypothermia in a surgical intensive care unit. BMC Anesthesiol. 2005, 5, 7.

CHAPTER 9

Post-Operative Benefits of Animal-Assisted Therapy in Pediatric Surgery: A Randomised Study

Valeria Calcaterra[1,2], Pierangelo Veggiotti[3,4], Clara Palestrini[5] , Valentina De Giorgis[4] , Roberto Raschetti[1] , Massimiliano Tumminelli[3] , Simonetta Mencherini[6] , Francesca Papotti[5] , Catherine Klersy[7] , Riccardo Albertini[8] , Selene Ostuni[9] , Gloria Pelizzo[9,10]

[1] *Department of the Mother and Child Health, Pediatric Unit, Fondazione IRCCS Policlinico San Matteo, Pavia, Italy,*

[2] *Department of Internal Medicine, University of Pavia, Pavia, Italy,*

[3] *Department of Child Neurology and Psychiatry C. Mondino National Neurological Institute, Pavia, Italy,*

[4] *Brain and Behaviour Department, University of Pavia, Pavia, Italy,*

[5] *Dipartimento di Scienze Veterinarie e Sanità Pubblica, Università degli Studi di Milano, Milano, Italy,*

[6] *Department of Anaesthesiology and Intensive Care, Fondazione IRCCS Policlinico San Matteo, Pavia, Italy,*

[7] *Biometry & Clinical Epidemiology, Scientific Direction, Fondazione IRCCS Policlinico San Matteo, Pavia, Italy,*

[8] *Laboratory of Clinical Chemistry, Fondazione IRCCS Policlinico San Matteo, Pavia, Italy,*

[9] *Pediatric Surgery Unit, Department of the Mother and Child Health, Fondazione IRCCS Policlinico San Matteo, Pavia, Italy,*

[10] *Department of ClinicalSurgical, Diagnostic and Pediatric Sciences University of Pavia, Pavia, Italy*

ABSTRACT

Background

Interest in animal-assisted therapy has been fuelled by studies supporting the many health benefits. The purpose of this study was to better understand the impact of an animal-assisted therapy program on children response to stress and pain in the immediate post-surgical period.

Patients and Methods

Forty children (3–17 years) were enrolled in the randomised open-label, controlled, pilot study. Patients were randomly assigned to the animal-assisted therapy-group (n = 20, who underwent a 20 min session with an animal-assisted therapy dog, after surgery) or the standard-group (n = 20, standard postoperative care). The study variables were determined in each patient, independently of the assigned group, by a researcher unblinded to the patient's group. The outcomes of the study were to define the neurological, cardiovascular and endocrinological impact of animal-assisted therapy in response to stress and pain. Electroencephalogram activity, heart rate, blood pressure, oxygen saturation, cerebral prefrontal oxygenation, salivary cortisol levels and the faces pain scale were considered as outcome measures.

Results

After entrance of the dog faster electroencephalogram diffuse beta-activity (> 14 Hz) was reported in all children of the animal-assisted therapy group; in the standard-group no beta-activity was recorded (100% vs 0%, $p<0.001$). During observation, some differences in the time profile between groups were observed for heart rate (test for interaction $p = 0.018$), oxygen saturation (test for interaction $p = 0.06$) and cerebral oxygenation (test for interaction $p = 0.09$). Systolic and diastolic blood pressure were influenced by animal-assisted therapy, though a higher variability in diastolic pressure was observed. Salivary cortisol levels did not show different behaviours over time between groups ($p=0.70$). Lower pain perception was noted in the animal-assisted group in comparison with the standard-group ($p = 0.01$).

Conclusion

Animal-assisted therapy facilitated rapid recovery in vigilance and activity after anaesthesia, modified pain perception and induced emotional prefrontal responses. An adaptative cardiovascular response was also present.

INTRODUCTION

The relationship between human beings and animals, especially dogs, has existed for thousands of years. Historically, animals have held an important role in this relationship as they provide company, stimulus and motivation. Animals are excellent company, since they do not discriminate or segregate anyone, that is, they are free of prejudice [1]. In spite of the long-lasting presence of companion animals in human life, the idea that interaction with animals may exert a positive effect on human health is rather recent [2].

The American Veterinary Medical Association classifies therapeutic animal assisted interventions (AAI) into three categories: animal assisted activities (AAA) that utilize companion animals; animal assisted therapy (AAT) that utilizes therapy animals and service animal programs (SAP) that utilize service animals. AAT in particular, is a goal-directed intervention in which an animal that meets specific criteria is an integral part of the treatment process. AAT is technically defined as the use of trained animals by trained health professionals to facilitate specific, measurable goals for individual patients for whom there is documented progress [3].

Interest in AAT has been fueled by studies supporting the many health benefits. AAT has proven a useful adjunct in a variety of settings including mental health facilitaties [4–10], nursing homes [11,12] and hospitals [13,14] where most studies have been performed with adult patients with variable interventions, goals, patient characteristics and patient needs. In these studies, AAT resulted in significant reductions in anxiety, agitation and fear. In children, AAT dogs decreased distress during painful medical procedures, promoted calmness in children with post-traumatic stress disorders and increased attention and positive behaviors in children with pervasive developmental disorders [10, 15–19].

Surgical procedures and hospitalization can be stressful for both children and their parents and they are associated with pain, helplessness, fear and boredom [19]. AAT has been shown to facilitate a child's ability to cope with hospitalization, but to date, no studies on AAT benefits in pediatric surgery have been reported.

Emotion regulation is the adaptive behavioural response to stress. Emotions promote biochemically-mediated neurologic responses to emotionally-based stimuli [20]. Neurophysiologic measurement based on electrophysiological techniques can detect a wide range of dynamics of the emotional state by directly accessing the fundamental structure in the brain from which an emotional state emerges [21,22]. Hormonal indicators, such as cortisol level, and autonomic measurements, for example heart rate, blood pressure, can objectively detect the physiological responses of the endocrinological and autonomic nervous systems related to stress [22,23].

The purpose of this study was to better understand the effects of an AAT program on neurological, cardiovascular and endocrinological responses to stress and pain in the immediate post-operative period in children undergoing surgical procedures.

MATERIALS AND METHODS

Ethics Statement

The study was performed according to the Declaration of Helsinki. The ethics committee of the Fondazione IRCCS Policlinico S. Matteo and Department of Internal Medicine, University of Pavia, approved the study protocol on April 11, 2013.

Animal-assisted therapy was also approved, on April 11, 2013, by the ethics committee of the Fondazione IRCCS Policlinico S. Matteo. No invasive intervention or drug sperimentation on the dog was performed; therefore the application of D.lgs. 116/92, European Directives 86/609/EE for the protection of animals used in scientific and experimental studies and 2010-63UE was not required. The dog owner provided consent for its use in the study.

Participant were recruited between September 01, 2013 and April 01, 2014.

The study was registered at ClinicalTrials.gov (Identifier: NCT02284100) after enrolment was initiated, because animal assisted-therapy was considered a complementary treatment. The authors confirm that all ongoing and the related trial for this drug/intervention are registered.

Study protocol

The study protocol and supporting CONSORT checklist are available as supporting information; see S1 CONSORT Checklist and S1 and S2 Protocol.

This was a randomised open-label, controlled, pilot study. The different arms consisted of an experimental group with a AAT session after a surgical procedure and a control group with standard care after surgery. The flow diagram of the progress through the phases of randomization is reported in Fig 1.

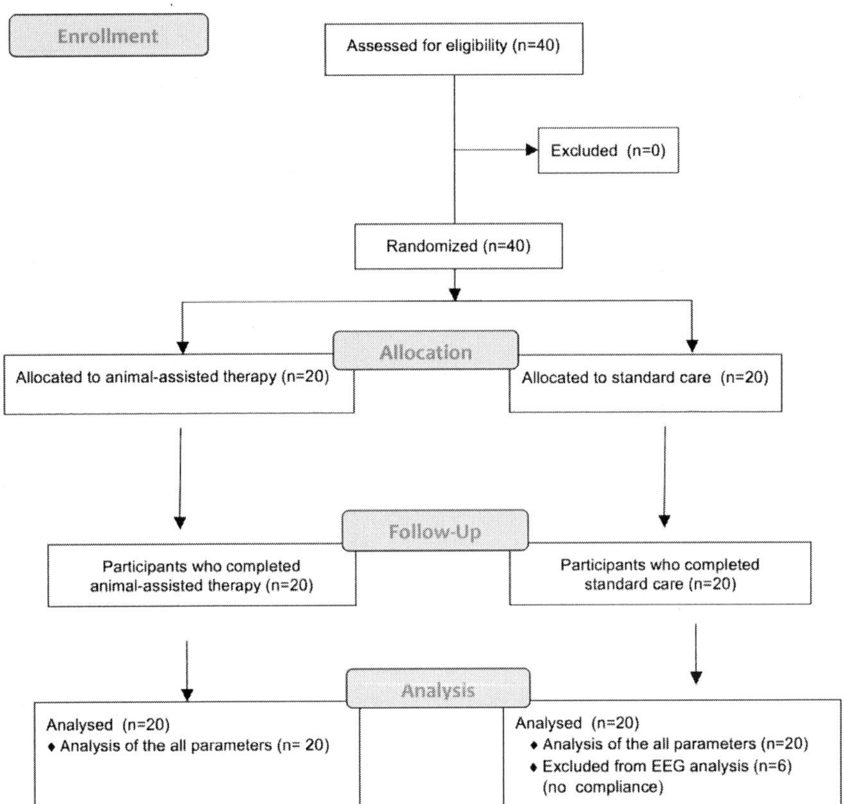

Figure 1. Flow diagram of the progress through the phases of the randomised trial of two groups (animal-assisted therapy and standard care after surgery).

The study variables were determined in each patient independently of the assigned group, before and after the experimental intervention, by a researcher unblinded to the patient's group.

The outcomes of the study were to define the impact of AAT on neurological (primary endpoint), cardiovascular (secondary endpoint 2) and endocrinological signs (secondary endpoint 3), in response to stress and pain in children undergoing surgical procedures.

As outcome measures we considered:

1. -. for neurological impact, the difference in the prevalence of beta (>14 Hz) electroencephalogram (EEG) activity between the intervention and control groups [24–27];
2. -. for cardiovascular impact, the difference in vital signs (cerebral prefrontal oxygenation (HbO_2) with near-infrared spectroscopy (NIRS), heart rate (HR), blood pressure (BP), oxygen saturation (SpO_2) between intervention and control groups;
3. -. for endocrinological impact, the difference in salivary cortisol levels, between the intervention and control groups.

The Wong-Baker Faces pain scale (FPS) was used to measure the child's self-reported pain.

After trial commencement, no changes to the methods or trial outcomes were made.

Participants

Forty immunocompetent children (both genders), aged 3 to 17 years, undergoing surgical procedures (including orchidopexy, inguinal or umbelical hernia repair, circumcision, varicocele treatment) were sequentially enrolled in this study.

A sample size of 20 patients per group for a total of 40 subjects was determined based dog availability. No data in the literature were available. We tentatively hypothesized a low prevalence of beta activity (primary endpoint) in the control group (with 2 different scenarios 5% and 10%), 80% power, alpha (2-tailed) = 0.10 (pilot study); the effect size (difference in prevalence between the 2 treatment groups) that could be elicited based on these assumptions was 31% and 34% respectively using a Chi2 test or 37% and 40% applying the continuity correction. The corresponding power obtainable using the Fisher exact test was 72% and 84% for the 2 scenarios, respectively Exclusion criteria included: allergy or fear of dogs, previous AAT experience, immunodeficiency, chronic illness, obesity and use of any medications.

In all subjects, surgery was performed between 8.30 am and 12 am and under general anaesthesia, at the Pediatric Surgery Unit, Fondazione IRCCS Policlinico S. Matteo, Pavia. Parental permission was obtained through a written and oral informed consent. Written assent by the patient was also obtained in children eight years of age and older before enrolment.

For the AAT session, a 7 year old Golden Retriever was employed as the therapy animal. Prior to the study, the dog underwent rigorous screening although she had previous experience in Animal Assisted Interventions and was already trained and prepared for this type of work. The dog was fully vaccinated, bathed regularly, screened for enteric pathogens, and treated for internal and external parasites on a monthly basis. The dog and handler met hospital policy for participating in animal-assisted therapy, including documentation of the dog's current vaccinations, controllability and temperament.

The dog's welfare was monitored by a veterinarian specialized specialist in "applied ethology and animal welfare", European college of animal welfare and behavioural medicine (ECAWBM) certification.

Follow-up

The participants were followed for the duration of the immediate postoperative period, an expected average of four hours.

Procedures and data collection

At admission, auxological examination of the children included measurement of height, weight and body mass index. Height measurement was performed using a Harpenden stadiometer, performed with patients in an upright position, without shoes, with their heels together, arms extended down the sides of the body and head positioned parallel to the floor. Weight was measured with the children barefooted and wearing light clothes, standing upright in the centre of the scale platform with their arms extended down the sides of the body. BMI was calculated as body weight in kilograms divided by body height squared in meters.

Pre-surgery, the children were randomly assigned to the experimental AAT-group where the dog was present during post-operative awakening (2 hours after surgery) or the standard-group (STAND-group) where children had standard post-operative medical care. Randomization was carried out using opaque envelopes, half of which contained a paper that said "AAT-group" or "STANDARD-group" and was based on a permuted blocks randomization.

The AAT-group underwent a 20 min session with the AAT dog. The therapy dog was specially prepared and chosen for the interactions, which were evaluated as suitable and registered together with the handler. The handler monitored the dog, tended to the dog's needs and answered questions about the dog.

The STAND-group underwent the same measurements over the same time frame, but without the dog being present.

EEG activity, cerebral prefrontal oxygenation, heart rate, blood pressure, oxygen saturation, salivary cortisol levels, faces pain scale were considered as indicators of neurological, cardiovascular and endocrinological response to stress and pain.

Data collection was performed in the following phases:

1. -. for all parameters, post-operatively baseline (T1), two hours after surgery at re-admission to the Unit. Pre dog-intervention in the AAT-group;
2. -. for all parameters, in the twenty minutes following T1 (T2). During the dog-intervention in the AAT-group. For the STAND-group, the child was asked to sit quietly;
3. -. only for salivary cortisol levels, between 11 pm and midnight (T3, the time when cortisol is normally at its lowest)

Electroencephalogram

Regarding the AAT-group patients, an EEG recording was obtained when the child was awake, 2 hours after surgery and before and during AAT intervention. In the STAND-group patients, and EEG recording was made 2 hours after surgery.

Any change in the normal physiological structure of the EEG, correlating the effect of the anaesthesia, post-operative stress of AAT intervention was recorded and monitored. In both cases, the protocol included an EEG recording tasting about 20 minutes, while awake, with open and closed eyes.

Vitals signs

HR, BP, SpO_2, HbO_2 were monitored (Dräger Primus) and recorded as follows:

1. -. T1 (every 5 minutes, for 10 minutes. Mean values were used for the statistical analysis);
2. -. T2 (every 5 minutes, for 20 minutes. Mean values were used for the statistical analysis).

Endocrinological parameters

Salivary cortisol levels were measured at T1, after T2 and T3. The saliva samples were collected using a standardized salivette and frozen at -20° until analysis in the laboratory. After thawing, salivary fluids were centrifugated to precipitate mucins and cortisol was assayed in the supernatant with a solid-phase radioimmunoassay, wherein 125-I labeled cortisol competes for a fixed time with cortisol in the biological sample (The Coat-A-Count Cortisol, Siemens, Los Angeles, CA).

Pain response

The faces pain scale was used to measure the child's self-reported pain at T1 and at the end of T2. This scale consisted of 6 cartoon faces with varying expressions ranging from very happy to very sad. The child rated the pain intensity on a scale, with point 0 being no pain and point 10 being the worst pain.

Analgesic treatment in the first 12 hours post-intervention was recorded.

ANESTHESIA PROTOCOL

Written consent was obtained from the parents of the children scheduled for surgical procedures. All subjects were in good physical condition. Anesthesia included propofol (2–2.5 mh/g) as a sedative-hypnotic agent and fentanyl (1–1.5 mcg/kg) as the analgesic. After laryngeal mask or tracheal tube positioning, patients underwent volume controlled mechanical ventilation with an inspired mixture of air and oxygen using a closed breathing system (fresh gas flow of 0.75 l min^{-1} oxygen and 1.5 l min^{-1} air during anesthesia) adjusted to achieve an end-tidal carbon dioxide of 32–35 mmHg. Anesthesia was maintained via administration of Sevoflurane gas. Sevoflurane was administered in a 0.75 to 1.25 MAC range. Twenty minutes before the end of the intervention, all patients received Paracetamol 15 mg/kg, as an analgesic. At the end of the operation, the patient was transferred from the operating theater to the recovery room and subsequently was re-admitted to the Unit.

Statistical analysis

Continuous variables were described as the mean and standard deviation (SD) or median and quartiles and categorical variables as counts and percentages. The prevalence of EEG beta activity was compared with the

Fisher exact test. To test the effects of AAT on vital signs and on endocrinological parameters over time in the two groups, regression models for repeated measures were used, including a main effect for group and time, as well as their interaction; the latter was used to assess treatment effect (comparison of profiles over time). We also included a term for the duration of anesthesia, which was unbalanced between groups. Model assumptions were checked by visual inspection of the residual vs. fitted plot. Distribution of pain scores at the end of the session in the two groups was compared with the Fisher exact test. Differences between groups and 95% confidence intervals (95%CI) at each time point were computed. A 2-sided p-value<0.05 was considered statistically significant. No multiple test correction was applied given the exploratory nature of the pilot study. All statistical analyses were performed using Stata 13.1 (StataCorp, College Station, TX, USA).

RESULTS

The 40 children (32 M and 8 F; mean age 7.97±3.17 years; range 3.8–16.3), were randomly assigned to 1 of the 2 groups: the AAT-group (n = 20) or the STAND-group (n = 20), Fig 1.

Demographic, auxological features at admission and vital signs pre experimental intervention are reported in Table 1. The STAND-group showed a longer anesthesia time and lower cerebral oxygenation in comparison with the AAT-group; no correlation between these two variables was found (p = 0.5).

Table 1. Demographic and auxological features of the children at admission and vital signs pre intervention.

	Animal-assisted therapy group (n = 20)	Standard group (n = 20)
Age (yr)	8.59±3.70	7.36±2.48
Gender (M/F)	15/5	17/3
Height (cm)	131.00±23.59	120.70±17.83
Weight (kg)	31.50±17.18	25.35±10.38
BMI (kg/m^2)	18.4±2.4	17.6±3.2
Anesthesia time (min)	45.0 (45.0–55.0) *	62.5 (50.0–70.0) *
Heart rate (bpm)	96.83±20.06	95.85±23.52
Systolic blood pressure (mmHg)	107.45±8.95	111.03±10.87
Diastolic blood pressure (mmHg)	65.55±7.37	66.20±7.99
Oxygen saturation (%)	98.65±1.04	98.47±0.94
Cerebral oxygenation-HbO_2 (%)	79.5 (74.5–87.25) *	77.0 (67.5–79.00) *

*data are expressed as median (interquartile range)

Primary endpoint

Electroencephalogram activity.

A complete EEG recording was obtained in 20 patients in the AAT-group and 14 patients in the STAND-group (in 6 STAND-children, no compliance was obtained).

In 11/14 (80%) of the STAND-subjects and in 20/20 (100%) of the AAT-group patients the presence of atypical alpha-like (10–14 Hz) activity in anterior derivations (frontal and temporal) was observed during post-operative wakening (Fisher exact test $p = 0.06$).

In the STAND-group the alpha-like activity was unchanged with no beta activity recorded. On the contrary, after entrance of the dog faster EEG diffuse beta activity (> 14 Hz) was reported in all children of the AAT-group (0% vs 100%, $p<0.001$).

Secondary endpoints

Vital signs data.

Differences in treatment effect at every step during the intervention are detailed in Table 2. Given a marked inbalance in anesthesia time between groups, all models were adjusted for this difference.

Some differences in the time profile between groups were observed for heart rate, oxygen saturation and cerebral oxygenation. Only the former reached statistical significance (test for interaction $p = 0.018$), with heart rate increasing immediately after the start of observation in the AAT group, with the highest difference at 10 minutes, and decreasing afterwards. The difference between groups in oxygen saturation was highest at the beginning of the observation period, due to a decrease in the STAND group (test for interaction $p = 0.06$). Finally cerebral oxygenation tended to increase more in the AAT group in the later part of the observation period (test for interaction $p = 0.09$). Systolic and diastolic blood pressure were influenced by AAT, though a higher variability in DBP was observed. (Fig 2).

Differences and 95%CI are derived from repeated measurements models (including main effects for time and groups and their interaction) adjusted for anesthesia time.

Table 2. Vital signs measured in treated and standard groups.

Time	Heart rate (bpm)	Systolic blood pressure (mmHg)	Diastolic blood pressure (mmHg)	Oxygen saturation (%)	Cerebral oxygenation-HbO_2 (%)
0	2.4 (-13.6 to 18.5)	-3.9 (-11.6 to 3.9)	-0.9 (-6.9 to 5.1)	-0.4 (-1.1 to 0.2)	7.1 (1.5 to 12.7)
1	7.4 (-4.0 to 18.9)	1.9 (-6.1 to 10.0)	-1.2 (-7.6 to 5.2)	2.0 (-3.4 to 7.5)	4.4 (-2.7 to 11.5)
5	6.6 (-2.9 to 16.0)	3.3 (-4.1 to 10.7)	-1.9 (-9.6 to 5.7)	2.6 (-2.9 to 8.1)	3.9 (-2.7 to 10.5)
10	15.7 (5.2 to 26.2)	1.5 (-6.0 to9.0)	3.5 (-4.4 to 11.4)	-0.3 (-1.0 to 0.5)	5.9 (0.3 to 11.4)
15	7.0 (-3.4 to 17.5)	2.1 (-6.1 to 10.3)	-2.1–8.5 to 4.2)	0.2 (-0.4 to 0.9)	5.0 (-1.0 to 10.9)
20	4.7 (-6.5 to 16.0)	2.4 (-5.9 to 10.8)	2.9 (-4.6 to 10.5)	-0.1 (-1.0 to 0.7)	6.6 (0.6 to 12.6)
Treatment effect (p for interaction)	**0.018**	**0.18**	**0.37**	**0.06**	**0.09**
Change over time within STAND (p-value)	*0.79*	*0.73*	*0.012*	*0.05*	*0.22*
Change over time within AAT (p-value)	*0.034*	*0.39*	*0.07*	*0.50*	*0.17*

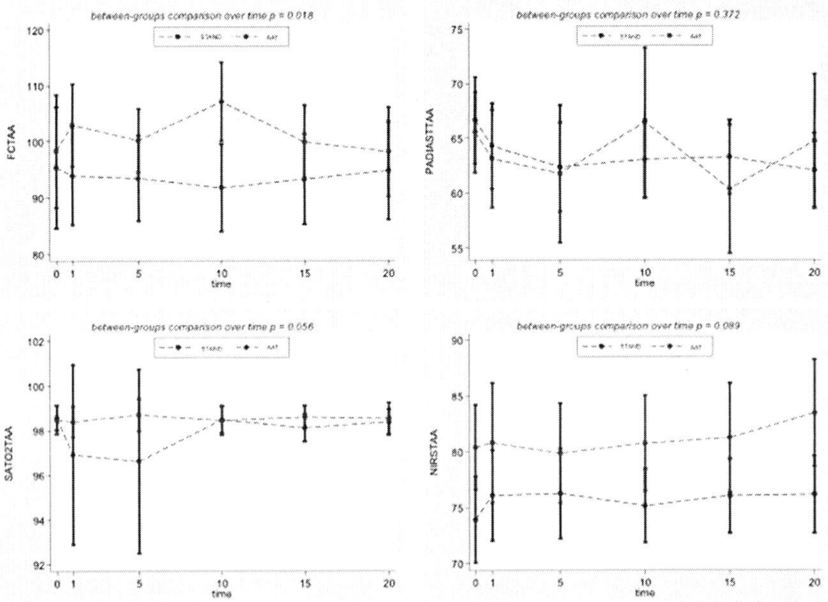

Figure 2. Comparison of the model-derived profiles for heart rate (HR), diastolic blood pressure (DBP), oxygen saturation (SpO_2), cerebral prefrontal oxygenation (HbO_2) in animal-assisted therapy (AAT) and standard (STAND) groups.

All models are adjusted for time of anesthesia

Endocrinological data.

Repeated measurement of salivary cortisol levels did not show differences over time between the AAT and STAND-groups (test for interaction, p = 0.70). At T1, cortisol concentrations in the AAT and STAND groups were respectively 10.98±9.49 vs 14.17±18.2 nmol/L (Difference -3.2 nmol/L, 95%CI -13.4 to 7.0, p = 0.53), at T2, 14.94±18.09 vs 13.29±11.23 nmol/L (Difference 1.6, 95%CI -9.0 to 12.3, p = 0.75) and at T3, 7.89±9.71 vs 8.92±7.89 nmol/L (Difference -0.9, 95%CI -7.2 to 5.3, p = 0.76))

Pain scale findings.

At T2, a lower perception of pain was noted in the AAT group in comparison with the STAND-group (p = 0.01), Table 3. Four cases in the AAT group had a reduction in pain from a level of six to zero.

Table 3. Faces pain scale in animal assisted therapy (AAT) and standard (STAND) groups (p = 0.01).

Pain levels	Number of patients (%)		
	AAT group (n = 20)	STAND group (n = 20)	Total
0	18 (66.67%)	9 (33.33%)	27 (100%)
2	2 (22.22%)	7 (77.78%)	9 (100%)
4	0 (0%)	3 (100%)	3 (100%)
6	0 (0%)	1 (100%)	1 (100%)

No differences in analgesic treatment in the first 12 hours post-intervention were recorded in the groups.

DISCUSSION

Surgery is one of the most stressful events a child may experience. The effects of AAT in pediatric surgery have not been described yet. We showed that AAT intervention during post-surgical procedures induces neurological and cardiovascular responses in the immediate post-operative period in children undergoing surgical procedures. Early vigilance recovery and activity after anaesthesia and activation of the emotional prefrontal area were noted in the AAT-group.

AAT has been utilized world-wide, as of 1960, and consists of animals with therapeutic training for patients with emotional, physical and mental diseases. In the presence of these animals, patient levels of anxiety and stress are reduced following painful procedures, physical activity is stimulated and interpersonal relationships, self-care and cardiovascular parameters are improved [19, 28–33].

The presence of atypical alpha-like (10–14 Hz) activity in anterior derivations (frontal and temporal) was observed during post-operative wakening in almost all of the STAND patients; this finding may be related to propofol treatment [34,35]. This activity, as reported in the literature, was constant throughout the duration of the postoperative recording, tending to fragment and disappear in the hours and days following anaesthesia. The 20 patients in the AAT-group, 2 hours after post-operative awakening, also showed the presence of atypical anterior alpha-like EEG activity. However, after the AAT intervention their EEG showed a quick and clear tendency to fragment the alpha-like activity up to its disappearance and replacement by quicker beta activity usually observed in awakening or during activities that require attention.

The more rapid disappearance of the propofol-induced atypical alpha-like anterior activity in patients in the AAT-group compared to the STAND-group is clear in our protocol; therefore, even though AAT stimulation is new and innovative (with few mentions in the literature), we can affirm that an early post-operative intervention with AAT stimulation could facilitate a rapid recovery of vigilance and activity after anaesthesia with propofol.

Hemodynamic changes during assessment of brain activity have been described [36]. The autonomic nervous system regulates involuntary body functions and is commonly evaluated by measuring systolic and diastolic blood pressure and heart rate reflex responses to physiological and pharmacological stimuli. In this study, the autonomic cardiovascular changes in HR and blood pressure could be considered as adaptative responses correlated to an early neurological response. Autonomic nervous system activity is an important component of human emotion. Mental processes influence bodily physiology, which in turn feeds back to influence thoughts and feelings. Afferent cardiovascular signals from arterial baroreceptors in the carotid sinuses are processed within the brain and contribute to this two-way communication with the body [22].

Several studies have investigated the pain-relieving effects of dog therapy in pediatric and adult patients and significant pain reduction has been reported reported [19,37]. EEG, although limited by its poor spatial resolution and high susceptibility to noise, has been widely used to investigate the brain dynamics relative to emotion as it enables the detection of immediate responses to emotional stimuli with excellent temporal resolution [21]. In the AAT-group the presence of beta activity was correlated with an increase in attention; these data could explain the finding of a higher threshold of pain sensitivity in these patients compared to the STAND-group, which instead had greater pain perception, presumably correlated to higher levels of drowsiness and apathy as demonstrated in the EEG findings.

The capacity to regulate emotions is particularly important during and after encountering a stressor. Emotional responses engage many different areas of the brain including parts of the limbic system and prefrontal cortex [38]. NIRS monitors hemodynamic activity in the brain measuring changes in oxygenated [HbO_2] and deoxygenated haemoglobin [HHb] concentrations in regional cerebral blood flow. Exposure to emotionally-laden stimuli is known to produce measurable changes in chromophore concentration (ie [HbO_2] and [HHb]). In this study, the prefrontal hemodynamic response as measured by NIRS, supports brain activity suggestive ofexecutive processing and a more positive activation of emotional areas during the AAT intervention.

In response to a stressor, the HPA axis (the major neuroendocrine system in mammals) regulates circulating levels of glucocorticoid hormones and provides rapid response and defense against stress. Cortisol is the main glucocorticoid, it is an essential hormone for normal homeostasis and is a useful indicator of short term stress in humans [23].

Brief dog therapy visits also result in a reduction in stress hormones [37]. Repeated measurement of salivary cortisol was employed to assess the endocrinological response on the HPA-axis during AAT. We used salivary cortisol because, unlike plasma, saliva collection is pain free for participants. With a delay of a few minutes, salivary cortisol represents an equivalent of the free, non-protein-bound cortisol in plasma [38]. A few studies have reported on the cortisol response to AAT. Barker et al measured serum cortisol in healthcare workers at baseline and one hour after a brief dog therapy visit or quiet rest and showed that a reduction in cortisol occurred with each intervention [4–7]. Viau et al [39] showed that the cortisol awakening response of children with autism is sensitive to the presence of service dogs, without any effect on the children's average diurnal cortisol levels. Beets et al reported benefits on stress modulation in male children with attachment issues [15,16]. We noted no different cortisol responses in the AAT and STAND-groups. The effect of AAT was probably not evident due to the high intra-individual biological variability of cortisol. However, the generally lower values found at T3 suggest a substantial integration of the HPA-axis after the surgical procedure. A possible explanation for the somewhat contradicting results might be the timing of the stressor, and according by different physiological stress systems might have an effect on emotion regulation [20]. To understand the real benefit of AAT on the stress response, an evaluation of other stress markers, such as epinephrine, norephinefrine, endorphin, oxitocin would be useful.

Notably, the use of AAT is not without risks [19] although no adverse outcomes resulted from this study. Without contact with oral secretions of the pets, the likelihood of transmission of infection from an immunized animal to an immunocompetent child is low [19]. Banks et al reported on 1690 patients visited by AAT dogs over a five year period and no zoonotic infections were reported [11,12]. AAT appears to be a therapeutic modality in which the benefits greatly outweigh the risks.

This pilot study was limited by the small sample size, thus further studies with a larger number of patients are mandatory to confirm the positive role of AAT after-surgery in children.

CONCLUSION

Therapy dogs offer a novel and useful complementary therapy for children undergoing surgical procedures. AAT facilitates rapid recovery of vigilance and activity after anaesthesia, modifies pain perception and induces an emotional prefrontal response. An adaptative cardiovascular response was also observed.

ACKNOWLEDGMENTS

We wish to acknowledge Debra Buttram, instructor in animal assisted interventions and President of Natura Animale; A.B.C. Burlo Onlus for supporting psychological research in our Pediatric Surgery Unit.

AUTHOR CONTRIBUTIONS

Conceived and designed the experiments: VC PV CP GP. Performed the experiments: VD MT RR FP SM SO. Analyzed the data: VC PV CK GP. Contributed reagents/materials/analysis tools: RA. Wrote the paper: VC PV VD GP.

REFERENCES

1. Miller J, Connor K Going to the dogs...for help. Nursing 2003;30: 65–67.

2. Wohlfarth R, Mutschler B, Beetz A, Kreuser F, Korsten-Reck U. Dogs motivate obese children for physical activity: key elements of a motivational theory of animal-assisted interventions. Front Psychol. 2013;4: 796. pmid:24194726
3. Kamioka H, Okada S, Tsutani K, Park H, Okuizumi H, Handa S, et al. Effectiveness of animal-assisted therapy: A systematic review of randomized controlled trials. Complement Ther Med. 2014; 22: 371–390. pmid:24731910
4. Barker SB, Dawson KS. The effects of animal-assisted therapy on anxiety ratings of hospitalized psychiatric patients. Psychiatr Serv. 1998;49: 797–801. pmid:9634160
5. Barker SB, Knisely JS, McCain NL, Best AM. Measuring stress and immune response in healthcare professionals following interaction with a therapy dog: a pilot study. Psychol Rep. 2005;96: 713–729. pmid:16050629
6. Barker SB, Pandurangi AK, Best AM. Effects of animal-assisted therapy on patients' anxiety, fear, and depression before ECT. J ECT 2003;19: 38–44. pmid:12621276
7. Barker SB, Wolen AR. The benefits of human-companion animal interaction: a review. J Vet Med Educ. 2008;35: 487–495. pmid:19228898
8. Bardill N, Hutchinson S. Animal-assisted therapy with hospitalized adolescents. J Child Adolesc Psychiatr Nurs. 1997;10: 17–24. pmid:9257630
9. Richeson NE. Effects of animal-assisted therapy on agitated behaviors and social interactions of older adults with dementia. Am J Alzheimers Dis Other Demen. 2003;18: 353–358. pmid:14682084
10. O'Haire ME. Animal assisted intervention for autism spectrum disorder: a systematic literature review. J Autism Dev Disord. 2013; 43: 1606–1622. pmid:23124442
11. Banks MR, Banks WA. The effects of animal-assisted therapy on loneliness in an elderly population in long-term care facilities. J Gerontol A Biol Sci Med Sci. 2002;57: M428–32. pmid:12084804
12. Banks MR, Willoughby LM, Banks WA. Animal-assisted therapy and loneliness in nursing homes: use of robotic versus living dogs. J Am Med Dir Assoc. 2008;9: 173–177. pmid:18294600
13. Barba BE. The positive influence of animals: animal-assisted therapy in acute care. Clin Nurse Spec. 1995;9: 199–202. pmid:7634227
14. Cole KM, Gawlinski A. Animal-assisted therapy: the human-animal bond. AACN Clin Issues. 2000,11: 139–149. pmid:11040560
15. Beetz A, Julius H, Turner D, Kotrschal K. Effects of social support by a dog on stress modulation in male children with insecure attachment. Front Psychol. 2012; 3:352. pmid:23162482
16. Beetz A, Uvnäs-Moberg K, Julius H, Kotrschal K. Psychosocial and psychophysiological effects of human-animal interactions: the possible role of oxytocin. Front Psychol. 2012; 3:234. pmid:22866043
17. Berry A, Borgi M, Francia N, Alleva E, Cirulli F. Use of assistance and therapy dogs for children with autism spectrum disorders: a critical review of the current evidence. J Altern Complement Med. 2013;19: 73–80. pmid:22978246
18. Martin F, Farnum J. Animal-assisted therapy for children with pervasive developmental disorders. West J Nurs Res. 2002;24: 657–670. pmid:12365766
19. Braun C, Stangler T, Narveson J, Pettingell S. Animal-assisted therapy as a pain relief intervention for children. Complement Ther Clin Pract. 2009;15: 105–109. pmid:19341990
20. Kinner VL, Het S, Wolf OT. Emotion regulation: exploring the impact of stress and sex Front Behav Neurosci. 2014;8: 397. pmid:25431554
21. Kim MK, Kim M, Oh E, Kim SP. A review on the computational methods for emotional state estimation from the human EEG. Comput Math Methods Med. 2013: 573734.
22. Makovac E, Garfinkel S, Bassi A, Basile B, Macaluso E, Cercignani M, et al. Effect of Parasympathetic Stimulation on Brain Activity During Appraisal of Fearful Expressions. Neuropsychopharmacology. 2015;Jan 12. [Epub ahead of print]
23. Spiga F, Walker JJ, Terry JR, Lightman SL. HPA Axis-Rhythms. Compr Physiol. 2014;4: 1273–1298. pmid:24944037

24. Bekkedal MY, Rossi J 3rd, Panksepp. Human brain EEG indices of emotions: delineating responses to affective vocalizations by measuring frontal theta event-related synchronization.Neurosci Biobehav Rev. 2011;35: 1959–1970. pmid:21596060
25. Harmon-Jones E, Winkielman P. Asymmetrical frontal cortical activity, affective valence, and motivational direction. In: Guilford Press editor. Social Neuroscience. New York; 2007. pp. 137–156.
26. Herrmann MJ, Ehlis AC, Fallgatter AJ. Prefrontal activation through task requirements of emotional induction measured with NIRS. Biol Psychol. 2003;64:2 55–263.
27. ausovec N, Jausovec K, Gerlic I. Differences in event-related and induced EEG patterns in the theta and alpha frequency bands related to human emotional intelligence. Neurosci. Lett. 2001;311: 93–96. pmid:11567786
28. Ernst LS. Animal-assisted therapy: paws with a cause. Nurs Manage. 2013;44: 16–19. pmid:24257450
29. Ernst LS. Animal-assisted therapy: using animals to promote healing. Nursing. 2012;42: 54–58. pmid:23014288
30. Fisher A, Wagner R, Newsome JT. Pet-assisted therapy. Lab Anim (NY). 2014;43: 122–123.
31. Hooker SD, Freeman LH, Stewart P. Pet therapy research: a historical review. Holist Nurs Pract. 2002;16: 17–23. pmid:12465214
32. Kaldy J. Alternative therapies for a new era of health care. Consult Pharm. 2013;28: 84–90. pmid:23395808
33. Marcus DA. The science behind animal-assisted therapy. Curr Pain Headache Rep. 2013;17: 322. pmid:23430707
34. Murphy M, Bruno MA, Riedner BA, Boveroux P, Noirhomme Q, Landsness EC, et al. Propofol anesthesia and sleep: a high-density EEG study. Sleep. 2011;34: 283–291. pmid:21358845
35. Ní Mhuircheartaigh R, Warnaby C, Rogers R, Jbabdi S, Tracey I. Slow-wave activity saturation and thalamocortical isolation during propofol anesthesia in humans. Sci Transl Med. 2013;5: 208ra148.
36. Corazza I, Barletta G, Guaraldi P, Cecere A, Calandra-Buonaura G, Altini E, et al. A new integrated instrumental approach to autonomic nervous system assessment. Comput Methods Programs Biomed. 2014;11: 267–277
37. Moghmi S, Kushki A, Guerguerian AM, Chau T. Characterizing emotional response to music in the prefrontal cortex using near infrared spectroscopy. Neurosci Lett. 2012;525: 7–11 pmid:22842396
38. Woodside DB, Winter K, Fisman S. Salivary cortisol in children: correlations with serum values and effect of psychotropic drug administration. Can J Psychiatry.1991;36: 746–748. pmid:1790521
39. .Viau R, Arsenault-Lapierre G, Fecteau S, Champagne N, Walker CD, Lupien S. Effect service dogs on salivary cortisol secretion in autistic children. Psychoneuroendocrinology 2010;35: 1187–93. pmid:20189722

CHAPTER 10

Specialty Grand Challenge in Veterinary Surgery and Anesthesiology

Christopher R. Byron

Department of Large Animal Clinical Sciences, Virginia-Maryland College of Veterinary Medicine, Blacksburg, VA, USA

INTRODUCTION

Truly incredible advances have been gained in the past few decades with regard to the clinical practice of veterinary surgery and anesthesiology. Novel investigational techniques and medical procedures have greatly enhanced our abilities to diagnose and treat animals. Use of minimally invasive diagnostic and procedural techniques, improvements in medical equipment, and the advent of molecular biological methods have advanced veterinary medical practice to a level well beyond what was possible just a few short years ago. However, tough challenges remain that will require novel approaches and new ways of thinking. Education of future clinician scientists is at the heart of this advancement; traditional instructional techniques may not meet the needs of future students. An increase in the flexibility and efficiency of veterinary medical education will be needed. In addition, we must have greater awareness of the interplay between human, animal, and environmental health. The importance of these relationships has recently been identified in the tenets of the One Health initiative; this movement has great relevance to the advancement of veterinary surgery and anesthesiology. Another critical component in the growth of these specialties will be concerted application of evidence-based clinical decision making. To that end, we will need to deliberately design high-quality randomized controlled clinical trials and to report the findings of these studies transparently and completely. In this editorial, I will make the case for the importance of these issues in veterinary surgery and anesthesiology and for the role of this specialty journal section in solving these problems. The purpose of the *Veterinary Surgery and Anesthesiology* section of *Frontiers in Veterinary Science* is to provide an efficient forum for dissemination and discussion of high-quality scientific information relevant to these and closely related disciplines. This is an exciting endeavor because it fosters engagement of authors, peer reviewers, editors, and readers in the process. Together, we can tackle these challenges!

Keywords: surgery, anesthesiology, education, One Health, evidence-based medicine, reporting guidelines, CONSORT, ARRIVE

A PARADIGM SHIFT IN VETERINARY SURGERY AND ANESTHESIA EDUCATION

Veterinary education is presently facing many simultaneous challenges. The costs of education for students are rising dramatically and, compounding this problem, recent economic analysis indicates that the net present value and return on investment of veterinary education are low (1). To address these and other issues facing the veterinary educational community, the North American Veterinary Medical Education Consortium determined strategic goals and recommendations including graduation of proficient career-ready veterinarians, use of competency-driven assessments, sharing of resources within the veterinary education community, increasing the economic viability of veterinary education, and increasing focus on innovation (2). Although there appears to be wide support for such measures, progress is slow and will require ongoing efforts (3).

Along with veterinary medical education in general, there has been a paradigm shift in veterinary surgery and anesthesia education. Trends are toward decreased use of live animals for teaching of procedures. As a result, there is great interest in use of alternate educational methods. Dedicated clinical skills laboratories are becoming more common at veterinary teaching institutions; these facilities incorporate innovative methods and custom models for use in surgical instruction. Validation is an important component of the process for refinement of these teaching methods. The goal is to improve the surgical education of veterinary students, interns, and surgical residents while decreasing unnecessary use of live animals. Recently, development of simulators (4) and use of video games (5) has been investigated for instruction and prediction of minimally invasive surgery skills of veterinarians and veterinary students. In addition, minimally invasive surgery can be useful for the instruction of traditional clinical skills. As an example, the use of laparoscopic guidance can improve student learning of rectal examination in horses (6). Such approaches are likely to become more common in traditional and minimally invasive surgical education. It is critical that descriptions and results of scientific analysis of these techniques be disseminated through primary literature publications. Use of scientific methodology for analysis of pedagogical approaches will result in improved educational methods through better and more efficient use of resources. Another benefit of this approach may be reduced cost of education, which will be increasingly important for economical delivery of instruction.

In addition to veterinary clinical education needs, there will be a growing requirement for training of scientists in relevant applied and basic science fields. With the pool of funding sources for support of scientific inquiry shrinking, discovery and training will continue to be a challenge. As well, recruitment of talented young clinician-scientists is difficult. New approaches to veterinary scientific exploration may increase available funding. In particular, more concerted efforts should be made to identify and evaluate naturally occurring animal diseases as models of human disease. In addition, veterinary scientific discovery would benefit from increased use of multidisciplinary and interdisciplinary teams of investigators working in multiple areas including veterinary and human clinical sciences, basic and applied biological sciences, chemistry, and materials engineering. Communication of such approaches to investigation through scientific journals is important, because areas in which further investigation is needed can be identified and such multidisciplinary approaches promoted.

Scientific journals play an integral role in supporting and advancing veterinary education. Peer review and dissemination of the descriptions of pedagogical techniques should be encouraged. Similarly, the efficacy of competency-driven assessments of skills (including surgery and anesthesia) will require validation. High-quality open-access publications are an efficient and cost-effective avenue for delivery of information to students and instructors. Description and validation of novel techniques that reduce the use of live animals in instruction and increase the economic and educational efficiency of veterinary learning can be effectively accomplished through publication in such a forum. And, journals such as Frontiers provide a high-quality open-access mechanism for dissemination of peer-reviewed manuscripts that are essential for the support of veterinary graduate students and investigators.

ONE HEALTH: VETERINARY SURGICAL MODELS OF HUMAN DISEASE

The above approaches to invigorating veterinary scientific education and discovery overlap with several tenets of the One Health Initiative. This movement is intended to foster co-equal collaborations between veterinary and human medical professionals to enhance the health of animals, humans, and the environment[1]. Veterinarians are uniquely positioned to contribute to these goals[2]. Prominent examples of topics for which this movement is expected to have high impact are zoonotic and vector-borne diseases, food production, the human–animal bond, and population growth. In the disciplines of veterinary surgery and anesthesiology, diagnosis and treatment of naturally occurring diseases often have direct applicability to One Health. Many animal diseases are excellent models of human disease; techniques for treatment of diseases in veterinary species can substantially aid treatment of diseases in humans, and vice versa. For example, femoropatellar joint osteoarthritis is a common malady that develops spontaneously in dogs. Experimentally induced and naturally occurring knee arthritis in dogs are considered excellent models of such diseases in humans (7) because diagnostic techniques, treatment methods, and outcome measures are similar in these species. Therefore, advances in osteoarthritis treatment for either humans or dogs can be of direct benefit to the other species. Horses are another veterinary species that commonly develop degenerative joint disease. Osteoarthritis is the most common cause of lameness in horses. Therefore, this is a potentially useful model for osteoarthritis in humans. As with dogs, the diagnostic methods and medical and surgical treatments for horses with arthritis are very similar to those used for humans. In addition, veterinarians have extensive clinical experience with this disease in horses (8). Such experiences, together with the commonality of the disease, afford the opportunity to use knowledge of equine osteoarthritis pathophysiology, biomarker detection, targets of therapy, and outcomes for the purposes of comparative rheumatology. Reliable experimental induction of osteoarthritis in horses, including use of non-terminal models (9), can also contribute greatly to such comparative approaches to human disease.

Recognition of animal diseases as models of human diseases is vitally important. However, an efficient method is needed to allow timely sharing of basic and clinical scientific knowledge. To that end, internet-based registries of diseases can be an effective way to disseminate such data; this could encourage multi-institutional studies, potentially increasing numbers of animals in such studies and broadening the geographic relevance of such work. Finally, a critical component of this comparative approach to modern medical advancement is an efficient scientific communication method.

Expedited communication of high-quality scientific discovery via open-access journals, such as Frontiers in Veterinary Science, is an efficient mechanism to support the progress of investigators in these endeavors.

EVIDENCE-BASED VETERINARY SURGERY AND ANESTHESIA

Evidence-based medicine seeks to apply evidence determined via use of the scientific method for optimization of clinical decision making. There have been calls for deliberate and purposeful application of this principle in veterinary practice (10). Success will depend on a transition from reliance on personal clinical experience and published small case series to use of evidence from large randomized controlled clinical trials. Unfortunately, the available scientific evidence in veterinary clinical disciplines is often of low quality and insufficient to inform clinical decision making with confidence (11). The strength of such evidence is a continuum from results determined through *in vitro* studies or statements of personal opinion (low quality) to meta-analyses of large randomized controlled trials (high quality). Historically, impediments to achievement of this gold standard have hindered application of evidence-based medicine to veterinary surgery and anesthesiology. In particular, low case numbers in published studies from single institutions, the lack of well-defined outcome measures, and absence of appropriate controls diminish the quality of evidence and its applicability to a geographically diverse veterinary practice. Increased multi-institutional collaborations among investigators would lead to an increase in the number of animals enrolled in clinical trials and may also increase the relevance of results to a geographically diverse population. In addition, such studies should be thoughtfully designed to include biologically relevant outcome measures and, when possible, this information should be collected by use of validated clinical outcome instruments (12, 13). These outcome instruments include questionnaires, standardized clinical scoring algorithms, and objective measures, such as force plate data (for assessment of lameness), physiologic parameters, and biomarkers. Further, study designs should include appropriate control populations and incorporate sound statistical methods. An additional method for enhancing collaboration in clinical scientific discovery and increasing the statistical power of clinical trials may be use of online registry databases as mentioned above, which could greatly aid the speed and quality of discovery.

Scientific publications have arole in evidence-based medicine. Dissemination of clinical evidence determined through well designed and carefully conducted studies provide the necessary scientific information for evidence-based decision making in veterinary practice. A high quality of manuscript preparation and meticulous review are vital in this process. Further, expedited and open-access publication ensure that clinically relevant evidence is disseminated in a timely fashion to the audience that most needs the information, regardless of their access to well-stocked medical libraries.

REPORTING GUIDELINES

An essential component of evidence-based medicine is meticulous reporting of the results of studies. Without this, findings cannot be adequately analyzed and considered by readers and therefore cannot be used for sound clinical decision making. Incomplete or inaccurate reporting of otherwise well-designed clinical trials therefore minimizes the impact of such important scientific knowledge. This can be a particularly vexing problem for veterinary surgery, as it is often difficult to accurately gage success and compare outcomes for various procedures because of sparse or incomplete information in the scientific literature. Likewise, these problems pose great difficulties for anesthesiologists; anesthetic complications for veterinary patients can be rare although important, and a lack of accurate and detailed reporting of clinical studies hampers use of such information to improve the quality and safety of patient care. Undoubtedly, shortcomings in reporting are typically unintentional and the result of the myriad choices that authors face when preparing manuscripts. Recently, there has been an effort to develop guidelines for standardized reporting of medical research. The EQUATOR network[3] is an organization of medical research stakeholders including investigators, journal editors, and peer reviewers that have developed reporting guidelines for over 250 specific types of studies. The stated mission of this group is to "achieve accurate, complete, and transparent reporting of all health research studies to support research reproducibility and usefulness ... to minimize avoidable waste of financial and human investments in health research projects." One of the most relevant of these guidelines to surgery and anesthesiology is the CONSORT (Consolidated Standards of Reporting Trials) statement[4]. This is an evidence-based list of recommendations for reporting of randomized trials, and includes checklists and flow diagrams to help authors with complete and transparent reporting of study findings. Employment of these items substantially increases the quality and utility of published work. Important components include recommendations regarding descriptions of trial design, subject recruitment, outcomes, sample size, blinding, statistical methods, study limitations, and interpretation. In addition, the ARRIVE (Animal Research: Reporting of *In Vivo* Experiments) guidelines[5] have been developed "to improve the design, analysis, and reporting of research using animals – maximizing information published and minimizing unnecessary studies." These recommendations are applicable to all areas of animal research, and provide valuable resources that can readily be used to improve reporting of *in vivo* surgery and anesthesiology research. Deliberate use of standardized guidelines during study design, manuscript preparation, and peer review should be encouraged, as it ultimately enhances study quality and serves to maximize impact for evidence-based decision making.

SUMMARY

There are many important challenges facing the veterinary profession. In particular, challenges in educational methods, application of the tenets of One Health, integration of evidence-based medicine, and careful integration of standardized study design and reporting guidelines are germane to the specialties of surgery and anesthesiology. The *Veterinary Surgery and Anesthesiology* specialty section of *Frontiers in Veterinary Science* will be an excellent forum for rapid dissemination of high-quality research findings in many facets of our profession. As well, it will be an open forum for discussion of the challenges facing us. As per the mission statement, the goals of this section are dissemination of findings from clinical research and basic science studies of naturally occurring and experimental disease with translational potential. The use of multidisciplinary and interdisciplinary approaches is encouraged. It is my sincere belief that this will be a dynamic and effective forum for solving these and other challenges that lie ahead. I invite you to actively engage in this scientific discussion and help forge the future of veterinary surgery and anesthesiology!

CONFLICT OF INTEREST STATEMENT

The author declares that the research was conducted in the absence of any commercial or financial relationships that could be construed as a potential conflict of interest.

FOOTNOTES

1. ^http://www.onehealthinitiative.com

2. ^www.avma.org/KB/Resources/Reference/Pages/One-Health94.aspx
3. ^http://www.equator-network.org
4. ^http://www.consort-statement.org
5. ^https://www.nc3rs.org.uk/arrive-guidelines

REFERENCES

1. Knippenberg R, Dicks MR, Bain B, Dow M. Estimating the financial return on a veterinary education. *J Am Vet Med Assoc* (2015) **246**:422–4.
2. AAVMC. *NAVMEC Report: Roadmap for Veterinary Medical Education in the 21st Century: Responsive, Collaborative, Flexible*. (2015). Available at: www.aavmc.org/data/files/navmec/navmec_roadmapreport_web_booklet.pdf
3. Andrews K, Chaddock M, Osburn BI. Baseline survey of progress by veterinary medical colleges in implementing recommendations from the North American Veterinary Medical Education Consortium. *J Am Vet Med Assoc* (2013) **243**:826–32.
4. Fransson BA, Ragle CA. Assessment of laparoscopic skills before and after simulation training with a canine abdominal model. *J Am Vet Med Assoc* (2010) **236**:1079–84.
5. Millard HA, Millard RP, Constable PD, Freeman LJ. Relationship among video gaming proficiency and spatial orientation, laparoscopic, and traditional surgical skills of third-year veterinary students. *J Am Vet Med Assoc* (2014) **244**:357–62.
6. Radcliffe RM, Manchester AC, Mohammed HO, Ortved K, Reesink HL, Schnabel LV, et al. Laparoscopic-guided compared to skilled instructor support for student rectal examination training using live horses in the veterinary curriculum. *Vet Surg* (2015) **44**:352–8.
7. Gregory MH, Capito N, Kuroki K, Stoker AM, Cook JL, Sherman SL. A review of translational animal models for knee osteoarthritis. *Arthritis* (2012) **2012**:764621.
8. McIlwraith CW, Frisbie DD, Kawcak CE. The horse as a model of naturally occurring osteoarthritis. *Bone Joint Res* (2012) **1**:297–309.
9. Boyce MK, Trumble TN, Carlson CS, Groschen DM, Merritt KA, Brown MP. Non-terminal animal model of post-traumatic osteoarthritis induced by acute joint injury. *Osteoarthritis Cartilage* (2013) **21**:746–55.
10. Schulz KS, Cook JL, Kapatkin AS, Brown DC. Evidence-based surgery: time for change. *Vet Surg* (2006) **35**:697–9.
11. Aragon CL, Budsberg SC. Applications of evidence-based medicine: cranial cruciate ligament injury repair in the dog. *Vet Surg*(2005) **34**:93–8.
12. Cook JL. Outcomes-based patient care in veterinary surgery: what is an outcome measure? *Vet Surg* (2007) **36**:187–9.
13. Cook JL. Canine orthopedic outcome measures program: where are we now? *Vet Surg* (2014) **43**:229–31.

CHAPTER 11

Experimental Animal Models of Arteriovenous Malformation: A Review

Jude Amal Raj and Marcus Stoodley *

The Australian School of Advanced Medicine, Macquarie University, NSW 2109, Australia

ABSTRACT

Arteriovenous malformations (AVMs) are congenital lesions that cause brain haemorrhage in children and young adults. Current treatment modalities include surgery, radiosurgery and embolization. These treatments are generally effective only for small AVMs. Over one third of AVMs cannot be treated safely and effectively with existing options. Several animal models have been developed with the aims of understanding AVM pathophysiology and improving treatment. No animal model perfectly mimics a human AVM. Each model has limitations and advantages. Models contribute to the understanding of AVMs and hopefully to the development of improved therapies. This paper reviews animal models of AVMs and their advantages and disadvantages.

Keywords: cerebral arteriovenous malformations; animal models

1. INTRODUCTION

Cerebral arteriovenous malformations (AVMs) are abnormal connections between arteries and veins that lead to the formation of a tangled collection of vessels referred to as a "nidus" [1,2,3]. The high-pressure shunt of arterial blood flowing through the fragile vessels in the nidus and the draining veins can cause intracranial haemorrhage resulting in death or disability [4,5]. The prevalence of this condition is 0.01%–0.5% of the population with presentation common in children and young adults [2,6]. In addition to haemorrhage, AVMs can present with epileptic seizures, chronic headaches, migraines and ischaemic neurologic deficits [7,8,9,10].

Current treatment options include surgical resection, endovascular occlusion and radiosurgery, or combinations of these [11]. Surgery is generally limited to small AVMs that are superficially located [12,13,14]. Large AVMs (more than 3 cm diameter) and those located in critical regions such as the thalamus and basal ganglia remain a challenge for effective treatment: There is a higher risk of morbidity following surgery in such cases [11,15,16]. Radiosurgery, although attractive as a relatively non-invasive treatment, is often not as effective as surgery [17,18]. There is also a 2–3 year delay to AVM occlusion after radiosurgery treatment and a success rate of approximately only 80% even for small lesions [14,19,20,21].

There is an urgent need for a new treatment for the over one third of AVMs that cannot be safely and effectively treated with current protocols [21,22]. Animal models are required to investigate the pathogenesis of AVM formation and their biological and haemodynamic characteristics. Development of a new effective treatment for AVMs is also likely to depend on the use of appropriate animal models. Improving responses of AVM tissue to radiation, and developing adjuvants such as radiosensitisers, could be best achieved using animal models. Potential new biological therapies such as gene therapy or vascular targeting to induce AVM thrombosis will require animal models for developing and trialling the therapeutic agents. Refining endovascular techniques is a further field where animal models are extremely useful.

Many animal models have been developed in the last three decades. Each has advantages and disadvantages and there is no model that is suitable for all purposes. This review discusses each of the research settings where models are used and discusses the advantages and disadvantages of each model. Standardising animal model use for each research indication would hopefully lead to uniformity in research methods and more efficient progress towards the goal of developing better treatments for AVM patients.

2. STUDY OF AVM HAEMODYNAMICS

Replicating AVM haemodynamics is necessary for the study of AVM pathophysiology and refinement of endovascular techniques. Most models include an arteriovenous shunt and a nidus.

The rete mirabile in swine has been used as an AVM model for its network of microarteries that are interconnected and resemble an AVM nidus, at least angiographically [23]. However its arterial-to-arterial structure does not mimic that of a human AVM, which is arterial to venous [23]. This limitation was addressed by creating a short-term arteriovenous shunt between the rete and the internal carotid artery and the cavernous sinus [23]. Although this model has morphological resemblances to human AVMs, there are several adverse effects on the animals such as proptosis, chemosis and sub-conjunctival haemorrhage. The model lasts for a maximum period of only a week by which time the connections have occluded spontaneously. As with most AVM models, the rete mirabile model is not inside the brain. The model however is particularly useful for refining endovascular techniques and trialling new embolic agents [23].

In an attempt to simplify the creation of the model, Massoud *et al.*, created an AVM model in the neck of swine, thus avoiding the adverse effects of the rete-cavernous fistula model. In order to increase blood flow in the fistula, three arteries are occluded in the neck region. This is followed by a side-to-side anastomosis forming a carotid-jugular fistula [24].

This model has been criticised for its use of microcatheter and detachable balloons, which are very expensive [25]. Qian *et al.* created a similar model in sheep, but did not occlude the arterial branches (Figure 1). The sheep rete mirabile is quite different to that in swine. The retia appear as two units, connected by bridging vessels in sheep and as one in swine. This proves to be an advantage for sheep over swine where it gives the possibility for researchers to study multiple AVMs [25].

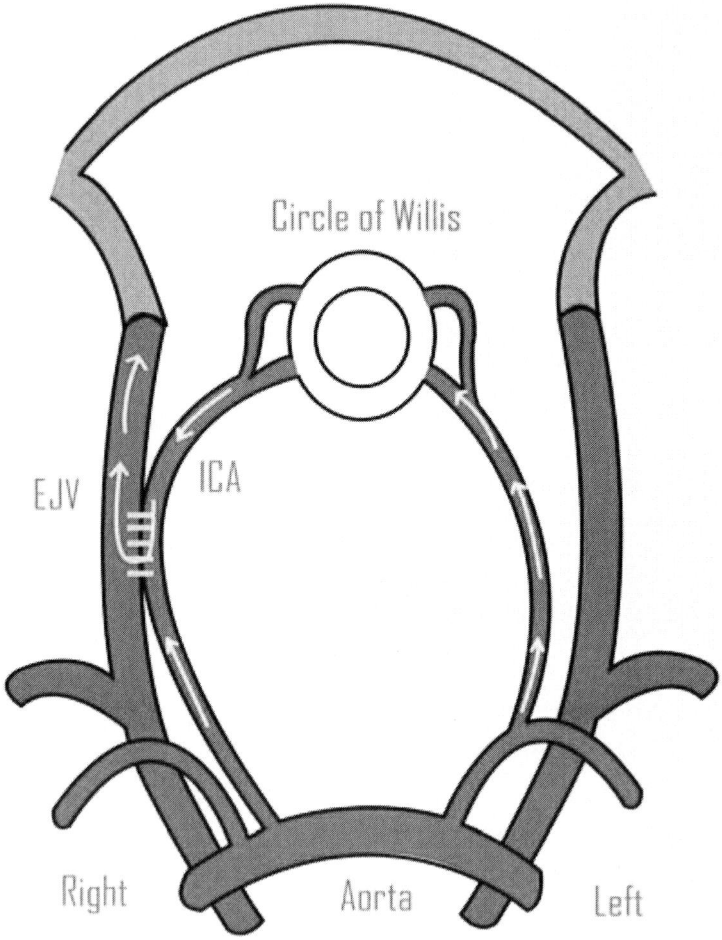

Figure 1. Animal model of cerebral AVM in sheep. Side-to-side anastomosis was carried out between the CCA and EJV (Right). The arrows show the direction of the circulatory flow after anastomosis. ICA: common carotid artery, EJV: external jugular vein. (Adapted from Qian *et al.*, 1990) [25].

Swine are preferred for their large neck vessels, which allow for faster and easier fistula creation. There are models in which external arteries need not be occluded [26,27]. An end-to-end anastomosis instead of a side-to-side anastomosis shows an increase in blood flow in the rete mirabile. Spontaneous occlusion does not occur as in previous models, which means these could be used for research that involves a longer duration [23,24,26,27]. Additional advantages with swine model are that the swine coagulation system is similar to humans and they are inexpensive [24].

Haemodynamic models are also used to study perfusion pressure breakthrough, the role of venous hypertension and the role of thrombosis in AVMs. Rat models are often used for research in these areas [28,29]. Various haemodynamic arrangements have been used, although the model usually is located in the neck region by creating an anastomosis between the common carotid artery and external jugular vein [6,30]. In these models, the common carotid artery acts as the feeding artery and the external jugular vein acts as the arterialized vein. In one model, the anastomosis is between the rostral carotid artery and the caudal jugular vein. This arrangement results in intracranial hypoperfusion, analogous to the "steal" phenomenon around AVMs, and has been used to study the effects of chronic ischaemia [31,32]. The carotid jugular fistula model by Morgan *et al.* was used to study hypoperfusion. An end-to-end anastomosis was performed between the common carotid artery and the external jugular vein (Figure 2). The involvement of both the intracranial arterial and venous systems with the extracranial venous systems is a feature of this model [32].

In other models, the anastomosis is a side-to-end connection between the common carotid artery and the rostral external jugular vein. In this arrangement arterial blood flows through the external jugular vein and its branches (the "nidus") and the outflow is then through the transverse sinus and the contralateral jugular vein (draining vein) (Figure 3) [33,34]. This model has been used to study the molecular and morphological changes in AVMs treated with radiosurgery [6,35]. Critically, the model produces an endothelial phenotype that resembles that of human AVMs; this is crucial when studying radiation effects because the endothelial response to radiation is different for different phenotypes. Blood flow has been shown to increase with time in this model, at least until 42 days post-surgery [6].

In the many AVM models that claim to have a blood flow similar to what is observed in human AVMs, the model is usually extracranial in location and not intracranial [24,36,37]. In response to this, an AVM model was developed in beagle dogs (Figure 4) [38]. An end-to-end anastomosis between the superficial temporal artery and the middle cerebral artery and an end to side anastomosis between the superficial temporal artery and the dorsal sagittal sinus were carried out. A muscle graft supplied by the superficial temporal artery was fixed in the ischaemic region of the brain. Ischaemia was found to be aggravated in the arteriovenous fistula region [38]. An advantage of this model is its intracranial location. This model has not been reported in any subsequent investigations.

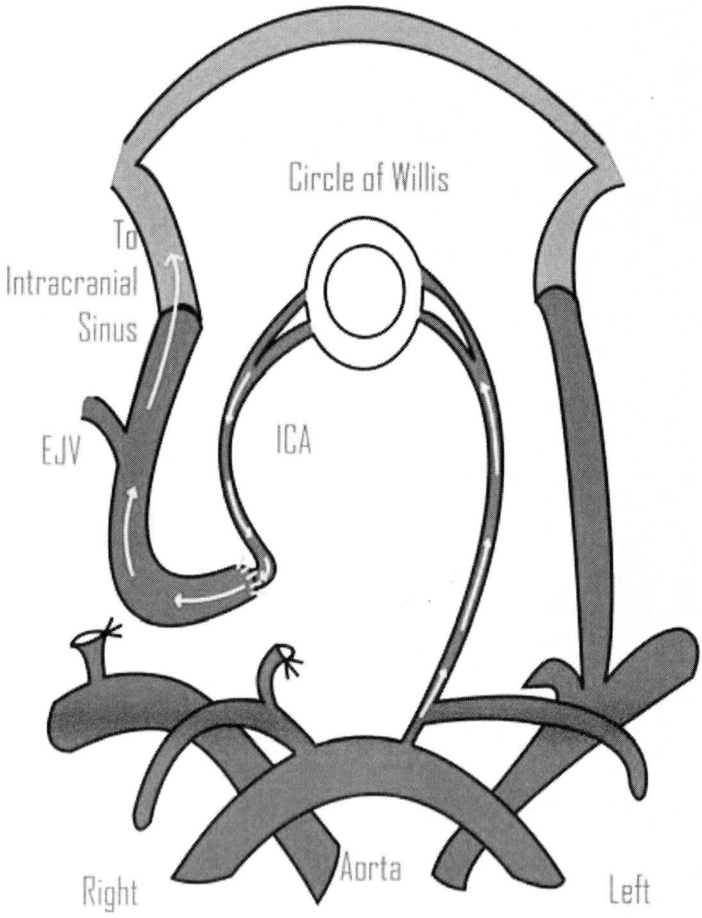

Figure 2. Schematic representation of animal model of cerebral AVM in rat. Carotid jugular fistula model that involves an end-to-end anastomosis between the common carotid artery and external jugular vein. ICA: internal carotid artery; EJV: external jugular vein. (Adapted from Morgan *et al.*, 1989) [32].

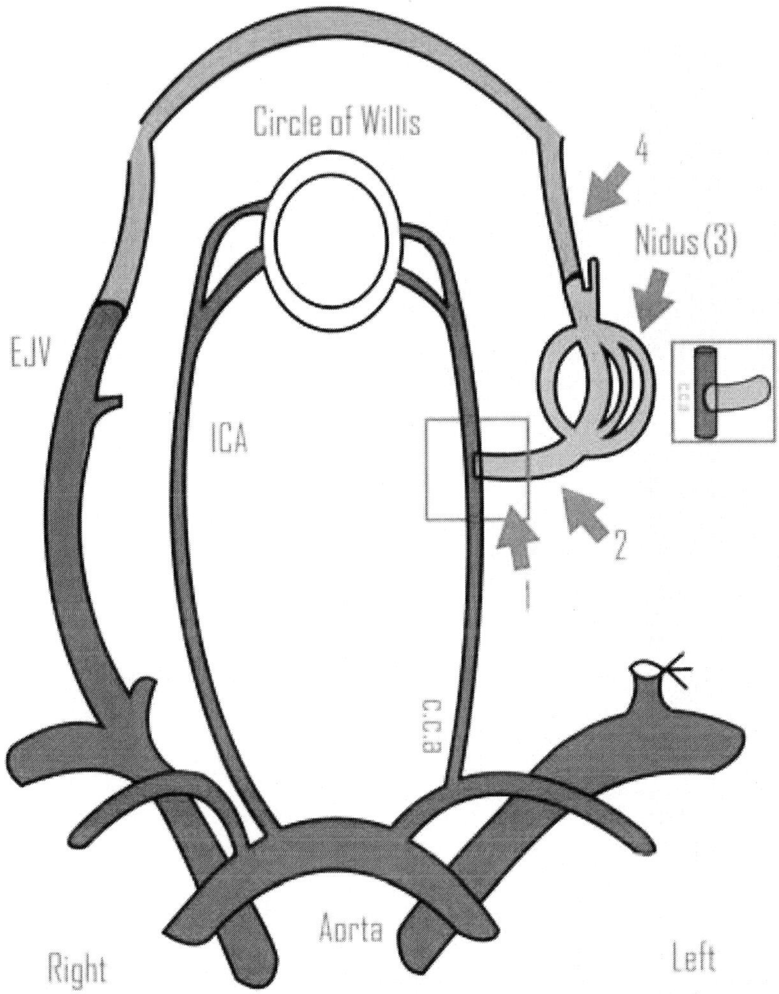

Figure 3. Schematic representation of an AVM model in rat. This is a carotid jugular, side-to-end anastomosis between the common carotid artery and external jugular vein. This is a very efficient model for radiosurgery related studies. CCA: common carotid artery, ICA: internal carotid artery, EJV: external jugular vein. Anastomosis (**1**) arterial feeder (**2**) nidus (**3**) draining vein (**4**).

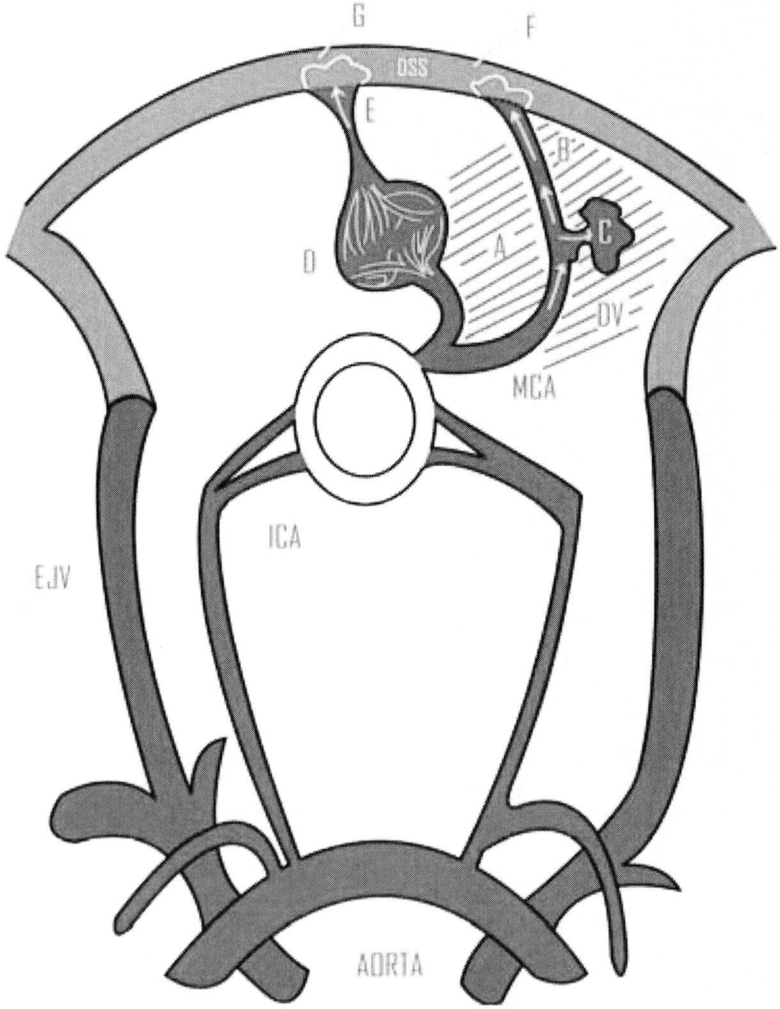

Figure 4. Animal model of cerebral AVM in beagle dogs. The dog model located intracranially has the benefit of the vascular capillary network around the AV shunt, resembling the human AVM more closely. DSS: Dorsal sagittal sinus MCA: Middle cerebral artery Area shaded with slanted lines: Relative ischaemia to stimulate angiogenesis. **A**: End-to-end anastomosis **B**: End-to-side anastomosis **C**: Muscle graft implanted in the ischaemic brain tissue and supplied by the bypass **D**: Nomal capillary network in the brain parenchyma **E**: Vein with normal drainage **F**:Arterialized blood **G**: Venous blood. (Adapted from Pietila *et al.*, 2000) [38].

A model was developed by Numazawa *et al.* to study hypoperfusion and impaired CO_2 reactivity. A femoral vein graft was used to connect a cortical branch of the middle cerebral artery and the superior sagittal sinus, thus creating an arteriovenous shunt. This model was shown to closely resemble the haemodynamics of AVM patients with hypoperfusion and decreased CO_2 reactivity [39].

The rete mirabile is a suitable model for the development of endovascular techniques. Surgery is not required for this arterio-arterial system. However for studies involving the pathophysiology of AVMs and their haemodynamic properties, an arterio-venous shunt is required. This swine model can be used for this purpose, however the arteriovenous model in rats has an advantage in terms of simpler animal care and maintenance and is far more economical than swine and sheep.

3. RADIOSURGERY

Radiosurgery is one of the treatment options for AVMs [40,41,42]. A single, focussed high dose of radiation is delivered to a target volume of AVM tissue [42,43]. Free radicals react with the DNA causing damage [44,45]. This is followed by endothelial proliferation, smooth muscle cell proliferation and thrombosis, which lead to AVM obliteration [43]. The exact mechanism is still unknown. AVM animal models have therefore been developed to study the short-term effects of radiosurgery.

De Salles *et al.* treated eight swine with 20, 30, 40, 50, 60, 70, 80 and 90 Gy dosages. The rete mirabile was selected as the target tissue. Neurologic dysfunctions, such as abnormal eye movement, were commonly observed in the animals. A decrease in vascularity was observed in animals that received 50 Gy or more dosage and complete obliteration of the rete mirabile was observed in the animal that received 90 Gy [46]. The rete mirabile, which is located in the cavernous sinus, is an arterioarterial system unlike an AVM that is an arteriovenous system [47,48,49]. This means that that model lacks the high-pressure blood flow in arterialized veins typical for an AVM.

Jahan *et al.* used the model by Massoud *et al.* to study the effects of radiosurgery on a swine model where a carotid-jugular fistula was created in the neck [50,51]. The preferred dosage was 40 Gy. Decreased vascularity was observed in treated animals when compared to those not treated at a three-month time point [50]. Histopathological changes post-radiation were studied. The study showed expressions of Type IV collagen to be very similar to those in resected specimens post-radiation from humans [50].

Radiosurgery is shown to cause vascular obliteration in several models. Intimal hyperplasia, necrosis, oedema and gliosis have been observed in the tissues surrounding the target volume [46]. In the rete mirabile model for instance, these could be pointed to as the main cause for several neurological deficits such as hemiparesis, loss of coordination, disruption in vision and seizures [46]. Mut *et al.* studied a carotid jugular fistula model in a rat to see the effects of radiosurgery in the tissues surrounding the target volume for a dosage of 25 Gy [43]. DNA strand breaks and the level of free radicals were analysed. There was significant apoptosis in irradiated groups irrespective of the presence of the fistula [43].

Great care is taken in the complete obliteration of AVMs during surgical excision. A small amount of tissue left behind is capable of AVM angiogenesis. This applies to the radiosurgery treatment modality as well. AVM angiogenesis after radiosurgery is highly dose dependent according to the corneal angiogenesis model study [52]. Resected AVM specimens from humans were implanted in the corneal tissues of rats. Dosages of 1.5 Gy, 3 Gy, 15 Gy and 30 Gy were administered to the animals. The latter two dosage groups showed a greater anti-angiogenesis effect than the former two [52].

Another major application radiosurgery models have is in the search for a potential vascular target for inducing thrombosis. In the field of cancer, vascular targeting has been successful in inducing thrombosis due to the molecular differences that distinguish tumor vessels from normal vessels [33]. This however requires the AVM vessels to differ significantly from normal vessels. Tu *et al.* have reported increased levels of ICAM, VCAM and E selectin in human AVM endothelium when compared to the endothelium in normal vessels, however the changes were not sufficiently discriminating to be used as a vascular target [53].

The carotid jugular anastomosis rat model was to see if radiosurgery induced molecular changes sufficient to discriminate them from normal vessels [53]. In this model, the external jugular vein was anastomosed to the common carotid artery. After six weeks when the fistulae were considered mature, AVM niduses were treated with radiosurgery (25 Gy). In this work, radiosurgery was given not as a treatment for the AVM but rather as a priming technique to explore the endothelial molecular changes that would come about as a result of radiosurgery [54]. Subsequent studies have used the model to demonstrate induced thrombosis with vascular targeting [33,35].

Swine and rat models are highly suitable to study the effects and efficacy of radiosurgery. The swine model is resource intensive and suffers from the fact that the "nidus" (rete mirabile) is purely arterial. The rat model is more practical and consists of arterialized veins that more closely resemble the molecular characteristics of human AVMs.

4. GENETIC STUDIES

There is a growing interest in gene therapy to treat large AVMs and those located in regions such as the thalamus, basal ganglia and brainstem that cannot be treated by surgery [55,56,57,58]. For this to be successful, a family of molecules or a particular gene responsible for a certain characteristic feature such as narrowing of the arteries needs to be identified [58,59]. Once identified, that particular family of molecules or the gene may be overexpressed to remove the AVM successfully [58]. AVMs are typically sporadic [60]. However, 2% of AVMs are familial, including patients with hereditary haemorrhagic telangiectasia (HHT) or Rendu-Osler Weber syndrome [60,61,62]. In HHT, abnormal malformations in different organs such as the lung, liver, brain and spine occur as a result of mutations in the endoglin and ACVRL1 genes [63,64]. The causal factors for this genetic disorder are unknown [65]. This poses the need for HHT animal models to better understand the pathophysiology of HHT and also to develop new therapeutic modalities. Animal model studies have successfully been able to identify certain genetic mutations and risk factors [2,66]. The TGF-β genes are a good example of those associated with non-sporadic AVMs. The TGF-β genes go through a loss of function mutation in endoglin and activin like kinase giving rise to the HHT (1 and 2) mutation resulting in AVM formation [65,67,68].

In the first transgenic mouse model study by Satomi *et al.*, ten endoglin heterozygous mice ($Eng^{+/-}$) were compared with 15 controls ($Eng^{+/+}$) mice. AVMs are believed to develop in the mice that are heterozygous at the endoglin locus [69]. Only 30% showed abnormalities that closely resemble those present in human AVMs. However, no abnormalities were observed in any of the control animals. Changes such as decreased arteriole constriction and increased relative dilatation were also observed. More research is required to determine if these changes actually contribute to the formation of AVMs in HHT patients [70]. In another mouse model that was stimulated with VEGF (vascular endothelial growth factor), 89% of the $Eng^{+/-}$ mice showed vascular abnormalities and none were observed in the controls [71]. The lung, liver and intestine showed dilation in some vessels with no external signs in one particular animal [70]. Another animal that had an ear telangiectasis showed an AVM nidus that resembled those in human and canine brain [70]. Despite the fact that $Eng^{+/-}$ and $Acvrl1^{+/-}$ mice are the best available HHT models at present, past studies do not show the HHT features at high frequency [69,72,73]. Although the two studies could be criticised as not being true AVM models and for their small sample size, they may contribute to research on the abnormal phenotype of AVMs in humans [66].

A transgenic arteriovenous fistula between the common carotid artery and external jugular vein with TAo from a donor mouse was created in 112 rats to investigate a possible gene therapy for AVMs. Grafting however would impose the challenge of not being able to investigate certain genetic changes in the fistula [58]. Limitations such as low lesion frequency in $Eng^{+/-}$ mice and lack of the human AVM phenotype in currently existing models call for more work to be carried out in this area of study. Two HHT 1 animal models were created to study the developmental and adult onset of cerebral AVMs [74]. *Eng* conditional knockout mouse lines were obtained by crossing *Eng*$^{2fl/2fl}$ with R26CreER, SM22α-Cre and LysM-Cre. The $Eng^{2fl/2fl}$ were crossbred with SM22αCre to study the effects of embryonic *Eng* deletion on the post-natal development of AVMs [74]. Almost 90% from the group *Eng*$^{2fl/2fl}$; SM22α-Cre showed the AVM phenotype. No lethality was observed until the fifth week [74]. In previous similar models involving a conditional knockout of the Alk1 gene, intracranial haemorrhage led to lethality by even post-natal day 5 and two weeks limiting the use of such models in studies related to new therapies [75,76]. The *Eng*$^{2fl/2fl}$; R26CreER were given a dose of tamoxifen for three consecutive days to induce *Eng* deletion. Those mice that were additionally injected with adeno-associated viral vector expressing vascular endothelial growth factor showed AVM phenotype in the form of lesions eight weeks after *Eng* deletion. No lethality was observed in the *Eng*$^{2fl/2fl}$; R26CreER group upto two months post-tamoxifen treatment [74].

Transgenic animal studies are useful for investigation of gene changes. Any potential gene therapy would need to be trialled on such models. The major drawback with these models is in identifying AVMs in individual animals.

5. CONCLUSIONS

Animal models have been developed because naturally occurring AVMs in animals are quite rare. There is no animal model that is perfect with all the necessary characteristics. The choice of model will depend on the purpose of the research. The rete mirabile is suitable for angiographic studies, but the lack of arterialized veins makes the model unsuitable for studies of AVM biology. Models with an arteriovenous shunt are useful for studying the biology of arterialized veins and for investigating the effects of cerebral hypoperfusion. Investigations using radiosurgery should use a model with arterialized veins and an identifiable model nidus. Transgenic models are appropriate for research related to HHT and for genetic studies.

ACKNOWLEDGMENTS

Our sincere thanks to Leena Swamy, Kaleidostrokes for her help with the figures.

AUTHOR CONTRIBUTIONS

Jude Amal Raj—Design, writing, revision of manuscript and presentation of figures. Marcus Stoodley—Concept, design, revision of manuscript and presentation of figures.

CONFLICTS OF INTEREST

The authors declare no conflict of interest.

REFERENCES

1. Halim, A.X.; Johnston, S.C.; Singh, V.; McCulloch, C.E.; Bennett, J.P.; Achrol, A.S.; Sidney, S.; Young, W.L. Longitudinal risk of intracranial hemorrhage in patients with arteriovenous malformation of the brain within a defined population. Stroke 2004, 35, 1697–1702.
2. Leblanc, G.G.; Golanov, E.; Awad, I.A.; Young, W.L.; Biology of Vascular Malformations of the Brain NINDS Workshop Collaborators. Biology of vascular malformations of the brain. Stroke 2009, 40, e694–e702.
3. Chen, W.; Choi, E.J.; McDougall, C.M.; Su, H. Brain arteriovenous malformation modeling, pathogenesis, and novel therapeutic targets. Transl. Stroke Res. 2014, 5, 316–329.
4. Achrol, A.S.; Guzman, R.; Varga, M.; Adler, J.R.; Steinberg, G.K.; Chang, S.D. Pathogenesis and radiobiology of brain arteriovenous malformations: Implications for risk stratification in natural history and posttreatment course. Neurosurg. Focus 2009.
5. Jeffree, R.L.; Stoodley, M.A. Postnatal development of arteriovenous malformations. Pediatr. Neurosurg. 2009, 45, 296–304.

6. Tu, J.; Karunanayaka, A.; Windsor, A.; Stoodley, M.A. Comparison of an animal model of arteriovenous malformation with human arteriovenous malformation. J. Clin. Neurosci. 2010, 17, 96–102.
7. Rosenkranz, M.; Regelsberger, J.; Zeumer, H.; Grzyska, U. Management of cerebral arteriovenous malformations associated with symptomatic congestive intracranial hypertension. Eur. Neurol. 2008, 59, 62–66.
8. Hofmeister, C.; Stapf, C.; Hartmann, A.; Sciacca, R.R.; Mansmann, U.; TerBrugge, K.; Lasjaunias, P.; Mohr, J.P.; Mast, H.; Meisel, J. Demographic, morphological, and clinical characteristics of 1289 patients with brain arteriovenous malformation. Stroke 2000, 31, 1307–1310.
9. Grzyska, U.; Fiehler, J. Pathophysiology and treatment of brain AVMs. Clin. Neuroradiol. 2009, 1, 82–90.
10. Friedlander, R.M. Clinical practice: Arteriovenous malformations of the brain. N. Engl. J. Med. 2007, 356, 2704–2712.
11. Klopfenstein, J.D.; Spetzler, R.F. Cerebral arteriovenous malformations: When is surgery indicated? Acta Neurochir.2005, 147, 693–695.
12. Pik, J.H.; Morgan, M.K. Microsurgery for small arteriovenous malformations of the brain: Results in 110 consecutive patients. Neurosurgery 2000, 47, 571–577.
13. Morgan, M.K.; Rochford, A.M.; Tsahtsarlis, A.; Little, N.; Faulder, K.C. Surgical risks associated with the management of Grade I and II brain arteriovenous malformations. Neurosurgery 2004, 54, 832–839.
14. Ferch, R.D.; Morgan, M.K. High-grade arteriovenous malformations and their management. J. Clin. Neurosci. 2002, 9, 37–40.
15. Morgan, M.K.; Drummond, K.J.; Grinnell, V.; Sorby, W. Surgery for cerebral arteriovenous malformation: Risks related to lenticulostriate arterial supply. J. Neurosurg. 1997, 86, 801–805.
16. McInerney, J.; Gould, D.A.; Birkmeyer, J.D.; Harbaugh, R.E. Decision analysis for small, asymptomatic intracranial arteriovenous malformations. Neurosurg. Focus 2001.
17. Karlsson, B.; Kihlstrom, L.; Lindquist, C.; Steiner, L. Gamma knife surgery for previously irradiated arteriovenous malformations. Neurosurgery 1998, 42, 1–6.
18. Steiner, W. Results of curative laser microsurgery of laryngeal carcinomas. Am. J. Otolaryngol. 1993, 14, 116–121.
19. Friedman, W.A.; Bova, F.J.; Bollampally, S.; Bradshaw, P. Analysis of factors predictive of success or complications in arteriovenous malformation radiosurgery. Neurosurgery 2003, 52, 296–308.
20. Friedman, W.A. Stereotactic radiosurgery of intracranial arteriovenous malformations. Neurosurg. Clin. N. Am. 2013, 24, 561–574.
21. Han, P.P.; Ponce, F.A.; Spetzler, R.F. Intention-to-treat analysis of Spetzler-Martin grades IV and V arteriovenous malformations: Natural history and treatment paradigm. J. Neurosurg. 2003, 98, 3–7.
22. Heros, R.C. Spetzler-Martin grades IV and V arteriovenous malformations. J. Neurosurg. 2003, 98, 1–2.
23. Chaloupka, J.C.; Vinuela, F.; Robert, J.; Duckwiler, G.R. An in vivo arteriovenous malformation model in swine: Preliminary feasibility and natural history study. Am. J. Neuroradiol. 1994, 15, 945–950.
24. Massoud, T.F.; Ji, C.; Vinuela, F.; Guglielmi, G.; Robert, J.; Duckwiler, G.R.; Gobin, Y.P. An experimental arteriovenous malformation model in swine: Anatomic basis and construction technique. Am. J. Neuroradiol. 1994, 15, 1537–1545.
25. Qian, Z.; Climent, S.; Maynar, M.; Uson-Garallo, J.; Lima-Rodrigues, M.A.; Calles, C.; Robertson, H.; Castaneda-Zuniga, W.R. A simplified arteriovenous malformation model in sheep: Feasibility study. Am. J. Neuroradiol. 1999, 20, 765–770.
26. Lv, M.M.; Fan, X.D.; Su, L.X. Is a swine model of arteriovenous malformation suitable for human extracranial arteriovenous malformation? A preliminary study. Cardiovasc. Intervent. Radiol. 2013, 36, 1364–1370.

27. Klisch, J.; Requejo, F.; Yin, L.; Eissner, B.; Schumacher, M. The two-in-one model: A new variation of the arteriovenous malformation model in swine. Neuroradiology 2001, 43, 393–397.
28. Lawton, M.T.; Jacobowitz, R.; Spetzler, R.F. Redefined role of angiogenesis in the pathogenesis of dural arteriovenous malformations. J. Neurosurg. 1997, 87, 267–274.
29. Herman, J.M.; Spetzler, R.F.; Bederson, J.B.; Kurbat, J.M.; Zabramski, J.M. Genesis of a dural arteriovenous malformation in a rat model. J. Neurosurg. 1995, 83, 539–545.
30. Yassari, R.; Sayama, T.; Jahromi, B.S.; Aihara, Y.; Stoodley, M.; Macdonald, R.L. Angiographic, hemodynamic and histological characterization of an arteriovenous fistula in rats. Acta Neurochir. 2004, 146, 495–504.
31. Sekhon, L.H.; Morgan, M.K.; Spence, I. Normal perfusion pressure breakthrough: The role of capillaries. J. Neurosurg.1997, 86, 519–524.
32. Morgan, M.K.; Johnston, I.; Besser, M.; Baines, D. Cerebral arteriovenous malformations, steal, and the hypertensive breakthrough threshold. J. Neurosurg. 1987, 66, 563–567.
33. Storer, K.; Tu, J.; Karunanayaka, A.; Smee, R.; Short, R.; Thorpe, P.; Stoodley, M. Coadministration of low-dose lipopolysaccharide and soluble tissue factor induces thrombosis after radiosurgery in an animal arteriovenous malformation model. Neurosurgery 2007, 61, 604–611.
34. Maruyama, K.; Kawahara, N.; Shin, M.; Tago, M.; Kishimoto, J.; Kurita, H.; Kawamoto, S.; Morita, A.; Kirino, T. The risk of hemorrhage after radiosurgery for cerebral arteriovenous malformations. N. Engl. J. Med. 2005, 352, 146–153.
35. Storer, K.P.; Tu, J.; Stoodley, M.A.; Smee, R.I. Expression of endothelial adhesion molecules after radiosurgery in an animal model of arteriovenous malformation. Neurosurgery 2010, 67, 976–983.
36. Morgan, W.R.; Majeski, J.A. Idiopathic arteriovenous renal vascular malformation treated by ex vivo repair. J. S. C. Med. Assoc. 1989, 85, 469–471.
37. Altschuler, E.; Lunsford, L.D.; Kondziolka, D.; Wu, A.; Maitz, A.H.; Sclabassi, R.; Martinez, A.J.; Flickinger, J.C. Radiobiologic models for radiosurgery. Neurosurg. Clin. N. Am. 1992, 3, 61–77.
38. Pietila, T.A.; Zabramski, J.M.; Thellier-Janko, A.; Duveneck, K.; Bichard, W.D.; Brock, M.; Spetzler, R.F. Animal model for cerebral arteriovenous malformation. Acta Neurochir. 2000, 142, 1231–1240.
39. Numazawa, S.; Sasaki, T.; Sato, S.; Watanabe, Y.; Watanabe, Z.; Kodama, N. Experimental model of intracranial arteriovenous shunting in the acute stage. Neurol. Med. Chir. 2005, 45, 288–293.
40. Pollock, B.E. Stereotactic radiosurgery for arteriovenous malformations. Neurosurg. Clin. N. Am. 1999, 10, 281–290.
41. Gobin, Y.P.; Laurent, A.; Merienne, L.; Schlienger, M.; Aymard, A.; Houdart, E.; Casasco, A.; Lefkopoulos, D.; George, B.; Merland, J.J. Treatment of brain arteriovenous malformations by embolization and radiosurgery. J. Neurosurg.1996, 85, 19–28.
42. Jahan, R.; Solberg, T.D.; Lee, D.; Medin, P.; Tateshima, S.; de Salles, A.; Sayre, J.; Vinters, H.V.; Vinuela, F. An arteriovenous malformation model for stereotactic radiosurgery research. Neurosurgery 2007, 61, 152–159.
43. Mut, M.; Oge, K.; Zorlu, F.; Undeger, U.; Erdem, S.; Ozcan, O.E. Effects of ionizing radiation on brain tissue surrounding arteriovenous malformations: An experimental study in a rat caroticojugular fistula model. Neurosurg. Rev. 2004, 27, 121–127.
44. Frankenberg-Schwager, M. Induction, repair and biological relevance of radiation-induced DNA lesions in eukaryotic cells. Radiat. Environ. Biophys. 1990, 29, 273–292.
45. Lunec, J. Free radicals: Their involvement in disease processes. Ann. Clin. Biochem. 1990, 27, 173–182.

46. De Salles, A.A.; Solberg, T.D.; Mischel, P.; Massoud, T.F.; Plasencia, A.; Goetsch, S.; de Souza, E.; Vinuela, F. Arteriovenous malformation animal model for radiosurgery: The rete mirabile. Am. J. Neuroradiol. 1996, 17, 1451–1458.
47. De Salles, A.A.; Manchola, I. CO2 reactivity in arteriovenous malformations of the brain: A transcranial Doppler ultrasound study. J. Neurosurg. 1994, 80, 624–630.
48. Spetzler, R.F.; Hargraves, R.W.; McCormick, P.W.; Zabramski, J.M.; Flom, R.A.; Zimmerman, R.S. Relationship of perfusion pressure and size to risk of hemorrhage from arteriovenous malformations. J. Neurosurg. 1992, 76, 918–923.
49. Arakawa, H.; Murayama, Y.; Davis, C.R.; Howard, D.L.; Baumgardner, W.L.; Marks, M.P.; Do, H.M. Endovascular embolization of the swine rete mirabile with Eudragit-E 100 polymer. Am. J. Neuroradiol. 2007, 28, 1191–1196.
50. Jahan, R.; Solberg, T.D.; Lee, D.; Medin, P.; Tateshima, S.; Sayre, J.; de Salles, A.; Vinters, H.V.; Vinuela, F. Stereotactic radiosurgery of the rete mirabile in swine: A longitudinal study of histopathological changes. Neurosurgery 2006, 58, 551–558.
51. Massoud, T.F.; Hademenos, G.J. Transvenous retrograde nidus sclerotherapy under controlled hypotension (TRENSH): A newly proposed treatment for brain arteriovenous malformations—Concepts and rationale. Neurosurgery 1999, 45, 351–365.
52. Kilic, K.; Konya, D.; Kurtkaya, O.; Sav, A.; Pamir, M.N.; Kilic, T. Inhibition of angiogenesis induced by cerebral arteriovenous malformations using gamma knife irradiation. J. Neurosurg. 2007, 106, 463–469.
53. Tu, J.; Stoodley, M.A.; Morgan, M.K.; Storer, K.P. Responses of arteriovenous malformations to radiosurgery: Ultrastructural changes. Neurosurgery 2006, 58, 749–758.
54. Reddy, R.; Duong, T.T.; Fairhall, J.M.; Smee, R.I.; Stoodley, M.A. Durable thrombosis in a rat model of arteriovenous malformation treated with radiosurgery and vascular targeting. J. Neurosurg. 2014, 120, 113–119.
55. Lawton, M.T.; Hamilton, M.G.; Spetzler, R.F. Multimodality treatment of deep arteriovenous malformations: Thalamus, basal ganglia, and brain stem. Neurosurgery 1995, 37, 29–36.
56. Hamilton, M.G.; Spetzler, R.F. The prospective application of a grading system for arteriovenous malformations. Neurosurgery 1994, 34, 2–7.
57. Heros, R.C.; Korosue, K.; Diebold, P.M. Surgical excision of cerebral arteriovenous malformations: Late results. Neurosurgery 1990, 26, 570–578.
58. Lawton, M.T.; Stewart, C.L.; Wulfstat, A.A.; Derugin, N.; Hashimoto, T.; Young, W.L. The transgenic arteriovenous fistula in the rat: An experimental model of gene therapy for brain arteriovenous malformations. Neurosurgery 2004, 54, 1463–1471.
59. Hashimoto, T.; Lam, T.; Boudreau, N.J.; Bollen, A.W.; Lawton, M.T.; Young, W.L. Abnormal balance in the angiopoietin-tie2 system in human brain arteriovenous malformations. Circ. Res. 2001, 89, 111–113.
60. Letteboer, T.G.; Mager, J.J.; Snijder, R.J.; Koeleman, B.P.; Lindhout, D.; Ploos van Amstel, J.K.; Westermann, C.J. Genotype-phenotype relationship in hereditary haemorrhagic telangiectasia. J. Med. Genet. 2006, 43, 371–377.
61. Matsubara, S.; Mandzia, J.L.; ter Brugge, K.; Willinsky, R.A.; Faughnan, M.E. Angiographic and clinical characteristics of patients with cerebral arteriovenous malformations associated with hereditary hemorrhagic telangiectasia. Am. J. Neuroradiol. 2000, 21, 1016–1020.
62. Willinsky, R.A.; Lasjaunias, P.; Terbrugge, K.; Burrows, P. Multiple cerebral arteriovenous malformations (AVMs): Review of our experience from 203 patients with cerebral vascular lesions. Neuroradiology 1990, 32, 207–210.
63. Govani, F.S.; Shovlin, C.L. Hereditary haemorrhagic telangiectasia: A clinical and scientific review. Eur. J. Hum. Genet.2009, 17, 860–871.
64. Du, R.; Hashimoto, T.; Tihan, T.; Young, W.L.; Perry, V.; Lawton, M.T. Growth and regression of arteriovenous malformations in a patient with hereditary hemorrhagic telangiectasia: Case report. J. Neurosurg. 2007, 106, 470–477.

65. Tual-Chalot, S.; Oh, S.P.; Arthur, H.M. Mouse models of hereditary hemorrhagic telangiectasia: Recent advances and future challenges. Front. Genet. 2015.
66. Young, W.L.; Yang, G.Y. Are there genetic influences on sporadic brain arteriovenous malformations? Stroke 2004, 35, 2740–2745.
67. Arthur, H.M.; Ure, J.; Smith, A.J.; Renforth, G.; Wilson, D.I.; Torsney, E.; Charlton, R.; Parums, D.V.; Jowett, T.; Marchuk, D.A.; et al. Endoglin, an ancillary TGFbeta receptor, is required for extraembryonic angiogenesis and plays a key role in heart development. Dev. Biol. 2000, 217, 42–53.
68. Marchuk, D.A.; Srinivasan, S.; Squire, T.L.; Zawistowski, J.S. Vascular morphogenesis: Tales of two syndromes. Hum. Mol. Genet. 2003, 12, R97–R112.
69. Bourdeau, A.; Dumont, D.J.; Letarte, M. A murine model of hereditary hemorrhagic telangiectasia. J. Clin. Investig.1999, 104, 1343–1351.
70. Satomi, J.; Mount, R.J.; Toporsian, M.; Paterson, A.D.; Wallace, M.C.; Harrison, R.V.; Letarte, M. Cerebral vascular abnormalities in a murine model of hereditary hemorrhagic telangiectasia. Stroke 2003, 34, 783–789.
71. Xu, B.; Wu, Y.Q.; Huey, M.; Arthur, H.M.; Marchuk, D.A.; Hashimoto, T.; Young, W.L.; Yang, G.Y. Vascular endothelial growth factor induces abnormal microvasculature in the endoglin heterozygous mouse brain. J. Cereb. Blood Flow Metab. 2004, 24, 237–244.
72. Torsney, E.; Charlton, R.; Diamond, A.G.; Burn, J.; Soames, J.V.; Arthur, H.M. Mouse model for hereditary hemorrhagic telangiectasia has a generalized vascular abnormality. Circulation 2003, 107, 1653–1657.
73. Srinivasan, S.; Hanes, M.A.; Dickens, T.; Porteous, M.E.; Oh, S.P.; Hale, L.P.; Marchuk, D.A. A mouse model for hereditary hemorrhagic telangiectasia (HHT) type 2. Hum. Mol. Genet. 2003, 12, 473–482.
74. Choi, E.J.; Chen, W.; Jun, K.; Arthur, H.M.; Young, W.L.; Su, H. Novel brain arteriovenous malformation mouse models for type 1 hereditary hemorrhagic telangiectasia. PLoS ONE 2014, 9, e88511.
75. Park, S.O.; Wankhede, M.; Lee, Y.J.; Choi, E.J.; Fliess, N.; Choe, S.W.; Oh, S.H.; Walter, G.; Raizada, M.K.; Sorg, B.S.; et al. Real-time imaging of de novo arteriovenous malformation in a mouse model of hereditary hemorrhagic telangiectasia. J. Clin. Investig. 2009, 119, 3487–3496.
76. Milton, I.; Ouyang, D.; Allen, C.J.; Yanasak, N.E.; Gossage, J.R.; Alleyne, C.H., Jr.; Seki, T. Age-dependent lethality in novel transgenic mouse models ofnervous system arteriovenous malformations. Stroke 2012, 43, 1432–1435.

CHAPTER 12

Adhesive Capsulitis in Eight Dogs: Diagnosis and Management

Brittany J. Carr, Sherman O. Canapp*, Debra A. Canapp, Lauri-Jo Gamble and David L. Dycus

Veterinary Orthopedic and Sports Medicine, Annapolis Junction, MD, USA

ABSTRACT

Objective: To describe clinical and diagnostic findings as well as management of adhesive capsulitis in dogs.

Background: Adhesive capsulitis, also known as frozen shoulder, is a syndrome defined by loss of range of motion of the shoulder and may be the end-stage manifestation of several primary conditions.

Evidentiary Value: This is a case series report of eight dogs with chronic forelimb lameness diagnosed with adhesive capsulitis.

Methods: Medical records (June 1, 2010–September 1, 2015) including, physical examination findings, radiographic findings, magnetic resonance imaging (MRI) findings, arthroscopy findings, and treatment plans were reviewed.

Results: All dogs presented with a chronic, grade III–VI/VI forelimb lameness. On orthopedic examination, all dogs had moderate to significant discomfort on shoulder extension and flexion and severe restriction of range of motion. Six of the eight dogs had evidence of bone remodeling and sclerosis in the affected shoulder on radiographs. Six of the dogs had an initial diagnostic ultrasound performed, which revealed evidence of fibrous scar tissue. Five dogs had MRI performed that revealed moderate shoulder effusion and enhancement of the synovial lining of the shoulder. Arthroscopy was performed in five of the eight patients. Three were noted to have significant contracture, adhesions, and fibrous scar tissue of the joint capsule. Severe inflammation was noted throughout the synovium of two patients. All eight patients tried conservative management consisting of oral medications and rehabilitation therapy. Five of the eight patients received extracorporeal shockwave therapy. Three patients received regenerative medicine treatment in the affected supraspinatus and shoulder. Regardless of the treatment elected, none of the dogs were reported to have significant improvement.

Conclusion: Adhesive capsulitis is an uncommon cause of chronic forelimb lameness. Further investigation is needed to describe the etiology and pathogenesis of adhesive capsulitis in dogs to evaluate the effectiveness of both non-surgical and surgical treatment modalities, establish treatment protocols, and evaluate short- and long-term clinical outcome of patients.

Application: Adhesive capsulitis should be considered in patients with chronic forelimb lameness and moderate to significant discomfort and restriction on shoulder range of motion.

Keywords: frozen shoulder, adhesive capsulitis, forelimb lameness, range of motion, fibrous scar tissue, synovitis

INTRODUCTION

Adhesive capsulitis, also known as frozen shoulder, is a syndrome appreciated in human sports medicine and defined by loss of range of motion of the shoulder, impairing the patient's ability to sleep, work, perform daily living activities, or desired recreational activities (1–5). Adhesive capsulitis may also be the end-stage manifestation of several primary conditions, including trauma, previous surgery, prolonged immobilization, endocrine disorders, or idiopathic causes (2, 6–12). Of patients affected by adhesive capsulitis, 70% are female and about 20–30% will develop the condition in the opposite shoulder (2). Only 20–30% of patients will report a history of minor trauma to the shoulder (6, 7), and it has been found to be more common in those with sedentary lifestyles (8). Regardless of the biological cause, adhesive capsulitis is defined as thickening and contracture of the joint capsule, which results in decreased intra-articular volume and capsular compliance so that glenohumeral motion is limited in all planes (2–4).

In humans, there are four stages associated with adhesive capsulitis, each based on the correlation of physical examination and arthroscopic examination of affected joint(s) and lasting about 6–9 months (1, 5). Stage 1 is characterized by pain with active and passive range of motion and a progressive loss of motion. Upon examination either under anesthesia or following intra-articular injection of local anesthetic, there is a significant improvement in range of motion to normal or minimal loss (13). Arthroscopic examination reveals a diffuse hypervascular glenohumeral synovitis, often most pronounced in the anterosuperior capsule (5). Histopathology reveals rare inflammatory cell infiltrates and findings consistent with a hypertrophic, hypervascular synovitis and a normal underlying capsule (14). Stage 2, also known as the "freezing stage," is characterized by chronic pain and progressive loss of range of motion. Examination after intra-articular injection of local anesthetic or scalene block reveals relief of pain with only partial improvement in range of motion (13). Arthroscopic examination reveals a diffuse, pedunculated synovitis and a tight capsule with a dense feel on the insertion of the arthroscope (5). Histopathology reveals hypertrophic, hypervascular synovitis with perivascular and subsynovial scar formation and capsular fibroplasia (14). Stage 3, also known as the "frozen stage," is characterized by minimal pain and significant shoulder stiffness. An intra-articular injection of local anesthetic or scalene block does not improvement in range of motion (13). Arthroscopic examination reveals a profound loss of capsular volume and remnants of fibrotic synovium that is not hypervascular (5). Histopathology reveals a dense, hypercellular collagenous tissue and a thin synovial layer without significant hypertrophy or hypervascularity (14). Stage 4, also known as the "thawing stage," is characterized by minimal pain and progressive improvement in range of motion, resulting in capsular remodeling. Since these patients rarely undergo surgery, there is little arthroscopic or histological data available for patients with stage 4 adhesive capsulitis (13).

To the authors' knowledge, there is only one paper reporting shoulder immobilization for 16 weeks in eight beagle dogs caused a loss of shoulder motion and focal capsular adhesion that resembled adhesive capsulitis seen in humans (15). Currently there is no other documentation of adhesive capsulitis in dogs in the literature. The purpose of this case series is to report clinical and diagnostic findings and management of adhesive capsulitis in dogs.

MATERIALS AND METHODS

Medical records (June 1, 2010–September 1, 2015) of eight dogs diagnosed with frozen shoulder were reviewed. Physical examination, diagnostic findings, arthroscopy findings, and treatment plans were reviewed. Inclusion criteria for patients were a diagnosis of frozen shoulder based on the presence of the following: history of unilateral thoracic limb lameness, severe restriction of shoulder extension (<110°) and flexion (>90°), and pain on manipulation of the shoulder. Patients were excluded from this study due to incomplete medical records or loss of follow-up.

Diagnostic Imaging

Routine lateral and caudocranial radiographs of the shoulders were performed in all eight dogs and evaluated by a board certified surgeon. Magnetic resonance imaging (MRI) was offered to all patients and elected in five of the eight patients. An MRI of the affected shoulder was performed using a 1.5-T MRI unit. Musculoskeletal Diagnostic Ultrasound of the Shoulder was also offered to all patients and elected in six of the eight patients. If elected, patients were placed in lateral recumbency and the shoulder was clipped and prepared routinely. A 15.4 MHz linear probe was used to acquire longitudinal and cross-sectional images of the supraspinatus, biceps, infraspinatus, and teres minor tendons. The echogenicity, fiber pattern, and size of each tendon was recorded and compared to the contralateral, unaffected shoulder. The shoulder joint capsule was also imaged in the same cross-sectional image to assess the supraspinatus and biceps tendons. The echogenicity, integrity, and thickness of the joint capsule was recorded and compared to the contralateral, unaffected shoulder. The caudal humeral head was also assessed on ultrasound to confirm the presence of osteochondrosis or osteochondritis dissecans by placing the transducer in a longitudinal fashion parallel to the caudal humeral head.

Shoulder Arthroscopy

A standard lateral portal was utilized for arthroscopic evaluation of the affected shoulder. Following aseptic preparation, a 22-gage needle was placed into the shoulder. Synovial fluid was aspirated using a 5-ml syringe to ensure the needle was in the joint. Approximately 5–10 ml of saline was then injected into the shoulder to distend the joint. A #15 blade was used to create a stab incision through the skin to allow for insertion of the cannula and trocar into the shoulder. The 1.9-mm arthroscope was then inserted into the cannula. Due to severe soft tissue contracture and joint space collapse, entry into the joint required fluoroscopic assistance in one patient. Another small 1-cm incision was made 3 cm caudal to the scope portal for insertion of instruments. Following arthroscopy, the incisions were closed routinely.

Regenerative Medicine Therapy

Regenerative medicine therapy consisting of adipose-derived progenitor cell (ADPC) and autologous conditioned serum (ACS) in the affected shoulder and an ultrasound-guided ADPC and platelet-rich plasma (PRP) injection in any concurrently affected tendons was offered to three patients.[1]

Adipose tissue was collected from the falciform ligament within the abdomen during a routine ventral midline celiotomy. Approximately 20 g of falciform adipose tissue was collected and placed into a sterile container with Delbecco's modified Eagle's medium (DMEM) supplemented with fetal bovine serum (FBS), penicillin, and streptomycin. Approximately 30 ml of blood was collected from the jugular vein using an 18-gage butterfly needle into a syringe with 5 ml of CPDA anti-coagulant. Both the adipose and the blood samples were then prepared for shipment. Adipose and blood samples were shipped in validated containers at 4°C to Virginia Tech's Regenerative Medicine Research Laboratory for processing. Adipose was processed using established techniques (33). The tissue was mechanically and enzymatically separated to release ADPCs. The cells were then washed in phosphate-buffered saline (PBS) and nucleated cells were counted and cultured in stem cell media at 37°C in a 5% CO_2 incubator with 95% humidity. Adherent cells were monitored daily for adherence, growth, and phenotype. Once 70% confluent, cells were detached from the flasks, washed in PBS, and suspended in autologous PRP. PRP was prepared from anti-coagulated blood via centrifugation to obtain a fourfold increase in platelets and 80% reduction in white blood cells over whole blood. Platelet and white blood cell counts were performed for each sample to confirm appropriate concentrations prior to use. The samples were then shipped back to VOSM for injection at 4°C using validated shipping containers. Ultrasound guidance was used to administer ADPC–PRP therapy with a fenestration technique (15, 17).

RESULTS

Population Sample

All patients included in this series were companion dogs. As described in Table 1, there were three mixed breed dogs, three Welsh Corgis, one German Shepherd Dog, and one Beagle. There were five male castrated dogs and three female spayed dogs. The mean age initial presentation was 8 years old (range 3–12 years old). All patients presented with a history of a chronic, unilateral forelimb lameness. One patient had a previous medical history of left medial patellar luxation, a left cranial cruciate ligament rupture and extracapsular suture repair, mitral valve insufficiency, and elevated liver enzymes. One patient had a previous medical history of hypothyroidism, which was controlled with medical management. The other six patients had no history of previous medical disease. Three patients had a history of previous trauma to the affected shoulder 12–18 months prior to presentation.

Table 1. Sample population, history, and physical examination.

Patient	Age (years)	Sex	Breed	Trauma	Lameness	Decreased ROM	Muscle atrophy	Other
1	3	MN	Welsh corgi	No	Chronic 4–5/6	Yes	Moderate	Increased abduction; pain left elbow
2	5	FS	Collie mix	No	Chronic 5–6/6	Yes	Severe	Swelling
3	10.5	FS	Maltipoo	Yes	Chronic 5–6/6	Yes	Mild	
4	11	FS	Beagle	No	Chronic 5–6/6	Yes	Severe	Mass cranial and medial to the shoulder
5	6	MN	German Shepherd	No	Chronic 3/6	Yes	Moderate	Increased abduction; "Loss of end feel" during the biceps stretch
6	7	MN	Mixed	No	Chronic 5/6	Yes	Severe	"Loss of end feel" during the biceps stretch
7	12	MN	Welsh corgi	Yes	Chronic 6/6	Yes	Severe	
8	8.5	MN	Welsh corgi	No	Chronic 5–6/6	Yes	Moderate	"Loss of end feel" during the biceps stretch

Initial Orthopedic Examination

All eight patients had an initial orthopedic examination performed by a board certified surgeon (see Table 1). All patients initially presented with a chronic, grade III–VI/VI unilateral forelimb lameness (5 left forelimb, 3 right forelimb) (Table 2). On initial orthopedic examination, all dogs had moderate to significant discomfort on shoulder extension and flexion and restricted range of motion in the affected shoulder. In all patients, shoulder extension was <110° and shoulder flexion was >90°. Four patients were too painful to allow for range of motion or abduction of the affected shoulder. Pain and spasm on direct palpation of the supraspinatus and biceps tendon of the affected limb was noted in two dogs. Pain in the medial compartment of the elbow of the affected limb was appreciated in two dogs. There was moderate to severe muscle atrophy of the shoulder musculature of the affected limb in six dogs and mild muscle atrophy noted in two dogs. Soft tissue swelling and effusion of the affected shoulder was appreciated in one patient. In one patient, the biceps muscle felt contracted and shortened and significantly limited elbow extension was noted. In three patients, there was pain, spasm, and empty end feel during biceps stretch (pain with no mechanical or anatomic resistance in shoulder flexion with concurrent elbow extension). The remainder of orthopedic examination was within normal limits in all patients. At the time of initial consult, all patients were categorized to be in stage 2 of the 4 stages associated with adhesive capsulitis.

Table 2. Lameness Numeric Rating Scale.

Grade of lameness	Description
1	Lameness is not perceptible at a walk but perceptible at a trot
2	Lameness is perceptible at a walk and apparent at the trot
3	Lameness is apparent at both a walk and trot
4	Lameness is apparent at a walk and severe to non-weight bearing at a trot
5	Non-weight bearing lameness at a walk and trot
6	Unable to rise and walk

Different diagnostics were performed on each dog, including radiographs of the affected shoulder in all eight patients, diagnostic ultrasound in six of them, and arthroscopy in five of the eight patients. Five dogs had MRI performed of the cervical spine and affected shoulder. Arthrocentesis of the affected shoulder was performed in four of the eight dogs and submitted for cytology, whereas histopathology of the shoulder synovium was performed in only one patient. Results are described in Table 3.

Radiographs

Routine lateral and caudocranial radiographs of the shoulders were performed in all eight dogs.[2]Radiographs were unremarkable in two of the eight dogs. Radiographs of the affected shoulder revealed severe soft tissue swelling and periosteal reaction of the distal scapula and humerus in one dog. Radiographs of the affected shoulder revealed sclerosis within the insertion of the capsule, teres, infraspinatus, and supraspinatus in four patients. Two patients were noted to have remodeling and sclerosis within the bicipital groove. Radiographs of the affected shoulder revealed severe remodeling, sclerosis, and collapse in the joint space in two patients (Figure 1). There was a large caudal glenoid fragment noted in one patient. Three patients were noted to have osteophytosis of the caudal humeral head with flattening of the glenoid cavity. Radiographs of the unaffected shoulder and both elbows were performed in all patients and did not have any significant findings.

Diagnostic Ultrasound

Six of the eight dogs had an initial diagnostic ultrasound of the shoulder that revealed evidence of fibrous scar tissue in all six dogs (Figures 2–4).[3] A diagnostic ultrasound of the affected shoulder in one patient revealed significant disruption of the supraspinatus fiber pattern, with shortened, irregular fibers at the insertion. In three of the six patients, diagnostic

Table 3. Diagnostics performed.

Patient	Radiographs	Ultrasound	MRI	Arthroscopy	Cytology and histopathology
1	Normal	Supraspinatus disruption; biceps tendon sheath effusion	Marked joint effusion and synovitis; damage to the supraspinatus tendon and medial glenohumeral ligament	Biceps tendinopathy; impingement of the biceps tendon; fraying and disruption of the subscapularis and of the MGL; disruption of joint capsule; grade III/IV cartilage erosion along the caudal humeral head and glenoid; RF treatment was performed	Mild mononuclear inflammation (76% large mononuclear cells and 24% small mononuclear cells) Negative culture
2	Severe soft tissue swelling; periosteal reaction of distal scapula and proximal	N/A	N/A	NA	Mononuclear Inflammation (large mononuclear cells predominate) Negative culture Lymphoplasmacytic synovitis with fibrosis
3	Sclerosis within the insertion of the teres, infraspinatus, and supraspinatus; large caudal glenoid fragment; bone spur off caudal humeral head; flattening of glenoid cavity	Fibrous scar tissue/adhesions	N/A	Collapsed joint capsule; severe capsular adhesion; grade IV/IV cartilage erosion of humeral head. MUA performed; abrasion arthroplasty performed, caudal glenoid fragment removed	N/A
4	Normal	Significant fibrous/periosteal reaction around the whole shoulder joint, including the biceps groove and infraspinatus insertion area	Marked shoulder effusion, severe inflammation of the shoulder joint capsule	N/A	Mononuclear inflammation (large mononuclear cells predominate) Negative culture
5	Remodeling and sclerosis within the bicipital groove; osteophyte off the caudal humeral head and glenoid cavity	Bilateral supraspinatus tendinopathy	N/A	Significant adhesions and fibrosis surrounding biceps tendon (debrided); severe supraspinatus bulge; severe synovitis; grade IV/IV cartilage lesions on humeral head. Abrasion arthroplasty performed. RF performed to all areas of inflammation and disruption	N/A
6	Sclerotic rim at the region of insertion of the capsule and tendons at the proximal humerus/ humeral head	Severe disruption and insertionopathy of the right biceps muscle/tendon unit; right supraspinatus tendinopathy; moderate disruption and insertionopathy of the left biceps tendon	Right shoulder muscular atrophy; severe biceps tendinopathy, mild shoulder effusion	Significant inflammation of the glenohumeral ligament, synovium, subscapularis tendon and joint capsule; mild bulge of the supraspinatus creating impingement of the biceps	Chronic sterile histiocytic inflammation, low-to-moderate grade chronic hemorrhage
7	Severe remodeling, sclerosis, and collapse in joint space; periosteal reaction along the proximal humerus and distal scapula	N/A	Severe joint inflammation and marked enhancement of the joint capsule	Shoulder joint collapsed; unable to identify normal anatomy; grade IV/IV cartilage erosion caudal humeral head. Abrasion arthroplasty performed and MUA performed	N/A
8	Severe osteoarthritis; irregular new bone proliferation in the region of the bicipital groove, less severe changes in the contralateral shoulder	Chronic biceps tendinopathy and bursitis., bilateral teres minor chronic tendinopathy	Right shoulder dysplasia with severe osteoarthritis, synovitis and joint capsule thickening, biceps tenosynovitis; mild supraspinatus, infraspinatus, and teres minor insertionopathy. Mild left shoulder dysplasia	N/A	N/A

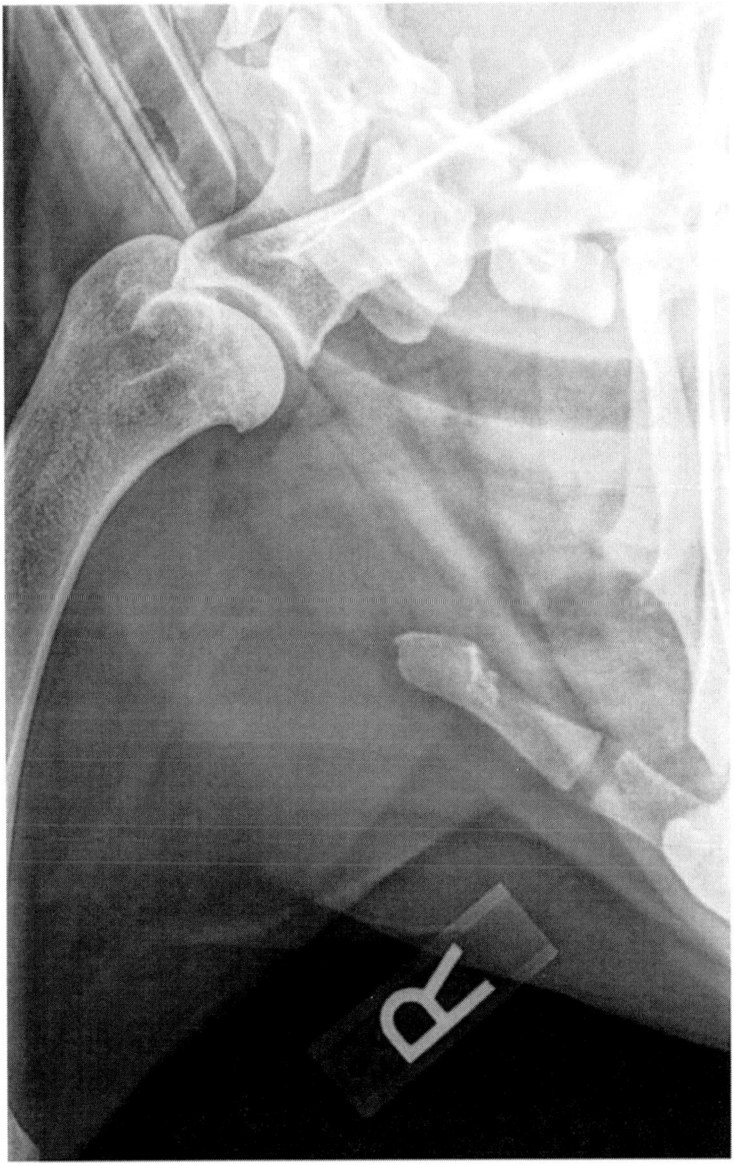

Figure 1. A right lateral shoulder radiograph of a patient diagnosed with adhesive capsulitis. There is a sclerotic rim at the region of insertion of the capsule and tendons at the proximal humerus and humeral head.

ultrasound of the affected shoulder showed an intact and overall normal supraspinatus insertion/origin. In the other three patients the supraspinatus was noted to have areas of hypoechoic and hyperechoic foci with a non-homogenous fiber pattern. In one patient, the supraspinatus appeared to impinge upon the biceps tendon. In one patient, there was a significant hyperechoic calcified foci impinging upon the biceps groove in addition to a hyperechoic periosteal reaction noted at the supraglenoid tubercle at the biceps origin. In three patients, significant fiber disruption was noted at the point of origin of the biceps, as well as large areas of hypoechogencity and periosteal reaction at the supraglenoid tubercle. There were changes in the surrounding connective tissue and infraspinatus and teres tendons, indicating fibrous scar tissue/adhesions in three patients. The joint capsule appeared hyperechoic, disrupted, and thickened in two patients. There was a moderate amount of shoulder joint effusion present in addition to free floating fibrous tissue noted in one patient.

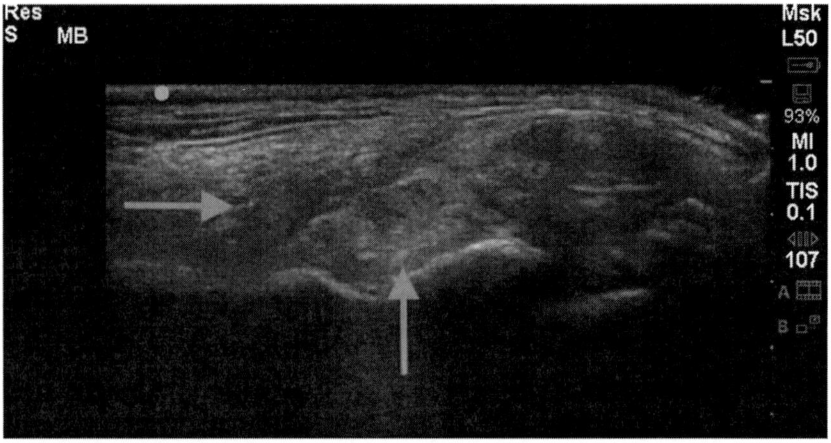

Figure 2. A longitudinal diagnostic ultrasound of the shoulder revealed evidence of significant fibrous scar tissue and disruption of the supraspinatus fiber pattern (red arrows).

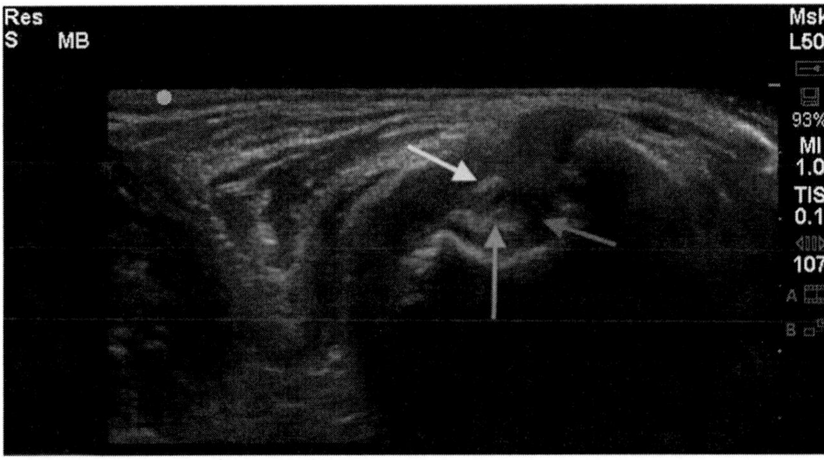

Figure 3. A cross-sectional diagnostic ultrasound of the shoulder revealed evidence of significant fibrous scar tissue (green arrow), shoulder effusion (blue arrow), and hyperechoic changes of the biceps (red arrow).

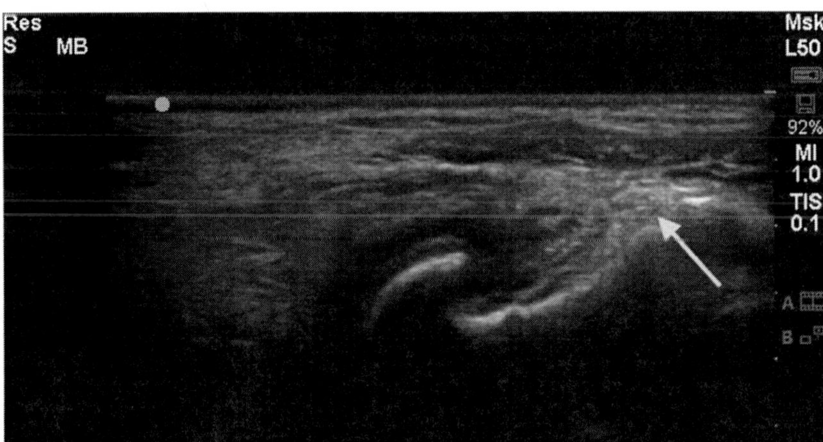

Figure 4. A longitudinal diagnostic ultrasound image of the shoulder revealed evidence of significant fibrous/periosteal reaction around the whole shoulder joint, including the infraspinatus insertion area (green arrow).

Magnetic Resonance Imaging

Magnetic resonance imaging was recommended in all patients to evaluate the soft tissue structures of the shoulder, particularly the joint capsule and supraspinatus tendon, which are commonly affected in human patients with adhesive capsulitis. Five of the eight patients elected to pursue MRI. Marked muscle atrophy of the shoulder musculature was noted in all five patients that had MRI of their cervical spine and affected shoulder. Moderate shoulder effusion and enhancement of the synovial lining of the shoulder, including the portion of the joint capsule that envelops the biceps tendon was appreciated in four of the five patients (Figure 5). Findings revealed a severe biceps tendinopathy with mild shoulder effusion in the last patient. One patient was noted to have damage to the supraspinatus tendon and medial glenohumeral ligament and another patient had signs of severe shoulder dysplasia with severe osteoarthritis, synovitis, and joint capsule thickening in addition to mild supraspinatus, infraspinatus, and teres minor insertionopathy. Mild enlargement of the axillary lymph node was noted in two of the patients.

Arthroscopy

Shoulder arthroscopy was performed in five of the eight patients. A standard lateral portal was utilized for arthroscopic evaluation of the affected shoulder. Following aseptic preparation, a 22-gage needle was placed into the shoulder. Synovial fluid was aspirated using a 5 ml syringe to ensure the needle was in the joint. Approximately 5–10 ml of saline was then injected into the shoulder to distend the joint. A #15 blade was used to create a stab incision through the skin to allow for insertion of the cannula and trochar into the shoulder. The 1.9 mm arthroscope was then inserted into the cannula. The shoulder was then assessed. A significant amount of scar tissue and adhesions were diffusely noted in three patients. The synovium was inflamed and disrupted in all patients, but severe inflammation was noted throughout the synovium of two patients (Figure 6). Moderate-to-severe inflammation of the biceps tendon and a moderate bulge of the supraspinatus, creating significant impingement of the biceps tendon was noted in one patient. Mild-to-moderate fraying and disruption of the subscapularis, medial glenohumeral ligament, and medial compartment/joint capsule was noted. Areas of grade III–IV/IV cartilage erosion on Modified Outerbridge Cartilage Scoring were noted along the caudal humeral head and glenoid in three patients (Figure 7). In these patients, abrasion arthroplasty was then performed. A large caudal glenoid slab fracture was noted and subsequently removed in one patient. In one patient, the entire joint space had collapsed, and the scapular-humeral joint appeared to have ankylosed through soft tissue contracture. In this patient, no normal anatomy could be identified at the time of arthroscopy. All incisions were routinely closed.

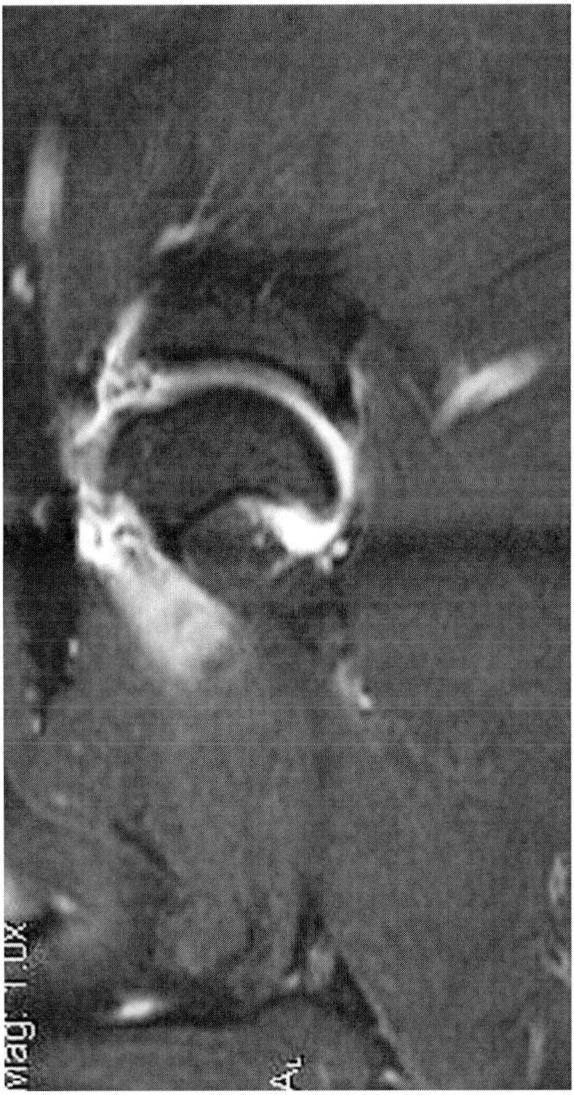

Figure 5. This is an MRI T2 sagittal view of a patient diagnosed with adhesive capsulitis. There is moderate shoulder effusion and enhancement of the synovial lining of the shoulder, including the portion of the joint capsule that envelops the biceps tendon was present.

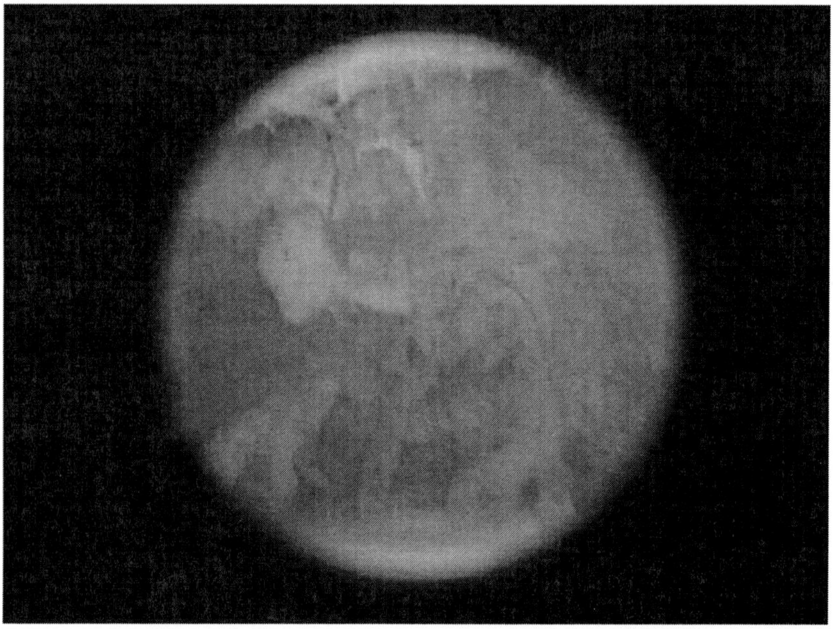

Figure 6. In all patients, arthroscopy of the affected shoulder revealed the synovium was inflamed and disrupted.

One patient had a second-look arthroscopic exploration of the affected shoulder 8 weeks following ADPC–PRP therapy, which revealed severe inflammation and tenosynovitis of the biceps tendon, with inflammation evident within the bicipital groove. The tendon had an "hour glass" appearance, and when probed, an area of irregularity and weakness in the fiber pattern was appreciated. Arthroscopic release of the tendon was performed. The previously noted bulge of the supraspinatus was significantly improved. The medial glenohumeral ligament and subscapularis were significantly improved, and areas of neovascularization /new blood supply were noted. Fibrocartilaganous growth was noted on the caudal humeral head and glenoid cavity. Significant synovitis was still noted.

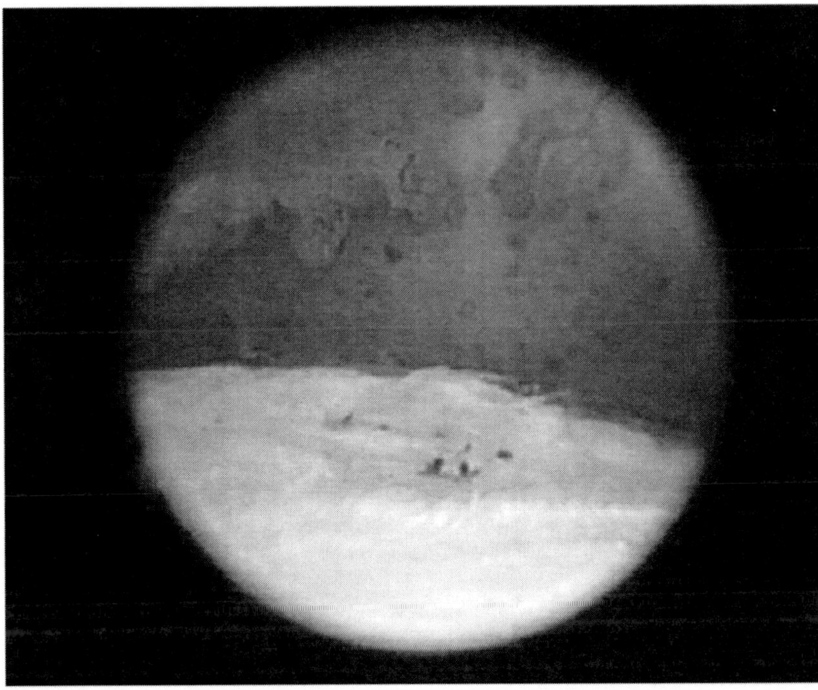

Figure 7. Areas of grade III/IV cartilage erosion on Modified Outerbridge Cartilage Scoring System were noted along the caudal humeral head and glenoid during arthroscopy in this patient diagnosed with adhesive capsulitis.

Cytology and Histopathology

Arthrocentesis of the affected shoulder was performed in four of the eight dogs and submitted for cytology. Cytology results were consistent with mononuclear inflammation (large mononuclear cells predominated) in three dogs and chronic sterile histiocytic inflammation with low grade to moderate and low-grade chronic hemorrhage in one dog. Histopathology of the shoulder synovium was performed in one patient and consistent with marked lymphoplasmacytic synovitis with fibrosis.

Conservative Management

All eight patients tried conservative management (Table 4). Seven patients were prescribed a non-steroidal anti-inflammatory medication (carprofen 2 mg/kg by mouth every 12 h, deracoxib 1–2 mg/kg by mouth every 24 h, or meloxicam 0.1 mg/kg by mouth every 24 h) and tramadol (2–4 mg/kg by mouth every 8 h). The other patient was prescribed codeine after a course of tramadol. Five dogs were prescribed gabapentin (5 mg/kg by mouth every 12 h). Four dogs were prescribed methocarbamol (15 mg/kg by mouth every 12 h). Three dogs were prescribed traumeel, a homeopathic, anti-inflammatory agent used to treat inflammation and pain (2 tablets by mouth every 8 h). All patients were directed to take a glucosamine/chondroitin sulfate joint supplement and omega-3 fatty acid supplement. Two patients were prescribed a course of prednisone (1 mg/kg by mouth every 24 h). Three patients received an intra-articular corticosteroid injection (methylprednisolone acetate, 20 mg). Seven patients participated in weekly rehabilitation, consisting of manual therapy (massage, stretching, passive range of motion, joint manipulation), therapeutic ultrasound/transcutaneous electrical nerve stimulation (TENS) combination therapy (frequency 2 MHz, intensity 0.5 pps, 5 min, duty 50%), laser/LED therapy to the carpi, elbows, supraspinatus, shoulders, scapula, and paraspinal musculature (dose: 5 J/cm^2), and magnetic field therapy.[4] Five of the eight dogs elected to try underwater treadmill hydrotherapy, but discontinued after two sessions due to worsening of lameness. Owners elected amputation in one patient after the patient did not respond to conservative management with medical therapy and rehabilitation therapy. Five of the eight dogs received three sessions of extracorporeal shockwave therapy (ESWT) at 2-week intervals (3.54 mJ applied at 0.14 mJ/mm^2, 240 shocks/min, 1000 shocks).[5] Slight improvement in lameness was noted in two of the five patients that received ESTW; however, the other three did not show significant improvement. Both patients that showed improvement originally presented with a grade V–VI/VI unilateral forelimb lameness, and after three sessions of ESWT at 2-week intervals, the patients had improved to a grade IV/VI unilateral forelimb lameness at a recheck examination 2 weeks after finishing ESWT.

Table 4. Treatments performed.

Patient	LLLT	Therapeutic ultrasound	ESWT	UWTM	IA steroids	Regenerative medicine	Amputation
1	Yes	Yes	Yes	Yes	Yes	IA ACS/ADSC IT PRP/ADSC	No
2	Yes	Yes	Yes	No	Yes	No	Yes
3	Yes	Yes	Yes	Yes	No	No	No
4	Yes	Yes	Yes	No	No	No	No
5	Yes	Yes		Yes		IA ACS/ADSC IT PRP/ADSC	
6	Yes	Yes	No	Yes	No	No	No
7	Yes	Yes		Yes	Yes	IA ACS/ADSC	
8	No	No	Yes	No	No	No	No

Stem Cell Therapy

Three patients received an intra-articular injection of ADPC and ACS in the affected shoulder and two of them received an ultrasound-guided ADPC and leukocyte-poor PRP injection in the affected supraspinatus (see text footnote 1) (15–17). A diagnostic ultrasound re-evaluation was performed at 4 and 8 weeks following regenerative medicine therapy. At 4 weeks, the supraspinatus tendon appeared to have a more organized fiber pattern and no impinge upon the biceps tendon as previously noted in two patients. However, both patients still had moderate shoulder joint and bicipital sheath effusion. At 8 weeks following regenerative medicine therapy, neither patient's lameness nor diagnostic ultrasound findings had improved. One patient elected to pursue a second ultrasound-guided ADPC–PRP injection in the affected supraspinatus and ADPC–ACS in the affected shoulder was performed. A recheck diagnostic ultrasound of the affected shoulder was performed 4 weeks later in this patient. Findings in the affected shoulder revealed chronic fibrous changes were still present. There also appeared to be an area of fibrous contracture developing within the infraspinatus and teres tendons, which were previously noted as normal. Eight weeks later, no improvement in lameness or diagnostic ultrasound findings was noted. A third ultrasound-guided ADPC–PRP injection was injected into the supraspinatus tendon bilaterally and ADPC–ACS were injected into the affected shoulder. A recheck diagnostic ultrasound of the shoulder was performed 8 weeks later, which revealed the supraspinatus fibers appeared more organized and homogeneous. However, a significant amount of hyperechoic chronic fibrous and periosteal changes noted along the biceps groove. The joint capsule appeared unorganized, hyperechoic, and thickened. The teres tendon and infraspinatus insertion point appeared remodeled and showed hyperechoic fibrous tissue similar to prior scans. The area identified as a firm area within the muscle belly of the infraspinatus on palpation and restricting full shoulder extension on manual palpation remained unchanged. Overall, no improvement was appreciated on diagnostic ultrasound in spite of regenerative medicine therapy in this patient.

Follow-Up

Three of the eight dogs were lost to follow up. A mean of 20.5 months follow-up was available (range 12–29 months). Five owners were asked to fill out a questionnaire and Canine Brief Pain Inventory at the time of re-evaluation. None of the five dogs were reported to have any improvement since their last visit at VOSM. One of the four dogs had died of an unrelated cause. Four of the five dogs were still said to be non-weight bearing the majority of the time in the affected limb. One dog was reported to have a mild weight bearing lameness, which worsens after 15 min of walking activity. None of the owners felt that any of the therapies or treatments attempted helped; however, two owners reported that their dog appeared to be subjectively more comfortable immediately following rehabilitation therapy. Three of the five owners felt that having a frozen shoulder significantly interfered with their dogs gait, general activities, and enjoyment of life.

DISCUSSION

Despite anatomical difference between human and canine shoulders, orthopedic examination findings in the eight patients appear to be similar to those of human patients with adhesive capsulitis. Human patients present with restricted ROM, pain on shoulder manipulation, and impaired function. Similarly, all eight of the patients initially presented with a chronic, grade III–VI/VI unilateral forelimb lameness and had moderate to significant discomfort on and restriction of shoulder extension and flexion. While intra-articular anesthesia is commonly used in humans at the time of diagnosis, this was not performed at the time of initial consult in the patients included in this study due to the need for sedation. However, all patients were sedated and/or anesthetized for either diagnostics or treatment, at which time shoulder extension and flexion were confirmed to be restricted.

Cytologic and histopathologic findings were consistent with findings appreciated in human patients with adhesive capsulitis in the four patients who pursued these diagnostics. Three of the eight patients had a history of previous trauma to the affected limb. This could suggest that this patient had adhesive capsulitis secondary to chronic inflammation and instability. However, while the other five patients had no history of trauma, all patients had a history of a chronic, unilateral forelimb lameness. It is unclear in these cases whether the adhesive diagnosis is primary or potentially secondary to a chronic, untreated shoulder pathology. In the authors' opinion, just as in humans, adhesive capsulitis may be the end-stage manifestation of several primary conditions, it seems possible that regardless of the etiology of the primary shoulder pathology, chronic, untreated shoulder pathology could lead to adhesive capsulitis (2). Further study is necessary to confirm this. Additionally, three of the eight patients in this study were Corgis. Future studies should be performed to assess if there is either a breed predilection or congenital condition that predisposes patients to developing adhesive capsulitis.

While adhesive capsulitis in humans is usually a clinical diagnosis made upon review of patient history and assessment of range of motion before and after injection of local anesthetic, diagnostic imaging is performed to rule out other differential diagnosis and fully evaluate the shoulder. Radiographs of the shoulder are performed to rule out other comorbidities, including fracture, dislocation, calcific tendonitis/bursitis, acromioclavicular arthrosis, and abnormal acromial morphology. CT would have been required to further evaluate the cartilage and boney structures of the joint. However, in this case series, MRI was recommended over CT to evaluate the soft tissue structures of the shoulder, which was the primary area of interest. MRI is routinely performed in humans with suspected frozen shoulder to evaluate the soft tissue structures of the shoulder, including the joint capsule, rotator cuff, biceps, glenohumeral, and coracohumeral ligaments. Shoulder effusion and enhancement of the synovial lining of the shoulder, including the portion of the joint capsule that envelops the biceps tendon, has been documented in human patients with frozen shoulder (18). One recent study also reported that 62% of human patients with idiopathic adhesive capsulitis were found to have concurrent supraspinatus lesions on MRI (19). Four of the five patients who had an MRI were found to have moderate shoulder effusion and enhancement of the synovial lining of the shoulder, which has also been documented in human patients with frozen shoulder (18). One patient was also found to have MRI findings consistent with supraspinatus tendinopathy in the affected limb and another was found to have signs of supraspinatus, infraspinatus, and teres minor insertionopathy, which have also been reported in humans.

Diagnostic ultrasound is also used to evaluate the soft tissue structures of the shoulder in patients with suspected frozen shoulder. Findings that have been seen in patients with adhesive capsulitis are enhanced vascularity and hypoechoic change around the rotator interval (20, 21). Dynamic sonography has been used to help diagnose adhesive capsulitis, which reveals a limited but continuous sliding movement of the supraspinatus tendon against the acromion of the scapula (22). Six of the eight dogs had an initial diagnostic ultrasound performed, all of which revealed evidence of effusion, thickening of the synovium, and fibrous scar tissue. Three dogs were noted to have changes consistent with supraspinatus and biceps tendinopathy. Two dogs had recheck diagnostic ultrasounds performed after receiving ultrasound-guided ADPC and PRP injection in the supraspinatus and ADPC and ACS in the affected shoulder. While the fiber pattern of the supraspinatus was noted to be more regular and homogeneous, moderate inflammation and fibrous changes were still appreciated.

Arthroscopy was performed in five of the eight patients. Just as in human patients with adhesive capsulitis (23), significant adhesions, fibrous scar tissue of the joint capsule, severe inflammation throughout the synovium, severe soft tissue contracture, and loss of capsular volume were found, all of which are very characteristic findings in human patients with adhesive capsulitis.

Treatment offered to human patients with adhesive capsulitis depends on the stage of the disease and clinical symptoms. Frozen shoulder was originally described as a self-limiting disease, and supervised neglect with analgesia, supervised physical therapy, and/or intra-articular corticosteroid injections were recommended (24). Recent studies report that 90% of cases resolve with conservative therapy, which often includes non-steroidal anti-inflammatory medications, intra articular corticosteroid injections, oral steroids, and physical therapy and rehabilitation (6). Intra-articular corticosteroid injections are commonly used and have been reported to have similar outcomes to physiotherapy alone and to more invasive measurements, such as manipulation and hydrodilation (25–27). Moreover, a double-blind, sham-controlled randomized study demonstrated that corticosteroid injections were better than a sham injection in improving shoulder pain at 6 weeks, and the effect was maintained for 12 weeks (28). Interestingly, one recent prospective study compared intensive physical therapy, including passive stretching and manual mobilization versus supportive therapy and exercises with pain limits, or "supervised neglect" and found that patients receiving intensive physical therapy did not do as well as those treated with supervised neglect (29).

In this case series, conservative management consisting of oral medications, intra-articular depomedrol injection, laser/LED therapy, massage therapy, therapeutic ultrasound, magnetic field therapy, and hydrotherapy did not appear to have a significant clinical effect. Of the treatment modalities performed, ESWT seemed to be the most effective treatment modality. Of the five that received ESWT, two showed clinical improvement in lameness. Those dogs originally presented with a grade V–VI/VI unilateral forelimb lameness, and after three sessions of ESWT at 2-week intervals, the patients had improved to a grade IV/VI unilateral forelimb lameness. Two patients received ultrasound-guided ADPC and PRP injection in the affected supraspinatus, and ADPC and ACS in the affected shoulder. While the supraspinatus fibers appeared more homogeneous and organized on 4-week recheck diagnostic ultrasound, suggesting that the supraspinatus tendinopathy had improved, both cases still had evidence of severe, chronic fibrous changes present within the joint and joint capsule. There are currently no reports of the use of stem cell or PRP therapy for adhesive capsulitis in humans.

While conservative management is successful in many of human cases, if there is little to no improvement, surgery may be indicated (7). A recent prospective, randomized study, which compared arthroscopic capsular release to a gentle home stretching program, demonstrated both treatment options to be effective treatment modalities (30). Goals of surgery are to decrease pain, improve ROM, improve function, and shorten natural history of disorder (1, 31–33). The mainstay of surgical treatment is to perform arthroscopic capsular release (1, 2, 31–33). Postoperative care regiments generally include starting physical therapy within 1–2 days of surgery, and patients are encouraged to participate in a physical therapy program for 8–10 weeks (1). Recent studies suggest more patients have pain free range of motion around 10 weeks postoperatively (33). Moreover, patients reported excellent outcome following an arthroscopic anterior capsule and rotator interval release (34). To the authors' knowledge, there are currently no reports documenting surgical treatment of this condition in dogs. Further study is needed to assess the efficacy of surgical management of adhesive capsulitis in dogs.

Adhesive capsulitis is an uncommon cause of chronic forelimb lameness. Due to the rare nature of this condition, this case series only describe eight patients. While the initial presentation of adhesive capsulitis in dogs is very similar to that in humans, the progression, response to therapy, and outcomes found in this study were not (35, 36). Recent studies performed in humans show that adhesive capsulitis often resolves with conservative management (6, 25–29, 36). However, none of the patients treated had complete resolution or significant improvement, regardless of the therapy elected. It is also important to note that while many of these patients had evidence on diagnostic imaging (diagnostic ultrasound or MRI) of concurrent supraspinatus, biceps, and/or infraspinatus tendinopathies, the progression, response to therapy, and outcomes are also not consistent with a primary supraspinatus, biceps, and/or infraspinatus tendinopathy. Further investigation is needed to further describe the etiology and pathogenesis of adhesive capsulitis in dogs. Future controlled, randomized, and blinded investigations with validated objective outcome measures are also warranted to evaluate the effectiveness of both non-surgical and surgical treatment modalities, establish treatment protocols, and evaluate short- and long-term clinical outcome of patients.

ETHICS STATEMENT

In accordance with AAALAC International Rules of Accreditation, this retrospective study was performed with the approval of the VOSM Research Committee and with owner consent. In this retrospective study, all dogs whose records were reviewed were client owned dogs. All clients volunteered their dog for the study and provided written consent as required by Veterinary Orthopedic and Sports Medicine Group for every study participant.

AUTHOR CONTRIBUTIONS

All authors have made substantial contributions to the conception or design of the work. All authors have participated in the acquisition, analysis, or interpretation of data for the work. All authors have contributed to the drafting the work and revising it. All authors have consented final approval of the version to be published and agreed to be accountable for all aspects of the work in ensuring that questions related to the accuracy or integrity of any part of the work are appropriately investigated and resolved.

CONFLICT OF INTEREST STATEMENT

The authors certify that they have no affiliations with or involvement in any organization or entity with any financial interest, or non-financial interest in the subject matter or materials discussed in this manuscript.

ABBREVIATIONS

ADPC, adipose-derived progenitor cell; PRP, platelet-rich plasma; ACS, autologous conditioned serum; ESTW, extracorporeal shockwave therapy.

FOOTNOTES

1. ^Regenerative Medicine Service, Marion DuPont Scott Equine Medical Center, Virginia-Maryland Regional College of Veterinary Medicine, Leesburg, VA, USA.
2. ^SoundEklin, Sound™, Carlsbad, CA, USA.
3. ^Hitachi-Aloka, Aloka, Wallingford, CO, USA.
4. ^The Dynatron Solaris 708, Dynatronics, Salt Lake City, UT, USA.
5. ^VersaTron 4 Paws Trode, PulseVet, Alpharetta, GA, USA.

REFERENCES

1. Chalmers PN, Sherman SL, Ghodadra N, Nicholson GP. Arthroscopic management of frozen shoulder. In: Flatow E, Colvin AC, editors. *AAOS Atlas of Essential Orthopaedic Procedures*. Rosemont, IL: American Academy of Orthopaedic Surgeons (2013). p. 153–7.
2. Binder AI, Bulgen DY, Hazleman BL, Roberts S. Frozen shoulder: a long-term prospective study. *Ann Rheum Dis* (1984) 43:361–4.
3. Warner JJ. Frozen shoulder: diagnosis and management. *J Am Acad Orthop Surg* (1997) 5:130–40.
4. Hannafin JA, Chiaia TA. Adhesive capsulitis: a treatment approach. *Clin Orthop Relat Res* (2000) 372:95–109.
5. Neviaser AS, Hannafin JA. Adhesive capsulitis: a review of current treatment. *Am J Sports Med* (2010) 38:2346–56.
6. Griggs SM, Ahn A, Green A. Idiopathic adhesive capsulitis: a prospective functional outcome study of nonoperative treatment. *J Bone Joint Surg Am* (2000) 82:1398–407.
7. Levine WN, Kashyap CP, Bak SF, Ahmad CS, Blaine TA, Bigliani LU. Nonoperative management of idiopathic adhesive capsulitis. *J Shoulder Elbow Surg* (2007) 16:569–73.
8. Neviaser RJ, Neviaser TJ. The frozen shoulder: diagnosis and management. *Clin Orthop Relat Res* (1987) 223:59–64.
9. Massoud SN, Pearse EO, Levy O, Copeland SA. Operative management of the frozen shoulder in patients with diabetes. *J Shoulder Elbow Surg* (2002) 11:609–13.
10. Ogilvie-Harris DJ, Myerthall S. The diabetic frozen shoulder: arthroscopic release. *Arthroscopy* (1997) 13:1–8.
11. Bowman CA, Jeffcoate WJ, Pattrick M, Doherty M. Bilateral adhesive capsulitis, oligoarthritis and proximal myopathy as presentation of hypothyroidism. *Br J Rheumatol* (1988) 27:62–4.
12. Wohlgethan JR. Frozen shoulder in hyperthyroidism. *Arthritis Rheum* (1987) 30:936–9.
13. Sheridan MA, Hannafin JA. Upper extremity: emphasis on frozen shoulder. *Orthop Clin North Am* (2006) 37:531–9.
14. Hannafin JA, DiCarlo ER, Wickiewicz TL, Warren RF. Adhesive capsulitis: capsular fibroplasias of the glenohumeral joint. *J Shoulder Elbow Surg* (1994) 3:5.
15. Rha DW, Park GY, Kim YK, Kim MT, Lee SC. Comparison of the therapeutic effects of ultrasound-guided platelet rich plasma injection and dry needling in rotator cuff disease: a randomized controlled trial. *Clin Rehabil* (2013) 27(2):113–22.
16. Atashi F, Jaconi MEE, Pittet-Cuenod B, Modarressi A. Autologous platelet-rich plasma: a biological supplement to enhance adipose-derived mesenchymal stem cell expansion. *Tissue Eng* (2015) 21:253–62.
17. Chiavaras MM, Jacobson JA. Ultrasound-guided tendon fenestration. *Semin Musculoskelet Radiol* (2013) 17(1):85–90.
18. Carrillon Y, Noel E, Fantino O, Perrin-Fayolle O, Tran-Minh VA. Magnetic resonance imaging findings in idiopathic adhesive capsulitis of the shoulder. *Rev Rhum Engl Ed* (1999) 66:201–6.
19. Yoo JC, Ahn JH, Lee YS, Koh KH. Magnetic resonance arthrographic findings of presumed stage-2 adhesive capsulitis: focus on combined rotator cuff pathology. *Orthopedics* (2009) 32:22.
20. Lee JC, Sykes C, Saifuddin A, Connel D. Adhesive capsulitis: sonographic changes in the rotator cuff interval with arthroscopic correlation. *Skeletal Radiol* (2005) 34:522–7.
21. Walmsley S, Osmotherly PG, Walker CJ, Rivett DA. Power doppler ultrasonography in the early diagnosis of primary/idiopathic adhesive capsulitis: an exploratory study. *J Manipulative Physiol Ther* (2013) 36:428–35.
22. Ryu KN, Lee SW, Rhee YG, Lim JH. Adhesive capsulitis of the shoulder joint: usefulness of dynamic sonography. *J Ultrasound Med* (1993) 12:445–9.
23. Hand GC, Athanasou NA, Matthews T, Carr AJ. The pathology of frozen shoulder. *J Bone Joint Surg Br* (2007) 89:928–32.

24. Hsu JE, Warrender WJ, Abboud JA. Current review of adhesive capsulitis. *J Shoulder Elbow Surg* (2011) 20:502–14.
25. Jacobs LG, Smith MG, Khan SA, Smith K, Joshi M. Manipulation or intra-articular steroids in the management of adhesive capsulitis of the shoulder? A prospective randomized trial. *J Shoulder Elbow Surg* (2009) 18:348–53.
26. Ryans I, Montgomery A, Galway R, Kernohan WG, McKane R. A randomized controlled trial of intra-articular triamcinolone and/or physiotherapy in shoulder capsulitis. *Rheumatology (Oxford)* (2005) 44:529–35.
27. Tveita EK, Tariq R, Sesseng S, Juel NG, Bautz-Holter E. Hydrodilatation, corticosteroids and adhesive capsulitis: a randomized controlled trial. *BMC Musculoskelet Disord* (2008) 9:53
28. Prestgaard T, Wormgoor ME, Haugen S, Harstad H, Mowinckel P, Brox JI. Ultrasound-guided intraarticular and rotator interval corticosteroid injections in adhesive capsulitis of the shoulder: a double-blind, sham-controlled randomized study. *Pain* (2015) 156:1683–91.
29. Diercks RL, Stevens M. Gentle thawing of the frozen shoulder: a prospective study of supervised neglect versus intensive physical therapy in seventy-seven patients with frozen shoulder syndrome followed up for two years. *J Shoulder Elbow Surg* (2004) 13:499–502.
30. Smitherman JA, Struk AM, Cricchio M, McFadden G, Dell RB, Horodyski M, et al. Arthroscopy and manipulation versus home therapy program in treatment of adhesive capsulitis of the shoulder: a prospective randomized study. *J Surg Orthop Adv* (2015) 24:69–74.
31. Le Lievre HM, Murrell GA. Long-term outcomes after arthroscopic capsular release for idiopathic adhesive capsulitis. *J Bone Joint Surg Am* (2012) 94:1208–16.
32. Holloway GB, Schenk T, Williams GR, Ramsey ML, Iannotti JP. Arthroscopic capsular release for the treatment of refractory postoperative or post-fracture shoulder stiffness. *J Bone Joint Surg Am* (2001) 83:1682–7.
33. Nicholson GP. Arthroscopic capsular release for stiff shoulders: effect of etiology on outcomes. *Arthroscopy* (2003) 19:40–9.
34. Ray RK, Patel A, Cox G, Cole A. Patient satisfaction and patient reported outcome measure data in arthroscopic release of frozen shoulder. *Orthop Rheumatol Open Access J* (2015) 1:555553.
35. Schollmeier G, Uhthoff HK, Sarkar K, Fukuhara K. Effects of immobilization on the capsule of the canine glenohumeral joint. A structural functional study. *Clin Orthop Relat Res* (1994) 304:37–42.
36. Zukerman JD, Rokito A. Frozen shoulder: a consensus definition. *J Shoulder Elbow Surg* (2011) 20:322–5.

CHAPTER 13

Nutritional Support of Dogs and Cats after Surgery or Illness

Ronald J. Corbee[1*], Wim J. S. Van Kerkhoven[2]

[1]*Department of Clinical Sciences of Companion Animals, Faculty of Veterinary Medicine, Utrecht University, Yalelaan, Utrecht, The Netherlands*

[2]*Viyo International NV, Ijzerenpoortkaai 3, Antwerpen, Belgium*

ABSTRACT

Nutritional support early during the postoperative period or after onset of illness decreases the mortality rate and shortens the duration of hospitalization of dogs and cats. The preferred feeding route is dependent on the condition of the patient. If there are no contraindications, every patient must receive nutritional support, at least consistent with the resting energy requirement (RER). Several nutrients may be beneficial during healing and recovery from illness or surgery, but further research is needed to empirically certify the effects of these nutrients in critically ill patients.

Keywords: Recovery, Food Supplementation, Nutrition, Nutraceuticals

1. INTRODUCTION

Nutritional support is important for animals during recovery from illness or surgery. Many animals will recover from mild illness or standard surgical procedures at home, but patients with more severe disease or conditions will be hospitalized during recovery. It is recommended to have a protocol for nutritional support of hospitalized patients since early nutritional support has been reported to improve outcome and to shorten hospitalization time in humans [1] [2] and dogs [3] . Proper absorption of fluids, energy, and essential nutrients is of primary importance during the first 14 days after trauma or onset of illness [4] . It is important to provide nutrients to an animal recovering from illness or surgery to ensure optimal functioning of the gastrointestinal and immune systems. Enteral feeding is preferable to parenteral feeding because intraluminal agents stimulate the gastrointestinal tract and prevent bacterial translocation [4] .

In the present paper we review the current knowledge regarding mechanisms involved in impaired food absorption after surgery or during serious illness, and propose a solution to restore these unfavorable mechanisms.

2. HYPOMETABOLISM VERSUS HYPERMETABOLISM

A healthy animal that has insufficient food available will gradually adapt its metabolism to that situation. Once glycogen stores are exhausted (typically 24 to 48 hours after cessation of food intake) the body will primarily draw on fat reserves. Proteins will also be broken down to meet energy requirements; this may begin within 2 hours but is most dramatic after 2 weeks of a total lack of food intake [5] . The metabolic rate slows down in these conditions as a result of a decrease in the conversion of thyroxine into triiodothyronine (insulin-dependent event), which allows the animal to conserve available energy [6] [7] . This state can be termed hypometabolism.

Stress response is the term for the hormonal and metabolic changes in an animal after illness, trauma, or surgery. Stress results in activation of the sympathetic nervous system, which results in alterations to the endocrine system, primarily insulin resistance and changes in secretion of pituitary gland hormones [8] [9] (Table 1). In addition, there are immunologic (production of cytokines and acute-phase reaction) and hematologic (neutrophilia and lymphocyte proliferation) changes [8] (Table 2).

In human medicine, it requires an average of 14 days for optimal recovery after illness or a surgical procedure. This 14-day recovery period can be categorized into 2 phases of metabolic alterations. Phase 1 of recovery is the first 24 to 48 hours of the recovery period. During this time, fluid therapy and the provision of nutrients to the intestinal tract are important [10] [11] . The second phase of recovery begins after day 3. This period includes the need for increased calorie administration; thus, the need for nutritional support is often stressed for patients after 3 days of malnutrition [12] . It has been suggested, [13] but not proven, that timely nutritional support with nutrients such as arginine, glutamine, taurine, long-chain polyunsaturated omega-3 fatty acids, and prebiotics can shorten the recovery period of cats and dogs after illness or surgery.

Table 1. Endocrine and hormonal reactions after a surgical procedure.

Endocrine gland	Hormones	Changes in secretion
Anterior pituitary	ACTH*	Elevated
	Growth hormone	Elevated
	TSH	Can be elevated or lowered
	FSH and LH	Can be elevated or lowered
Posterior pituitary	AVP	Elevated
Adrenal cortex	Cortisol	Elevated
	Aldosterone	Elevated
Pancreas	Insulin	Elevated
Thyroid	Thyroxine, triiodothyronine	Lowered

*ACTH, adrenocorticotropic hormone (corticotropin); AVP, arginine vasopressin; FSH, follicle stimulating hormone; LH, luteinising hormone; TSH, thyroid stimulating hormone; Based on Desborough and Hall. Stress response after surgery is characterized by an increased secretion of pituitary hormones and by an activation of the sympathetic nervous system. These changes in pituitary secretion have a secondary effect on the hormonal secretion of the affected organs. These so-called target-organs and their hormonal changes are mentioned in this table. (Adapted from Desborough JP, Hall GM. Endocrine response to surgery. *Anesth Rev* 1993; 10:131-148. Reprinted with permission).

Table 2. Systemic reactions after a surgical procedure.

Sympathetic nervous system activation	
Endocrine stress response	Pituitary hormone secretion
	Insulin resistance
Immunological and hematological changes	Cytokin production
	Acute phase reaction
	Neutrophil leucocytosis
	Lymphocytes proliferation

Stress response is the name given to the hormonal and metabolic changes occurring after injury or trauma. It's part of a systemic wound reaction and comprises a wide range of endocrine, immunological and hematological changes. This table focuses on the systemic reactions that occur after surgery. (Adapted from Desborough J.P. The stress response to trauma and surgery; British Journal of Anaesthesia 2000; 85: 109-117. Reprinted with permission).

In general, phase 1 begins during surgery and lasts for 24 to 48 hours after surgery. Even before an animal awakes from anesthesia, vital needs must be met. Therefore, because the body must function with less energy and less oxygen, the relative metabolic rate, cardiac flow, and oxygen consumption are reduced (Figure 1). During phase 1, it is the maintenance of the

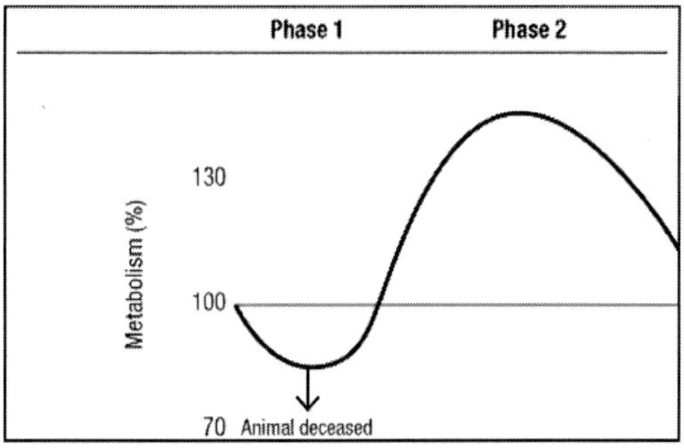

	Phase 1	Phase 2
Blood sugar concentration	↑	↑
Body temperature	↓	↑
Cardiac flow	↓	↑
Free fatty acids	↑	↑
Gluconeogenesis	↑	↑
Insulin	↓	↑
Insulin resistance	↑	↑
Lipolysis	↑	↑
Nitrogen balance	—	↓
Oxygen consumption	↓	↑
Protein degradation	—	↑
Stress hormones (catecholamine, glucagon and cortisol)	↑	↑

Figure 1. Changes in metabolism after a surgical procedure (top) and alterations in various factors (bottom).

circulation of bodily fluids through transfusion or other means that is of foremost importance, rather than the supply of energy. It is important during this phase that the intestines be provided with nutrients and fluids because of reduced blood flow through the gastrointestinal tract owing to the release of cortisol and catecholamines [8] [10] [11] . Phase 1 should focus on having an animal eat and drink as soon as possible after surgery [8] .

Next comes phase 2 and its energy requirement, in which metabolic rate, nitrogen excretion, and oxygen consumption are increased. Supplying alimentation in line with the severity of the invasiveness of the surgery becomes important when transitioning from phase 1 to phase 2 [8] [10] [11] . During phase 2, the goal is to provide an animal with sufficient energy (calories) (Table 3).

However, enteral feeding is not recommended when there is excessive hemorrhage into the gastrointestinal tract; a volvulus, severe constipation, or any other gastrointestinal tract blockage; or excessive vomiting [14]

Phase 1 is the first 24 to 48 hours of the recovery period during which the body must function with less energy and less oxygen; thus, the relative metabolic rate, cardiac flow, and oxygen consumption are reduced. Phase 2 of recovery begins on approximately day 3 and lasts until approximately day 14 after surgery. During phase 2, the metabolic rate, nitrogen excretion, and oxygen consumption are increased. ↑ = Increased. ↓ = Decreased. — = No change (Adapted from: Otani J. Feeding tube practices before and after gastrointestinal surgery. Infusion Therapy News 2003; 3 - 5. Yoshikawa K. Feeding tube techniques and carbohydrate metabolism Journal of clinical and experimental medicine-special issue: 9-14, (1996). Reprinted with permission).

Table 3. 2-phase system.

Recuperation phase	Support	Recuperation time
Phase 1 (critical start-up) 0 h - 48 h	Pet needs to be stimulated to eat: feed the gut (immune function stimulation) + intake of water + all essential recuperation nutrients.	Average 14 days
Phase 2 (Follow-up) From day 3	Arginine, Taurine, L-Glutamine, Omega 6 & 3 fatty acids + Normal diet (Energy)	

The animal needs an average of 14 days for an optimal recovery after illness or a surgical procedure. These 14 days are divided in 2 phases. During phase 2 the goal is to provide the animal with sufficient energy (calories), phase 1 focusses on having the animal eat and drink again as soon as possible and deliver the so-called recuperation nutrients: glutamine, taurine, arginine and omega 6 & 3 fatty acids. For an optimal recovery these recuperation nutrients will also have to be provided during phase 2 and so for 14 days.

The animal needs an average of 14 days for an optimal recovery after illness or a surgical procedure. These 14 days are divided in 2 phases. During phase 2 the goal is to provide the animal with sufficient energy (calories), phase 1 focusses on having the animal eat and drink again as soon as possible and deliver the so-called recuperation nutrients: glutamine, taurine, arginine and omega 6 & 3 fatty acids. For an optimal recovery these recuperation nutrients will also have to be provided during phase 2 and so for 14 days.

[15] . In these situations, the patient should be stabilized and then should receive parenteral nutritional support.

A state of hyper-metabolism exists during phase 2 [10] [11] [16] . Because of the effects of inflammatory cytokines in the catabolic state; protein will be catabolized (for energy and gluconeogenesis), used for the synthesis of acute-phase proteins and immunoglobulins, and needed for wound healing.

Given this hyper-metabolic state, it is important for early nutritional support to provide energy and protein to promote recovery. It is also essential to consider the supplementation of recovery nutrients because many foods are deficient in energy and protein, compared with the amounts of those nutrients required for optimal recovery [10] [11] . Nutritional support (arginine, taurine, l-glutamine, omega-6 fatty acids, and omega-3 fatty acids) for at least 3 days is generally needed to change from a catabolic to an anabolic state. Subsequent adaptation to full recovery may require several weeks. During phase 2, there continues to be an increased need for protein (> 12 g/MJ of ME [> 5 g/100 kcal of ME] in dogs and >18 g/MJ of ME [> 7.5 g/100 kcal of ME] in cats) and fat (>12 g/MJ of ME [> 5g/100 kcal of ME], unless fat restriction is desired (4.8 to 8.4 g/MJ of ME [2.0 to 3.5 g/100 kcal of ME) [7] .

3. ANOREXIA VERSUS HYPOREXIA

Sick animals often ingest an insufficient amount of nutrients because of the primary disease or because of admission to a clinic (e.g. change in diet or feeding routine). Before nutritional support is initiated, a patient must be hemodynamically stable and have adequate pain relief. In addition, clinicians should distinguish between anorexia and hyporexia. Hyporexia refers to a decreased appetite, rather than a total loss of appetite [14] . In the case of anorexia, forced feeding will not stimulate spontaneous food intake. However, enteral feeding or parenteral feeding should be provided to anorectic patients [5] [17] . It should be mentioned that treating the underlying disease also contributes to improving spontaneous food intake.

For hyporectic patients, stimulation of spontaneous food intake may be sufficient and diminishes the need for other assisted feeding techniques, which are more invasive and stressful. This should not lead to a one-more-day syndrome during which nutritional support is not initiated because of the expectancy that an animal will starteating at some point (e.g. the next day). If stimulation of spontaneous food intake results in an insufficient effect (i.e. does not result in an adequate amount of food intake) during phase 1, other assisted feeding techniques should be implemented to ensure the administration of sufficient amounts of nutrients during phase 2.

4. CACHEXIA AND SARCOPENIA

Cachexia has been defined as loss of lean body mass. Cachexia affects a large proportion of dogs and cats with congestive heart failure, chronic kidney disease, cancer, and a variety of other chronic diseases. By contrast, sarcopenia is loss of lean body mass that occurs with aging in the absence of disease. Both conditions are associated with increases in morbidity and mortality rates [18] . Fat is the predominant fuel source in healthy, anorectic dogs and cats, whereas cachectic dogs and cats use both fat and amino acids as fuel sources. This latter situation contributes to loss of lean body mass and possible decreases in immune function, wound healing, and survival rate. Nutritional assessment guidelines [19] have been formulated that include methods for early detection of muscle mass to enable clinicians to take preventive measures for animals in a pre-cachectic state. During the catabolic or hyper-metabolic recovery phase, feeding patients foods with increased energy density and increased protein concentration will help prevent additional loss of lean body mass.

5. IMPORTANCE OF NUTRITIONAL SUPPORT FOR WOUND HEALING AND FIGHTING INFECTION

It is vital that animals begin to eat and drink as soon as possible during the first 48 hours after trauma, surgery, or illness. Malnutrition leads to an increased mortality rate in humans because of decreasing potassium, phosphorus, and magnesium concentrations and reductions in immune function [1] [2] . Food intake provides nutrients (amino acids, fatty acids, vitamins, and minerals) that help optimize the immune response, tissue repair, and tissue synthesis as well as the metabolism of administered medications [1] -[3] . One problem that often arises is that patients refuse to eat during phase 1, usually as a consequence of catabolic stress. Almost all hospitalized animals have a negative energy balance, and almost half of the animals have that negative energy balance as a result of food refusal [20] . Animals in negative energy balance have a deficit of essential nutrients. As a consequence, palatability of nutritional products that promote recovery must be extremely high because the appetite of nearly all animals is dramatically reduced during the recovery [21] -[25] .

In addition, an optimally functioning immune system is crucial for recovery from illness and after surgery. It has been estimated that approximately 60% to 70% of the immune system function of cats and dogs is in the intestines; therefore, the intestinal structure (mucosa) must remain intact for proper immune development [15] . As mentioned previously, the first 14 days are typically extremely important, especially for the ingestion of fluids and essential nutrients via enteral alimentation. As such, nutrients ingested during this period promote optimal gastrointestinal tract function.

6. BACTERIAL TRANSLOCATION

Bacterial translocation is a phenomenon in which intestinal bacteria migrate from the intestines through the intestinal wall into the mesenteric lymph nodes and internal organs where they are not typically found. Surgical procedures can be associated with unexpected inflammation, serious infections, and resultant compromised organ function [1] -[4] . One possible cause is bacterial translocation (causing gastrointestinal tract–associated sepsis), and this phenomenon may be attributable to 1 or several causes. Malnutrition of the gastrointestinal tract (i.e. the gastrointestinal tract is deprived of intraluminal nutrients) decreases cell turnover and mucosal mucus production, which thereby decreases the barrier function and surface of the gastrointestinal tract and results in malabsorption of nutrients. The synthesis of immunoglobulins is often also decreased because of protein malnutrition, thus decreasing gastrointestinal tract–barrier function (gastrointestinal tract–associated lymphoid tissue) [26] . Therefore, support of the gastrointestinal tract with intraluminal nutrients will be beneficial in all anorectic animals.

The absence of intraluminal nutrients also reduces blood flow to the gastrointestinal tract. Parenteral nutrition does not deliver intraluminal nutrients and thus might predispose an animal to bacterial translocation. The postsurgical mortality rate decreased in laboratory rats with artificially induced peritonitis when the rats received enteral nutrition, compared with the mortality rate when rats received parenteral nutrition (Table 4) [27] . Several studies [4] [27] -[29] in humans have found that there are fewer (20% less) infections after surgery when enteral (peroral) feeding is used, as opposed to the number of infections after surgery when parenteral (IV) feeding is used. Furthermore, enteral feeding soon after surgery supports recovery of gastrointestinal tract motility and improvement of the general condition in humans [28] . It is likely that there may be a similar situation in veteri nary practice. Reduced blood flow to the gastrointestinal tract provokes damage to the epithelial cells and there

Table 4. Chances of postoperative survival for different forms of food administration in laboratory rats after laparotomy.

	N*	Survival rate (%)
Oral feeding	15	80.0
Enteral (tube) feeding	13	61.5
Intravenous feeding	10	20.0

*N = number of animals in treatment (Adapted from Fukushima R. Counteraction against bacterial translocation (1) the nutrient canal. *J Jpn Surg Soc* 2007; 108: 206-210. Reprinted with permission).

fore increases permeability of the gastrointestinal tract [17] . Bacterial overgrowth, inflammatory bowel disease, and other gastrointestinal diseases cause a loss of barrier function and therefore are possible causes of bacterial translocation. Bacterial translocation has been detected in cats [30] and humans [31] [32] ; however, bacterial translocation has not been proven in dogs with pancreatitis [33] [34] . Damage to the gastrointestinal tract is usually associated with bacterial translocation [30] [35] . A study [36] in dogs with gastric dilatation-volvulus revealed no bacterial translocation; however, bacterial translocation can occur in dogs and cats when the mucosal barrier is damaged by an underlying disease or by treatment with medications (e.g. chemotherapeutics, immunosuppressive drugs, and certain antimicrobials) that influence the gastrointestinal barrier and gastrointestinal tract motility [30] [37] . It should be mentioned that bacterial translocation does not occur when food is withheld from healthy dogs or healthy dogs are fed parenterally [38] . Most dogs and cats with an underlying illness are anorectic. A nonfunctional gastrointestinal tract is an indication for parenteral feeding, and animals with a nonfunctional gastrointestinal tract are especially at risk for bacterial translocation.

Investigators in 1 study [14] compared early-stage enteral nutrition and total parenteral nutrition with regard to the improvement of the nitrogen balance in humans after surgery for esophageal cancer. Those investigators found that it required 8 days after surgery for the nitrogen balance to normalize when enteral nutrition was provided, whereas it required 14 days after surgery for the nitrogen balance to normalize when parenteral nutrition was provided (Figure 2) [16] . It appears that there is a similar time frame for domestic animals [39] [40] .

7. STIMULATION OF SPONTANEOUS FOOD INTAKE

Spontaneous food intake is a sign of recovery, so it is important to offer food and to stimulate spontaneous food intake daily to monitor return of appetite. Hospitalization often is a stressful event for an animal; thus, it is important to recreate an animal's home environment as much as possible during hospitalization to encourage spontaneous food intake. A calm and relaxed atmosphere, appropriate and tender care, and a light cycle (No. of hours of light vs No. of hours of darkness) consistent with the time of year will provide ideal conditions for encouraging feed intake. Some medications, including opiates, antimicrobials, diuretics, immunosuppressives, and chemotherapeutics, can have inhibitory effects on food intake [41] . It is often preferable to initially provide a hospitalized animal with the food that it typically receives at home, rather than immediately feeding it the diet that will eventually be prescribed. This is certainly the case for animals with neophobia (suspicion of unknown foods). In our experience, aversion to diets administered during hospitalization is common, and this makes it more difficult to change to therapeutic diets. Furthermore, the authors do not recommend offering several foods at the same time.

A palatable food increases the chances of spontaneous food intake. Most critical care diets [A,B,C] are extremely flavorful. In addition, food can be made more attractive by increasing the moisture content. Higher

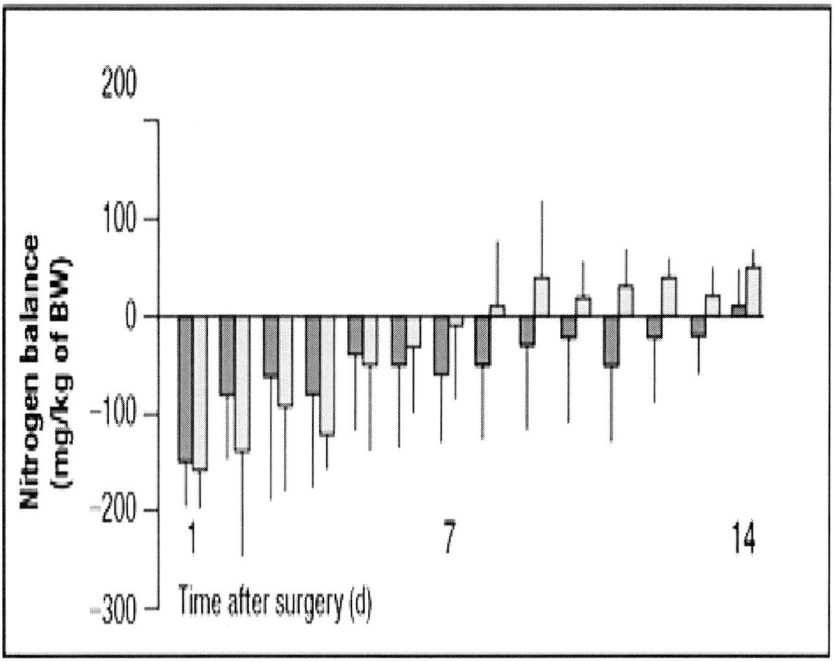

Figure 2. Postoperative nitrogen balance for different forms of food administration in humans.

Mean ± SD nitrogen balance in humans receiving nutrients via total parenteral nutrition (dark gray bars; n = 30) or enteral feeding (light gray bars; 20) after surgery. Notice that the postoperative nitrogen balance was better for enteral feeding than for parenteral feeding. BW = Body weight. A study compared the use of enteral feeding in an early stage with total parenteral feeding in order to improve the nitrogen balance in humans after a surgical procedure in case of an esophageal cancer. It was proven that the nitrogen balance is normalized after 8 days post-surgery in case of enteral feeding. 14 days were needed in case of total parenteral feeding (Adapted from Omata J, Fukatsu K, Maeshima Y, et al. Enteral nutrition rapidly reverses total parenteral nutrition-induced impairment. Clin Nutr 2009; 28: 668-673. Reprinted with permission).

moisture content also contributes to rehydration of patients and causes the nutrients to leave the stomach more quickly, making it less likely that the food will be vomited or regurgitated [42] . Canned foods have high moisture content and often higher fat and protein contents, which also enhance palatability. However, foods with a high fat content will slow down gastric emptying [42] . Still, such foods are often preferred, unless there are contraindications for a high fat or protein content.

Cats are often strongly attached to the texture of food. If cats were fed dry food since they were kittens, the chances are they will spontaneously eat dry food rather than canned alternatives. Adding sugar can improve the flavor of dog food, but cats have no taste receptors for sweetness [43] . Artificial sugars (e.g. xylitol) can lead to hypoglycemia in cats and dogs and should be avoided [17] .

Heating food to body temperature also has a positive impact on food intake, as does offering fresh food. Food that is not eaten within a short period after it is provided will likely not be eaten later by that an animal. Spontaneous food intake can be encouraged by the addition of water or meat stock to dry food, thereby releasing appetizing aromas.

Appetite-enhancing drugs such as cyproheptadine, mirtazapine, corticosteroids, and diazepam can be used; however, the authors do not recommend the use of these products in hospitalized patients because of their adverse effects. A product [D] has recently been marketed that stimulates the appetite of patients with hyporexia. That product contains flavoring agents as well as nutraceuticals and may be provided to hyporectic and anorectic animals via enteral feeding.

8. FEEDING METHODS

It is important to provide nutrients through the most natural route. Enteral provision of nutrients is preferred to parenteral provision of nutrients. The structure of intestinal villi was compared in rats after enteral and parenteral feeding for 14 days [15] [26] [27] . These studies revealed that there was atrophy of the intestinal villi within 3 days after onset of total parenteral nutrition. The villi of the small intestine became smaller and thinner, and the enzymatic activity decrease. In contrast, the intestinal mucosa structure was maintained in rats provided parenteral nutrition (Figure 3). However, when spontaneous food intake is not successful stimulated, tube feeding or parenteral nutrition should be initiated immediately.

Nutritional support should be administered in accordance with the RER of an animal on the basis of the current body weight. The RER can be calculated by use of the following equations: RER = (body weight in kg)$^{0.75}$ × 293 kJ of ME/d and RER = (body weight in kg)$^{0.75}$ × 70 kcal of ME/d. The volume of food should be attained gradually over a period of 2 to 4 days. Should an animal not voluntary consume the necessary volume of food (e.g. poor appetite or condition that prevents eating), then tube feeding should be initiated. Parenteral feeding is indicated for a nonfunctional gastrointestinal tract, the inability to place an enteral feeding tube, or excessive vomiting. Parenteral feeding requires an intensive care setting, and the preparation and handling of products used for parenteral feed should be performed under strict aseptic conditions. Additional information about enter

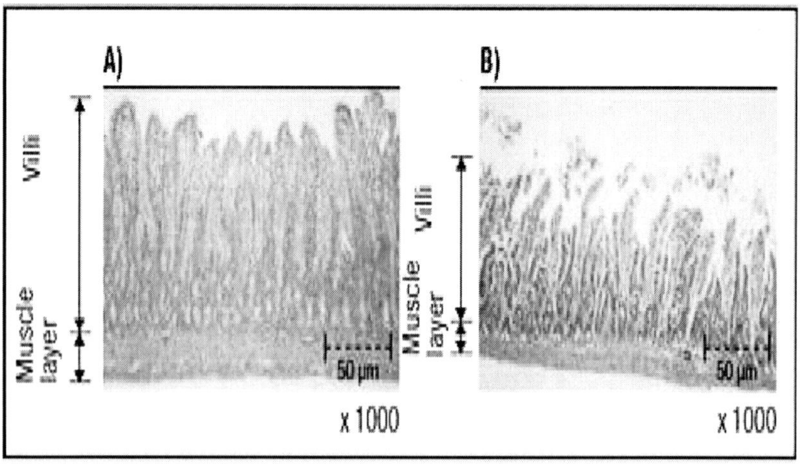

Figure 3. The structure of intestinal villi in the rat after 14 days grain feeding versus parenteral feeding in laboratory rats.

Photomicrographs of tissue sections of the small intestines obtained from rats fed grain for 14 days (A) or provided nutrients via total parenteral nutrition for 14 days (B). Notice that the intestinal villi atrophy and are smaller and thinner in rats after 14 days of total parenteral nutrition. Hematoxylin and Eosin stain; bar = 50 µm. A study in rats showed that villus atrophy occurred after parenteral feeding. The same study proved that within 3 days after parenteral feeding intestinal villi in the small intestine became thinner and smaller and that there was a decreased enzymatic activity. Conclusion is that parenteral feeding caused a negative effect on the structure of the intestinal mucosa (Adapted from Okuma T. Clinical nutrition in keywords: enteral tube feeding, in Proceedings. Nutr Soc 2007; 66: 299-306. Reprinted with permission).

al and parenteral feeding techniques have been reported elsewhere [5] [13] [41] .

9. NUTRITIONAL SUPPORT DURING RECOVERY

As mentioned previously, timely nutritional support with nutrients such as arginine, glutamine, taurine, longchain polyunsaturated omega-3 fatty acids, and prebiotics may possibly shorten the recovery period of cats and dogs after illness or surgery. Thus, it is important to provide these nutrients to hospitalized animals.

Amino acids

In humans, arginine at a dose of 1.12 g/MJ of ME (0.47 g/100 kcal of ME) enhances the T-cell response and increases the number of T-helper cells in trauma (including surgery) and infection [44] [45] . The mean dose in critical care diets for dogs and cats is 1.20 to 1.77 g/MJ of ME (0.50 to 0.74 g/100 kcal of ME). Arginine promotes nitrogen retention, protein turnover, conversion of ammonia to urea, and wound healing [46] . Rats receiving supplemental arginine have an increased ability to synthesize acute-phase proteins [47] and improved rates of wound healing, nitrogen retention, and growth [48] [49] . Investigators in another study [50] on rats found effects of arginine supplementation on recovery of damage to the intestinal tract as well as prevention of bacterial translocation. Arginine improves cardiac output and blood flow through the gastrointestinal tract via increased production of nitric oxide; such improvements result in better wound healing [51] . Arginine is an essential amino acid for dogs and cats. The dietary arginine concentration should be sufficiently high (> 0.46 g/MJ of ME) for parenteral feeding, and many human products do not contain that amount of arginine [15] .

Glutamine is one of the most important amino acids. It supports health of intestinal cells and improves immunity [52] . However, glutamine is not fully synthesized in stressed animals, including stress induced by surgery [53] . Glutamine supports intestinal activity and the immune system by stimulating the synthesis of specific acute-phase proteins and shortens the recovery period after surgery. Rapidly dividing cells (eg, fibroblasts, lymphocytes, and intestinal epithelial cells) have high glutaminase activity and therefore metabolize much of the available glutamine. Glutamine also serves as the fuel for enterocytes, and it enhances insulin sensitivity by increasing plasma concentrations of glucagon-like peptide-1 [54] . Glutamine also contributes to the synthesis of glutathione (a crucial antioxidant in the gastrointestinal mucosa), aids in the prevention of bacterial translocation, and improves function of the gastrointestinal tract and systemic immune system [50] . Glutamine also enhances de novo synthesis of arginine in the kidneys via the citrulline pathway [55] . Numerous studies [56] -[62] , [E] have indicated the effect of glutamine on wound healing and repair. To the authors' knowledge, the specific effective dose for dogs and cats has not been determined; however, > 1.2 g/MJ of ME (> 0.5 g/100 kcal of ME) is supported [58] . Typical critical care diets contain glutamine at a dose of 2.65 g/MJ of ME (1.11 g/100 kcal of ME).

Taurine promotes the healthy development and function of the immune system [63] . Taurine helps fight cellular attack from oxidative stress and plays a major role in bile conjugation, retinal function, and energy metabolism of the myocardium. Taurine is also involved in calcium exchange mechanisms [63] [64] and reduces blood pressure in humans [65] . Taurine deficiency causes dilatative cardiomyopathy in cats and dogs [66] . To the authors' knowledge, the specific effective dose of taurine for prevention of disease in dogs and cats has not been determined. Typical critical care diets contain taurine at doses of 0.30 g/MJ of ME (0.13 g/100 kcal of ME).

Fatty acids

The use of EPA and DHA offers potential health benefits. Not all omega-3 long-chain polyunsaturated fatty acids are as effective as EPA and DHA, so it is important to know the concentration of EPA and DHA in product that contain omega-3 fatty acids. In addition, those products should contain adequate amounts of antioxidants. Typically, the use of cod liver oil is not recommended because some of those preparations also contain high amounts of vitamins A and D. Furthermore, the products should not be contaminated with heavy metals (eg, mercury). [67] Eicosapentaenoic acid and DHA exert anti-inflammatory effects by altering eicosanoid synthesis and cytokine expression and reducing the expression of tumor necrosis factorand interleukin 1, which play major roles in anorexia and cachexia. They also contribute to a decrease in the production of leukotriene B_4 and prostaglandin E_2, which play important roles in inflammation and pain. Mitigating pain and inflammation will improve recovery time and accelerate the healing process [68] [69] . Omega-3 fatty acids decrease muscle loss in dogs with chronic heart failure and improve appetite [70] . Omega-3 fatty acids also are beneficial in animals with atopic dermatitis,

[71] osteoarthritis, [72] [73] cancer, [74] errors in lipid metabolism, [75] cardiac disease, [76] [77] kidney disease, [78] and ulcerative colitis [79] . The therapeutic use of long-chain omega-3 fatty acids and fish oils has recently been reviewed [80] . Dosage recommendations for some clinical conditions range from 0.25 to 0.66 g of EPA and DHA/MJ of ME (0.10 to 0.28 g/100 kcal of ME) for dogs and 0.25 to 1.81 g of EPA and DHA/MJ of ME (0.10 to 0.76 g/100 kcal of ME) for cats [80] . Typical critical care diets contain 0.42 to 1.41 g of EPA and DHA/MJ of ME (0.18 to 0.59 g/100 kcal of ME). Because of possible adverse effects of EPA and DHA on blood clotting, EPA and DHA should not be used in patients with coagulation problems [80] . Currently, there is no evidence to support the use of EPA and DHA in all hospitalized patients. Further research is needed on the use of omega-3 fatty acids in hospitalized patients.

Prebiotics

Prebiotics such as inulin and fructo-oligosaccharides are believed to contribute to health of the gastrointestinal tract. Prebiotics are non-digestible food ingredients that selectively stimulate the growth and activity of 1 or more types of bacteria (including Bifidobacterium spp and Lactobacillus spp) in the colon, thereby improving health of the animal [81] . Fermentation of prebiotics causes a decrease in the pH in the colonic contents, which promotes growth of probiotic bacteria; the decrease in pH also has antimicrobial effects on pathogens. Major effects are improvements in bowel function, increases in mineral absorption, alterations in lipid metabolism, reductions in ammonia absorption, and decreases in insulin concentrations [82] .

Prebiotics may be beneficial for the management or prevention of diarrhea in hospitalized patients that are receiving medications. Diarrhea is a frequent adverse effect of many medications, including NSAIDs, digitalis, thyroid gland hormones, magnesium-containing compounds, lactulose, anthelmintics, chemotherapeutics, topical insecticides, and antimicrobials. The most common adverse effect of orally administered antimicrobials is a change in the population of intestinal microflora (bacteria and yeast) that typically coexist in a beneficial balance in the gastrointestinal tract. This problem is also evident with parenterally administered antimicrobials. Depending on the antimicrobial used, there may be overgrowth of certain types of organisms, whereas other types of organisms may be suppressed or eliminated. One result of this imbalance can be diarrhea. A well-balanced intestinal flora creates a barrier that prevents the intrusion of harmful bacteria, promotes the transit of food through the intestines, stimulates the immune system, neutralizes the production of harmful elements, and ferments non-digestible fiber to form valuable short-chain fatty acids (e.g.

butyric acid). These short-chain fatty acids are used as a substrate by colonocytes to maintain the integrity of the gastrointestinal tract, facilitate the absorption of water and electrolytes, and thus minimize the risk of diarrhea. Short-chain fatty acids are a subgroup of fatty acids that have fewer than 6 carbon atoms. They include acetic acid, propionic acid, isobutyric acid, butyric acid, isovaleric acid, valeric acid, caproic acid, lactic acid, and succinic acid. Similar to mediumchain fatty acids, short-chain fatty acids are transported directly into the portal vein during lipid digestion. This is in contrast to long-chain fatty acids, which are packaged into chylomicrons and enter lymphatic capillaries before being admixed with blood in the subclavian vein [83] . Therefore, short-chain fatty acids are readily absorbed by colonocytes. To the authors' knowledge, the specific effective dose of prebiotics for dogs and cats has not been determined. It has been suggested [84] -[86] that 4 to 8 g of fructo-oligosaccharides and inulin/d are effective.

Nutrition assessment determines whether there is a need for nutritional support. If it is determined that there is insufficient intake relative to need (293× (bodyweight) 0.75 in kJ/ME per day) then the (medical) cause needs to be eliminated where possible. A distinction is then made between hyporexia (decreased appetite, appetite is present but not optimal) and anorexia (complete loss of appetite). In hyporexia, a start can be made in improving the environment, palatability etc. In anorexia, a distinction must be made between patients with a functioning gastrointestinal tract (GI-tract) and patients with a non-functioning GI-tract or patients with excessive vomiting. Patients with a functioning GI-tract can be supported with the various types of enteral tubes. The choice depends on the expected duration of the anorexia. Patients with a non-functioning GI-tract or patients with excessive vomiting can be parentally supported. The choice of centrally (through a jugular catheter) or peripherally (through a peripheral catheter) depends on the expected duration of the dysfunction and/or vomiting and on what is possible for the clinic. If insufficient enteral support can be provided then partial parenteral support can be used. If the GI-tract is functioning again and/or excessive vomiting has ceased then the parenteral feeding can be phased out and replaced by enteral feeding (Figure 4).

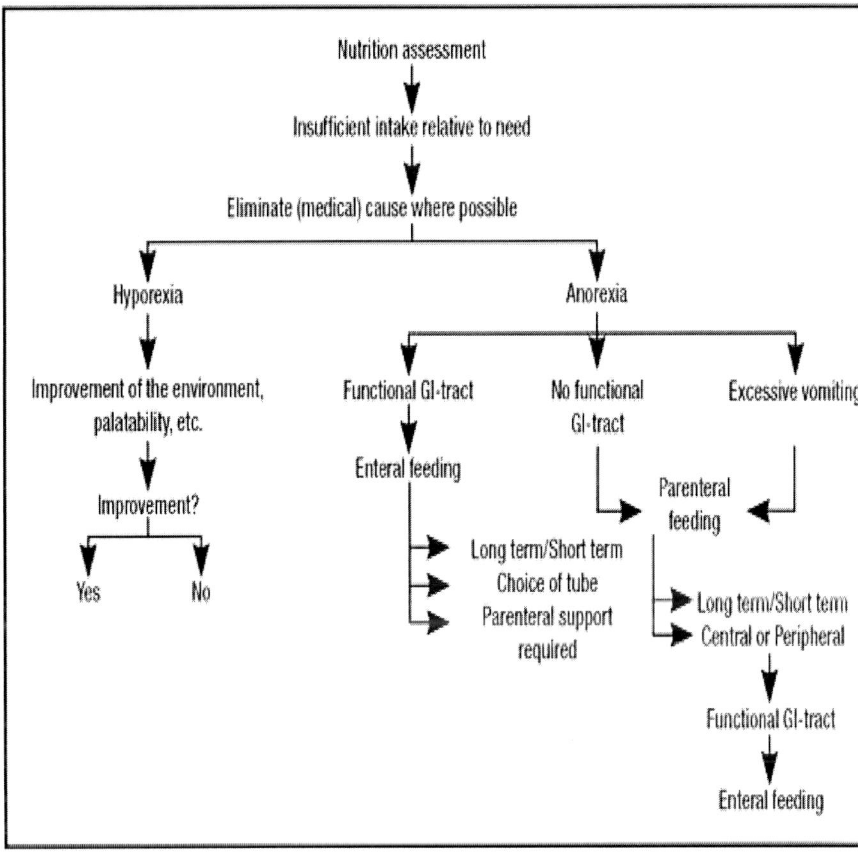

Figure 4. Decision tree for nutritional support.

10. CONCLUSION

After surgery or during serious illness several mechanisms disturb the nutritional situation resulting in delayed Nutrition assessment determines whether there is a need for nutritional support. If it is determined that there is insufficient intake relative to need (293× (bodyweight) 0.75 in kJ/ME per day) then the (medical) cause needs to be eliminated where possible. A distinction is then made between hyporexia (decreased appetite, appetite is present but not optimal) and anorexia (complete loss of appetite). In hyporexia, a start can be made in improving the environment, palatability etc. In anorexia, a distinction must be made between patients with a functioning gastrointestinal tract (GI-tract) and patients with a non-functioning GI-tract or patients with excessive vomiting. Patients with a functioning GI-tract can be supported with the various types of enteral tubes. The choice depends on the expected duration of the anorexia. Patients with a non-functioning GI-tract or patients with excessive vomiting can be parentally supported. The choice of centrally (through a jugular catheter) or peripherally (through a peripheral catheter) depends on the expected duration of the dysfunction and/or vomiting and on what is possible for the clinic. If insufficient enteral support can be provided then partial parenteral support can be used. If the GI-tract is functioning again and/or excessive vomiting has ceased then the parenteral feeding can be phased out and replaced by enteral feeding.

Ronald J. Corbee, Wim J. S. Van Kerkhoven recovery. There is evidence that nutritional support early during the postoperative period or after onset of illness decreases the mortality rate and shortens the duration of hospitalization for dogs and cats. The preferred feeding route will depend on the condition of the patient. In case there are no clinical contraindications, patients should receive nutritional support, at least consistent with the resting energy requirement. Whereas the addition of food supplements containing nutrients belonging to several groups may be beneficial during healing and recovery from illness or surgery, further studies are needed to certify the favorable effects of these.

REFERENCES

1. Chambrier, C. and Sztark, F. (2012) Société Francophone de nutrition clinique, et métabolisme (SFNEP), Société fran- çaise d'anesthésie et réanimation (SFAR). French clinical guidelines on perioperative nutrition. Update of the 1994 consensus conference on perioperative artificial nutrition for elective surgery in adults. Journal of Visceral Surgery, 149, e325-e336.
2. Serón-Arbeloa, C., Puzo-Foncillas, J., Garcés-Gimenez, T., et al. (2011) A Retrospective Study about the Influence of Early Nutritional Support on Mortality and Nosocomial Infection in the Critical Care Setting. Clinical Nutrition, 30, 346-350.
3. Liu, D.T., Brown, D.C. and Silverstein, D.C. (2012) Early Nutritional Support Is Associated with Decreased Length of Hospitalization in Dogs with Septic Peritonitis: A Retrospective Study of 45 Cases (2000-2009). Journal of Veterinary Emergency and Critical Care, 22, 453-459.
4. Chung, C.K., Whitney, R., Thompson, C.M., et al. (2013) Experience with an Enteral-Based Nutritional Support Regimen in Critically Ill Trauma Patients. Journal of the American College of Surgeons, 217, 1108-1117.
5. Saker, K.E. and Remillard, R.L. (2010) Critical Care Nutrition and Enteral-Assisted Feeding. In: Hand, M.S., Ed., Small Animal Clinical Nutrition, 5th Edition, Mark Morris Institute, Topeka, 439-476.
6. Daley, B.J. and Bistrian, B.R. (1994) Nutritional Assessment. In: Zaloga, G.P., Ed., Nutrition in Critical Care, St Louis, Mosby, 9-33.

7. Walton, R.S., Wingfield, W.E. and Ogilvie, G.K. (1996) Energy Expenditure in 104 Postoperative and Traumatically Injured Dogs with Indirect Calorimetry. Journal of Veterinary Emergency and Critical Care, 6, 71-79.
8. Desborough, J.P. (2000) The Stress Response to Trauma and Surgery. British Journal of Anaesthesia, 85, 109-117.
9. .P. and Hall, G.M. (1993) Endocrine Response to Surgery. Anesthesia Review, 10, 131-148.
10. Otani, J. (2003) Feeding Tube Practices before and after Gastrointestinal Surgery. Infusion Therapy News, 2, 3-5. [Citation Time(s):5]
11. Yoshikawa, K. (1996) Feeding Tube Techniques and Carbohydrate Metabolism. Journal of Experimental & Clinical Medicine, Special Issue, 9-14. [Citation Time(s):5]
12. WSAVA Global Nutrition Committee (2013) Feeding Guide for Hospitalized Dogs and Cats. www.wsava.org/sites/default/files/hospitalized%20 patient%20feeding%20guide.pdf
13. Larsen, J.A. and Perea, S.C. (2011) Enteral Nutrition and Tube Feeding and Parenteral Nutrition. In: Fascetti, A.J. and Delaney, S.J., Eds., Applied Veterinary Clinical Nutrition, West Sussex, Blackwell, Chichester, 329-373.
14. Seike, J., Tangoku, A., Yuas, Y., Okitsu, H., Kawakami, Y. and Sumitomo, M. (2011) The Effect of Nutritional Support on the Immune Function in the Acute Postoperative Period after Esophageal Cancer Surgery: Total Parenteral Nutrition versus Enteral Nutrition. The Journal of Medical Investigation, 58, 75-80.
15. Omata, J., Fukatsu, K., Maeshima, Y., Moriya, T., Murakoshi, S., Noguchi, M., Okamoto, K., Fukazawa, S., Saitoh, D., Mochizuki, H., Yamamoto, J. and Hase, K. (2009) Enteral Nutrition Rapidly Reverses Total Parenteral Nutrition-Induced Impairment of Hepatic Immunity in a Murine Model. Clinical Nutrition, 28, 668-673.
16. Delaney, S.J. (2006) Management of Anorexia in Dogs and Cats. Veterinary Clinics of North America: Small Animal Practice, 36, 1243-1249.
17. Freeman, L.M. (2012) Cachexia and Sarcopenia: Emerging Syndromes of Importance in Dogs and Cats. Journal of Veterinary Internal Medicine, 26, 3-17.
18. WSAVA Nutritional Assessment Guidelines Task Force Members (2011) WSAVA Nutritional Assessment Guidelines. Journal of Feline Medicine & Surgery, 13, 516-525.
19. Remillard, L. (2002) Nutritional Support in Critical Care Patients. Veterinary Clinics of North America: Small Animal Practice, 32, 1145-1164.
20. Vadurel, A. and Gogny, M. (1997) L'odorat du chien: Aspects physiologique set facteurs de variation (Sense of Smell in Dogs: Physiological Aspects and Factors of Variation). Point Veterinaire, 28, 1037-1044.
21. Kienzle, E., Meyer, H. and Schnieder, R. (1991) Investigations on Palatability, Digestibility and Tolerance of Low Digestible Food Components in Cats. Journal of Nutrition, 121, S56-S57.
22. Smith, J.C., Rashottea, M.E., Austina, T. and Griffin, R.W. (1984) Fine-Grained Measures of Dogs' Eating Behavior in Single-Pan and Two-Pan Tests. Neuroscience & Biobehavioral Reviews, 8, 243-251.
23. Stein, L.J. (2001) Chemosensory Insights. Petfood Industry 2001, 4-6.
24. Thomson, D.M.H. (1986) The Meaning of Flavour. In: Birch, G.G. and Lindley, M.G., Eds., Developments in Food Flavours, Elsevier Applied Science, London, 1-21.
25. Fukatsu, K. (2007) Bacterial Translocation and the Patient's Condition. Journal of Japan Surgical Society, 108, 138- 142.
26. Fukushima, R. (2007) Counteraction against Bacterial Translocation (1) the Nutrient Canal. ournal of Japan Surgical Society, 108, 206-210.
27. Carlson, G.L. and Dark, P. (2010) Acute Intestinal Failure. Current Opinion in Critical Care, 16, 347-352.
28. Nakamura, T. (2004) Nutrient Canal Management for ER and ICU Patients. Infusion Therapy News, 3, 3-5.

29. Marks, S.L., Cook, A.K., Griffey, S., Kass, P.H. and Rogers, Q.R. (1997) Dietary Modulation of Methotrexate-Induced Enteritis in Cats. American Journal of Veterinary Research, 58, 989-996.
30. Wells, C.L. and Erlandsen, S.L. (1996) Bacterial Translocation: Intestinal Epithelial Permeability. In: Rombeau, J.L. and Takala, J., Eds., Update in Intensive Care and Emergency Medicine: Gut Dysfunction in Critical Illness, SpringerVerlag, Berlin, 131-145.
31. MacFie, J. (2004) Current Status of Bacterial Translocation as a Cause of Surgical Sepsis. British Medical Bulletin, 71, 1-11.
32. Kazantsev, G.B., Hecht, D.W., Rao, R., Fedorak, I.J., Gattuso, P., Thompson, K., Djuricin, G. and Prinz, R.A. (1994) Plasmid Labeling Confirms Bacterial Translocation in Pancreatitis. American Journal of Surgery, 167, 201-207.
33. Liu, Q., Djuricin, G., Nathan, C., Gattuso, P., Weinstein, R.A. and Prinz, R.A. (2000) The Effect of Interleukin-6 on Bacterial Translocation in Acute Canine Pancreatitis. International Journal of Gastrointestinal Cancer, 27, 157-165.
34. Gennari, R. and Alexander, J.W. (1996) Effects of Hyperoxia on Bacterial Translocation and Mortality during GutDerived Sepsis. JAMA Surgery, 131, 57-62.
35. Winkler, K.P., Greenfield, C.L. and Schaeffer, D.J. (2003) Bacteremia and Bacterial Translocation in the Naturally Occurring Canine Gastric Dilatation-Volvulus Patient. Journal of the American Animal Hospital Association, 39, 361- 368.
36. Wells, C.L., Maddaus, M.A. and Simmons, R.L. (1988) Proposed Mechanisms for the Translocation of Intestinal Bacteria. Clinical Infectious Diseases, 10, 958-979.
37. Dahlinger, J., Marks, S.L. and Hirsh, D.C. (1997) Prevalence and Identity of Translocating Bacteria in Healthy Dogs. Journal of Veterinary Internal Medicine, 11, 319-322.
38. Hullár, I., Fekete, S., Andrásofszky, E., Szöcs, Z. and Berkényi, T. (2001) Factors Influencing the Food Preferences of Cats. Journal of Animal Physiology and Animal Nutrition, 85, 205-211.
39. Ohno, K. (2008) Clinical Adoption of Enteral Feeding—Nutritional Support for Severely Ill Patients: Enteral Feeding or Intravenous Nutrition. Proceedings of 29th Annual Meeting of Japan Society of Clinical Veterinary Medicine, Tokyo, 2008, 211-213.
40. Hand, M.S., Thatcher, C.D., Remillard, R.L., Roudebush, P. and Novotny, B.J. (2011) Small Animal Clinical Nutrition. 5th Edition, Mark Morris Institute.
41. Sachdeva, P., Kantor, S., Knight, L.C., Maurer, A.H., Fisher, R.S. and Parkman, H.P. (2013) Use of a High Caloric Liquid Meal as an Alternative to a Solid Meal for Gastric Emptying Scintigraphy. Digestive Diseases and Sciences, 58, 2001-2006.http://dx.doi.org/10.1007/s10620-013-2665-2
42. Jiang, P., Josue, J., Li, X., Glaser, D., Li, W.H., Brand, J.G., Margolskee, R.F., Reed, D.R. and Beauchamp, G.K. (2012) Major Taste Loss in Carnivorous Mammals. Proceedings of the National Academy of Sciences of the United States of America, 109, 4956-4961.
43. Bower, R.H., Cerra, F.B., Bershadsky, B., et al. (1995) Early Enteral Feeding of a Formula (Impact Registered Trademark) Supplemented with Arginine, Nucleotides, and Fish Oil in Intensive Care Unit Patients: Results of a Multicenter, Prospective, Randomized, Clinical Trial. Critical Care Medicine, 23, 436-449.
44. Goffschlich, M.M., Jenkins, M., Warden, G.D., Baumer, T., Havens, P., Snook, J.T. and Alexander, J.W. (1990) Differential Effects of Three Dietary Regimens on Selected Outcome Variables in Burn Patients. JPEN Journal of Parenteral and Enteral Nutrition, 14, 225-236.
45. Evoy, D., Lieberman, M.D., Fahey III, T.J. and Daly, J.M. (1998) Immunonutrition: The Role of Arginine. Nutrition, 14, 611-617.
46. León, P., Redmond, H.P., Stein, T.P., Shou, J., Schluter, M.D., Kelly, C., Lanza-Jacoby, S. and Daly, J.M. (1991) Arginine Supplementation Improves Histone and Acute-Phase Protein Synthesis during Gram-Negative Sepsis in the Rat. JPEN Journal of Parenteral and Enteral Nutrition, 15, 503-508.

47. Seifter, E., Rettura, G., Barbul, A. and Levenson, S.M. (1978) Arginine: An Essential Amino Acid for Injured Rats. Surgery, 84, 224-230.
48. Pui, Y.M. and Fisher, H. (1979) Factorial Supplementation with Arginine and Glycine on Nitrogen Retention and Body Weight Gain in the Traumatized Rat. Journal of Nutrition, 109, 240-246.
49. Duggan, C., Gannon, J. and Walker, W.A. (2002) Protective Nutrients and Functional Foods for the Gastrointestinal Tract. American Journal of Clinical Nutrition, 75, 789-808. [Citation Time(s):1]
50. Prins, H.A., Houdijk, A.P., van Lambalgen, A.A., Teerlink, T., Meijer, S., Thijs, L.G. and van Leeuwen, P.A.M. (1998) Paradoxical Changes in Organ Blood Flow after Arginase Infusion in the Non-Stressed Rat. Shock, 9, 422-427.
51. Campos, F.G., Waitzberg, D.L., Logulo, A.F., Mucerino, D.R. and Habr-Gama, A. (1996) The Role of Glutamine in Nutrition in Clinical Practice [in Portuguese]. Arquivos de Gastroenterologia, 33, 86-92.
52. O'Flaherty, L. and Bouchier-Hayes, D.J. (1999) Immunonutrition and Surgical Practice. Proceedings of the Nutrition Society, 58, 831-837.
53. Greenfield, J.R., Farooqi, I.S., Keogh, J.M., Henning, E., Habib, A.M., Blackwood, A., Reimann, F., Holst, J.J. and Gribble, F.M. (2009) Oral Glutamine Increases Circulating Glucagon-Like Peptide 1, Glucagon, and Insulin Concentrations in Lean, Obese, and Type 2 Diabetic Subjects. American Journal of Clinical Nutrition, 89, 106-113.
54. Houdijk, A.P., van Leeuwen, P.A., Teerlink, T., Flinkerbusch, E.L., Boermeester, M.A., Sauerwein, H.P. and Wesdorp, R.I.C. (1994) Glutamine-Enriched Enteral Diet Increases Renal Arginine Production. JPEN Journal of Parenteral and Enteral Nutrition, 18, 422-426.
55. Klimberg, V.S., Salloum, R.M., Kasper, M., et al. (1990) Oral Glutamine Accelerates Healing of the Small Intestine and Improves Outcome after Whole Abdominal Radiation. JAMA Surgery, 125, 1040-1045.
56. Kozar, R.A., Schultz, S.G., Bick, R.J., Poindexter, B.J., DeSoignie, R. and Moore, F.A. (2004) Enteral Glutamine but Not Alanine Maintains Small Bowel Barrier Function after Ischemia/Reperfusion Injury in Rats. Shock, 21, 433-437.
57. Souba, W.W., Roughneen, P.T., Goldwater, D.L., Williams, J.C. and Rowlands, B.J. (1987) Postoperative Alterations in Interorgan Glutamine Exchange in Enterectomized Dogs. Journal of Surgical Research, 42, 117-125.
58. Wischmeyer, P.E. (2006) Glutamine: Role in Gut Protection in Critical Illness. Current Opinion in Clinical Nutrition & Metabolic Care, 9, 607-612.
59. O'Dwyer, S.T., Smith, R.J., Hwang, T.L. and Wilmore, D.W. (1989) Maintenance of Small Bowel Mucosa with Glutamine-Enriched Parenteral Nutrition. JPEN Journal of Parenteral and Enteral Nutrition, 13, 579-585.
60. Ligthart-Melis, G.C., van de Poll, M.C.G., Dejong, C.H.C., Boelens, P.G., Deutz, N.E.P. and van Leeuwen, P.A.M. (2007) The Route of Administration (Enteral or Parenteral) Affects the Conversion of Isotopically Labeled L-[2-^{15}N] Glutamine into Citrulline and Arginine in Humans. JPEN Journal of Parenteral and Enteral Nutrition, 31, 343-350.
61. Yang, R., Martin-Hawver, L., Woodall, C., Thomas, A., Qureshi, N., Morrison, D. and Van Way, C. (2007) Administration of Glutamine after Hemorrhagic Shock Restores Cellular Energy, Reduces Cell Apoptosis and Damage, and Increases Survival. JPEN Journal of Parenteral and Enteral Nutrition, 31, 94-100.
62. Redmond, H.P., Stapleton, P.P., Neary, P. and Bouchier-Hayes, D. (1998) Immunonutrition: The Role of Taurine. Nutrition, 14, 599-604.
63. Lee, J.H., Jarreau, T., Prasad, A., Lavie, C., O'Keefe, J. and Ventura, H. (2011) Nutritional Assessment in Heart Failure Patients. Congestive Heart Failure, 17, 199-203.
64. Wójcik, O.P., Koenig, K.L., Zeleniuch-Jacquotte, A., Costa, M. and Chen, Y. (2010) The Potential Protective Effects of Taurine on Coronary Heart Disease. Atherosclerosis, 208, 19-25.

65. Sanderson, S.L. (2006) Taurine and Carnitine in Canine Cardiomyopathy. Veterinary Clinics of North America: Small Animal Practice, 36, 1325-1343.
66. Lenox, C.E. and Bauer, J.E. (2013) Potential Adverse Effects of Omega-3 Fatty Acids in Dogs and Cats. Journal of Veterinary Internal Medicine, 27, 217-226.
67. Johnson III, J.A., Griswold, J.A. and Muakkassa, F.F. (1993) Essential Fatty Acids Influence Survival in Sepsis. Journal of Trauma, 35, 128-131.
68. Kollman, K.A., Lein, E.L. and Vanderhoof, J.A. (1999) Dietary Lipids Influence Adaptation after Massive Bowel Resection. Journal of Pediatric Gastroenterology & Nutrition, 28, 41-45.
69. Freeman, L.M. (1998) Interventional Nutrition for Cardiac Disease. Clinical Techniques in Small Animal Practice, 13, 232-237.
70. Watson, T.D.G. (1998) Diet and Skin Disease in Dogs and Cats. Journal of Nutrition, 128, 2783S-2789S.
71. Corbee, R.J., Barnier, M.M.C., Tryfonidou, M.A., et al. (2013) The Effect of Dietary Long-Chain Omega-3 Fatty Acids Supplementation on Owner's Perception of Behavior and Locomotion in Cats with Naturally Occurring Osteoarthritis. Journal of Animal Physiology and Animal Nutritionr, 97, 846-853.
72. Roush, J.K., Dodd, C.E., Fritsch, D.A., et al. (2010) Multicenter Veterinary Practice Assessment of the Effects of Omega-3 Fatty Acids on Osteoarthritis in Dogs. Journal of the American Veterinary Medical Association, 236, 59-66.
73. Atasoy, B.M., Deniz, M., Dane, F., Özen, Z., Turan, P., Ercan, F., Çerikçioğlu, N., Aral, C., Akgün, Z., Abacioğlu, U. and Yeğen, B.Ç. (2010) Prophylactic Feeding with Immune-Enhanced Diet Ameliorates Chemoradiation-Induced Gastrointestinal Injury in Rats. International Journal of Radiation Biology, 86, 867-879.
74. Lasztity, N., Hamvas, J., Biro, L., Németh, É., Marosvölgyi, T., Decsi, T., Pap, Á. And Antal, M. (2005) Effect of Enterally Administered n-3 Polyunsaturated Fatty Acids in Acute Pancreatitis: A Prospective Randomized Clinical Trial. Clinical Nutrition, 24, 198-205
75. Smith, C.E., Freeman, L.M., Rush, J.E., Cunningham, S.M. and Biourge, V. (2007) Omega-3 Fatty Acids in Boxer Dogs with Arrhythmogenic Right Ventricular Cardiomyopathy. Journal of Veterinary Internal Medicine, 21, 265-273.
76. Freeman, L.M. (2010) Beneficial Effects of Omega-3 Fatty Acids in Cardiovascular Disease. Journal of Small Animal Practice, 51, 462-470.
77. Brown, S.A., Brown, C.A., Crowell, W.A., Barsanti, J.A., Allen, T., Cowell, C. and Finco, D.R. (1998) Beneficial Effects of Chronic Administration of Dietary Omega-3 Polyunsaturated Fatty Acids in Dogs with Renal Insufficiency. Journal of Laboratory and Clinical Medicine, 131, 447-455.
78. Trepanier, L. (2009) Idiopathic Inflammatory Bowel Disease in Cats. Rational Treatment Selection. Journal of Feline Medicine & Surgery, 11, 32-38.
79. Bauer, J.E. (2011) Therapeutic Use of Fish Oils in Companion Animals. Journal of the American Veterinary Medical Association, 239, 1441-1451.
80. Gibson, G.R., Beatty, E.R., Wang, X. and Cummings, J.H. (1995) Selective Stimulation of Bifidobacteria in the Human Colon by Oligofructose and Inulin. Gastroenterology, 108, 975-982.
81. Jackson, K.G., Taylor, G.R., Clohessy, A.M. and Williams, C.M. (1999) The Effect of the Daily Intake of Inulin on Fasting Lipid, Insulin and Glucose Concentrations in Middle-Aged Men and Women. British Journal of Nutrition, 82, 23-30.
82. St-Onge, M.P. and Jones, P.J. (2002) Physiological Effects of Medium-Chain Triglycerides: Potential Agents in the Prevention of Obesity. Journal of Nutrition, 132, 329-332. [Citation Time(s):1]
83. Kleessen, B., Sykura, B., Zunft, H.J. and Blaut, M. (1997) Effects of Inulin and Lactose on Fecal Microflora, Microbial Activity and Bowel Habit in Elderly Constipated Persons. American Journal of Clinical Nutrition, 65, 1397-1402.
84. Roberfroid, M.B. and Delzeene, N.M. (1998) Dietary Fructans. Annual Review of Nutrition, 18, 117-143.

85. Roberfroid, M. (1993) Dietary Fiber, Inulin, and Oligofructose: A Review Comparing Their Physiological Effects. Critical Reviews in Food Science and Nutrition, 33, 103-148.

86. Parts of this manuscript has been published in Dutch Corbee, R.J. and van Kerkhoven, W. (2012) Nutritionele ondersteuning van de opgenomen patiënt. Tijdschrift voor Diergeneeskunde, 137, 384-390.

87. [A] Prescription diet a/d, Hill's Pet Nutrition Inc, Topeka, Kan.

88. [B] Royal Canin Veterinary Diet Recovery RS, Royal Canin USA Inc., St Charles, Mo.

89. [C] Eukanuba Max Cal, Iams, P&G, Henderson, NC.

90. [D] Viyo Recuperation, Viyo International N.V., Antwerp, Belgium.

CHAPTER 14

Bleeding Diathesis in Fawn Hooded Rats—Possible Implications for Invasive Procedures and Refinement Strategies

Manon W. H. Schaap [1], Hugo van Oostrom [2], Saskia S. Arndt [1,*] and Ludo J. Hellebrekers [1]

[1]*Division of Animal Welfare and Laboratory Animal Science, Department of Animals in Science and Society, Faculty of Veterinary Medicine, Utrecht University, Yalelaan 2, 3584 CM Utrecht, The Netherlands*

[2]*School of Veterinary Sciences, Bristol University, Langford, Bristol BS40 5DU, UK*

SIMPLE SUMMARY

The Fawn Hooded (FH) rat population is commonly used in biomedical research. It is widely acknowledged that the FH rat has a bleeding disorder, which may lead to abundant bleeding. However, the clinical consequences of this bleeding disorder have not been described in current literature. During our study, surgical procedures on FH rats led to an unanticipated loss of animals due to abundant bleeding. Adjustments made to minimize the impact of this disorder in animals undergoing invasive procedures are described. It is strongly recommended to take the bleeding diathesis into account when performing invasive procedures in FH rats and to apply the described refinements.

ABSTRACT

The Fawn hooded (FH) rat is commonly used in biomedical research. It is widely acknowledged that the FH rat has a bleeding disorder; leading to abundant bleedings. Although this bleeding disorder is investigated to model the storage pool defect; its impact on commonly performed invasive laboratory procedures has not yet been described. Our research group experienced clinically significant consequences of this bleeding disorder following invasive procedures (including intraperitoneal injections and neurocranial surgery) in the Rjlbm: FH stock. The clinical consequences of the surgical and anesthetic protocols applied; are described including the subsequent procedural refinements applied to minimize the impact of this disorder. It is strongly recommended to take the bleeding diathesis into account when performing invasive procedures in FH rats and to apply the suggested refinement of procedures.

Keywords: fawn hooded rat; invasive procedures; bleeding disorder; anesthesia; surgery; refinement

1. INTRODUCTION

The Fawn hooded (FH) rat is widely used in biomedical research, e.g., as a model for various psychiatric disorders [1,2,3,4], (spontaneous) hypertension-associated renal failure [5,6], and blood factor studies (*i.e.*, atherosclerosis and coagulation [7,8]). Consequently, these FH rats are commonly subjected to invasive procedures, varying from injections to surgical interventions.

Several inbred substrains and outbred stocks of the FH rat are used, which differ genetically, physiologically and behaviorally [4,9]. It has been described that the platelets of FH rats have a defective secretion of coagulation mediators, which may lead to abundant bleeding in several inbred substrains [8,10,11]. Although this bleeding disorder of several FH populations has been investigated in the context of translational research (*i.e.*, to model the storage pool defect [8,12]), its (potential) impact on commonly performed invasive procedures has not been described in the literature. Recently, our research group was confronted with the clinical consequences of the bleeding disorder of one specific FH outbred stock, *i.e.*, the Rjlbm: FH, following invasive procedures. In accordance with the principle of the 3Rs (refinement, reduction, replacement) of Russell and Burch [13], we feel that it is important to publish this experience and to describe our refinement strategy.

2. MATERIALS AND METHODS

In a recent study [14], we performed neurocranial surgery on adult male FH rats (Elevage Janvier, Le Genet St. Isle, France; denomination: Rjlbm: FH) for permanent implantation of epidural electrodes. All procedures were approved by the Animal Experiments Committee of the Academic Biomedical Centre, Utrecht, The Netherlands (DEC-DGK number 2011.I.08.079). The Animal Experiments Committee based its decision on "De Wet op de Dierproeven" (the Dutch "Experiments on Animals Act", 1996) and on the "Dierproevenbesluit" (the Dutch "animal experiments decree", 1996). Both documents are available online at http://wetten.overheid.nl. Further, all animal experiments followed the "Principles of Laboratory Animal Care" and refer to the Guidelines for the Care and Use of Mammals in Neuroscience and Behavioural Research (National Research Council 2003).

Rats were housed individually in clear 1500 U Eurostandard Type IV S cages, measuring 48 × 37.5 × 21 cm and were provided with bedding material (Aspen chips), *ad lib* access to food (CRM, Expanded, Special Diets Services Witham, United Kingdom) and water and carton houses (Rat Corner House, Bio Services B.V., Uden, The Netherlands). The environment was temperature—(21 ± 2 °C) and humidity (47% ± 3%) controlled with an inversed 12:12 h light-dark cycle. After an acclimatization period of three weeks [14], the animals underwent surgery for permanent implantation of epidural electrodes. The anesthetic protocol used was designed to combine adequate surgical analgesia with the possibility to antagonize anesthesia and thus promote fast recovery. Anesthesia was induced with an intraperitoneal (i.p.) injection of 0.25 mg/kg fentanyl (Fentanyl Janssen, Janssen-Cilag B.V., Tilburg, The Netherlands, 0.05 mg/mL fentanyl citrate) and 0.15 mg/kg dexmedetomidine (Dexdomitor®, Pfizer Animal Health B.V., Capelle a/d IJssel, The Netherlands, 0.5 mg/mL dexmedetomidine hydrochloride). After loss of the pedal reflex, the animal was endotracheally intubated, and anesthesia was maintained with isoflurane in 100% O_2. The animals were provided with 8 mL of saline subcutaneously (s.c.) to support normal fluid balance and eye ointment to prevent drying of the cornea (Ophtosan eye ointment, AST farma B.V., Oudewater, The Netherlands, 10,000 IE vitamin A palmitate per gram). Subsequently, the animal was positioned in the stereotactic apparatus (model 963, Ultra Precise Small Animal Stereotaxic, David Kopf Instruments, Tujunga, CA, USA). Body temperature was monitored using a rectal probe thermometer and maintained at 37–38 °C with an adjustable electrically heated mattress. In addition to clinical assessment (*i.e.*, pedal reflexes), respiratory rate, heart rate, in- and expired CO_2 and SpO_2 were monitored continuously for assessment of anesthetic depth and anesthetic drug administration was adjusted appropriately. Following the skin incision, but prior to detachment of the periostium from the neurocranium, 3 mg/kg lidocaine solution (Alfacaine 2% plus adrenaline, Alfasan B.V., Woerden, The Netherlands, 20 mg/mL lidocaine hydrochloride and 0.01 mg/mL adrenaline) was applied to the periostium. Five small wired stainless steel screws (tip diameter 0.6 mm) were implanted epidurally over the vertex (4.5 mm caudal to bregma, 1 mm right from midline), primary somatosensory cortex (2.5 mm caudal to bregma, 2.5 mm right from midline), anterior cingulate cortex (1.5 mm rostral to bregma, 0.5 mm lateral from midline), and left and right frontal sinus (10 mm rostral to bregma, 1 mm lateral from midline). All electrodes were wired to a receptacle and fixed to the skull with antibiotic bone cement (Simplex™ P bone cement with tobramycin, Stryker Nederland B.V., Waardenburg, The Netherlands, 0.5 g tobramycin per 20 g cement powder). The skin was closed in a single layer around the receptacle. Subsequently, anesthesia was antagonized with 0.05 mg/kg buprenorphine (i.p., Buprecare, AST Farma B.V., Oudewater, The Netherlands, 0.3 mg/mL buprenorphine) and 0.6 mg/kg atipamezole (i.p., Antisedan, Pfizer Animal Health B.V., Capelle a/d IJssel, The Netherlands, 5 mg/mL atipamezole hydrochloride). Anesthesia was antagonized in the animal's home cage outside the surgery room in a separate room under red light conditions with extra oxygen supplied by a face-mask. After return of purposeful locomotion, the rat was returned to the animal housing room. Postoperative analgesia was provided with 0.05 mg/kg buprenorphine at 12 h intervals for 3 days after surgery and 0.2 mg/kg of meloxicam (s.c., Metacam, Boehringer Ingelheim B.V., Alkmaar, The Netherlands, 5 mg/mL) at 24 h intervals for 2 days after surgery.

3. OBSERVATIONS

This surgical and anesthetic protocol used was proven successful in earlier studies conducted in our lab, using in total over 200 adult male rats from different rat populations, including Wistar WU (Harlan Netherlands B.V. and Charles River, Germany) [15,16], Wistar Kyoto (WKY/KyoRj, Elevage Janvier, Le Genet St. Isle, France) [14] and Lister hooded (Charles River, Germany; manuscript in preparation). However, when applying this surgical and anesthetic protocol in the currently described 11 Rjlbm: FH rats, the bleeding disorder described in the literature for other FH substrains (from which the Rjlbm: FH is descending), proved to be clinically problematic in the Rjlbm stock. One animal died during the surgery due to a prolonged bleeding time after drilling holes in the neurocranium, and three animals were euthanized within two days after the surgery, because of multiple bleedings and severe weight loss.

During the surgery several remarkable observations were made. First off, several rats showed a blue discoloration of the scrotum, most likely due to a hemoscrotum, which developed in minutes after the i.p. injection to induce anesthesia. Notably, to our knowledge, hemoscrotums following an i.p. injection were never described in other rat strains or stocks. Although not quantified, compared to other rat populations used in our lab (e.g., Wistar WU, WKY/KyoRj and Lister hooded) the FH rats displayed prolonged bleeding times after incisions (skin and periosteum) and drilling of the holes in the neurocranium. Finally, qualitatively greater and prolonged post-surgical bleeding was observed in FH rats, compared to other rat populations used in our lab.

4. REFINEMENT OF THE SURGICAL AND ANESTHETIC PROTOCOL

Based upon those observations, the surgical and anesthetic protocols for the Rjlbm: FH rats were refined. Firstly, i.p. injections for anesthesia induction were replaced by s.c. injections in the neck area. Secondly, dexmedetomidine was excluded from the protocol based upon its potentially hypertensive side effect [17]. Instead, anesthesia was induced with 3 mg/kg midazolam (Midazolam, Actavis B.V., Baarn, The Netherlands, 5 mg/mL midazolam), 0.25 mg/kg fentanyl (Fentanyl Janssen, Janssen-Cilag B.V., Tilburg, The Netherlands, 0.05 mg/mL fentanyl citrate) and 10 mg/kg ketamine (Narketan, Vétoquinol S.A., Lure Cedex, France, 115.34 mg/mL ketamine hydrochloride). Thirdly, to minimize bleeding during the surgery, after detachment of the periostium, bleeding from the neurocranium was coagulated with 30% H_2O_2. Finally, to support the postoperative nutritional restoration, until three days after surgery, the rats received nutritional support using Convalescence Support (Royal Canin B.V., The Netherlands), containing a high level of energy and protein concentration. Otherwise, the surgical and other aspects of anesthetic protocols were unchanged.

The refined surgical and anesthetic protocols were used in six FH rats. No further bleeding diatheses occurred. One rat immediately died after induction of anesthesia (presumed anesthesia-related problems), before the start of the surgical procedure. The remaining five rats recovered without any complications.

Further refinement of the protocol could be considered during future studies. To completely avoid injections, a protocol using volatile induction and maintenance and oral administration of analgesics (e.g., [18]) administered pre-operatively could be an option. Post-operative analgesia can also be provided by the exclusive per os route, further minimizing the need of injections.

5. CONCLUSIONS AND RECOMMENDATION

In conclusion, the bleeding disorder of the adult male Rjlbm: FH rat, as experienced in our laboratory, may have a substantial impact when performing invasive procedures. Although invasive procedures in this specific stock can be performed, the bleeding diathesis is potentially clinically problematic and life threatening and demands special attention. This is the first report on the clinical implications of the FH's bleeding disorder, using the Rjlbm: FH stock. To what degree other FH substrains and stocks are clinically affected by the bleeding disorder during invasive procedures in the laboratory is yet unknown. However, it is strongly recommended to take the bleeding diathesis into account when performing invasive procedures in any FH population. Applying refinement strategies, as described here, will contribute to the reduction of the number of animals and their discomfort, in accordance with the principle of the 3Rs [13] regarding laboratory animals.

ACKNOWLEDGEMENTS

We would like to thank Annemarie Baars, José Lozeman- van 't Klooster, and Peter Hesseling for their technical support during the procedures.

AUTHOR CONTRIBUTIONS

Manon W.H. Schaap: Primary investigator, conducting the practical work, and writing the manuscript. Hugo van Oostom: Refined the anesthetic and surgical protocol, and reviewed the manuscript. Saskia S. Arndt: Supervised the whole process, reviewed the manuscript. Ludo J.

Hellebrekers: Refined the anesthetic and surgical protocol, reviewed the manuscript, and supervised the whole process.

CONFLICTS OF INTEREST

The authors declare no conflict of interest.

REFERENCES

1. Hensler, J.G.; Hodge, C.W.; Overstreet, D.H. Reduced 5-HT3 receptor binding and lower baseline plus maze anxiety in the alcohol-preferring inbred fawn-hooded rat. Pharmacol. Biochem. Behav. 2004, 77, 281–289.
2. Rezvani, A.H.; Parsian, A.; Overstreet, D.H. The fawn-hooded (FH/wjd) rat: A genetic animal model of comorbid depression and alcoholism. Psychiatr. Genet. 2002, 12, 1–16.
3. Martin, S.; Lawrence, A.J.; van den Buuse, M. Prepulse inhibition in fawn-hooded rats: Increased sensitivity to 5-HT1A receptor stimulation. Eur. Neuropsy chopharmacol. 2004, 14, 373–379.
4. Overstreet, D.H.; Rezvani, A.H. Behavioral differences between two inbred strains of fawn-hooded rat: A model of serotonin dysfunction. Psychopharmacology 1996, 128, 328–330.
5. Van Dokkum, R.P.E.; Alonso-Galicia, M.; Provoost, A.P.; Jacob, H.J.; Roman, R.J. Impaired autoregulation of renal blood flow in the fawn-hooded rat. Am. J. Physiol. 1999, 276, 189–196.
6. Burke, M.; Pabbidi, M.; Fan, F.; Ge, Y.; Liu, R.; Williams, J.M.; Sarkis, A.; Lazar, J.; Jacob, H.; Roman, R.J.; et al. Genetic basis of the impaired renal myogenic response in FHH rats. Am. J. Physiol. Ren. Physiol. 2013, 304, 565–577.
7. Raymond, S.L.; Dodds, W.J. Characterization of the fawn-hooded rat as a model for hemostatic studies. Thromb. Diath. Haemorrh. 1975, 33, 361–369.
8. Datta, Y.H. Genetic mapping and characterization of the bleeding disorder in the fawn-hooded hypertensive rat. Thromb. Haemost. 2003, 89, 1031–1042.
9. Provoost, A.P.; Shiozawa, M.; van Dokkum, R.P.E.; Jacob, H.J. Phenotype and genotype variability in four strains of fawn-hooded (FH) rats. Clin. Exp. Pharmacol. Physiol. 1998, 25. Abstract 33.
10. Kirchmaier, C.M.; Meyer, M.; Spangenberg, P.; Heller, R.; Haroske, D.; Breddin, H.K.; Till, U. Platelet membrane defects in fawn hooded bleeder rats. Thromb. Res. 1990, 57, 353–360.
11. Tschopp, T.B.; Zucker, M.B. Hereditary defect in platelet function in rats. Blood 1972, 40, 217–226.
12. Witters, P.; Hoylaerts, M.; Freson, K.; de Vos, R.; van Pelt, J.; Nevens, F.; van Geet, C.; Cassiman, D. ADP-degrading enzymes inhibit platelet activation in bile duct-ligated rats. J. Thromb. Haemost. 2010, 8, 360–368.
13. Russell, W.M.S.; Burch, R.L. The Principles of Humane Experimental Technique; Methuen: London, UK, 1959.
14. Schaap, M.W.H.; van Oostrom, H.; Doornenbal, A.; van't Klooster, J.; Baars, J.M.; Arndt, S.S.; Hellebrekers, L.J. Nociception and conditioned fear in rats: Strains matter. PLOS ONE 2013, 8.
15. Koot, S.; Koukou, M.; Baars, J.M.; Hesseling, P.; van't Klooster, J.; Joëls, M.; van den Bos, R. Corticosterone and decision-making in male wistar rats: The effect of

corticosterone application in the infralimbic and orbitofrontal cortex. Front. Behav. Neurosci. 2014, 8, 127.

16. Schaap, M.W.H.; van Oostrom, H.; Doornenbal, A.; Baars, J.M.; Arndt, S.S.; Hellebrekers, L.J. Predictability of painful stimulation modulates the somatosensory-evoked potential in the rat. PLOS ONE 2013, 8.
17. Gerlach, A.T.; Dasta, J.F. Dexmedetomidine: An updated review. Ann. Pharmacother. 2007, 41, 245–252.
18. Flecknell, P.A.; Roughan, J.V.; Stewart, R. Use of oral buprenorphine ("buprenorphine jello") for postoperative analgesia in rats—A clinical trial. Lab. Anim. 1999, 33, 169–174.

CHAPTER 15

A Large Animal Survival Model to Evaluate Bariatric Surgery Mechanisms

Vlad V. Simianu[1*], Jonathan G. Sham[1], Andrew S. Wright[1], Skye D. Stewart[1], Mouhamad Alloosh[2], Michael Sturek[2], David E. Cummings[3], David R. Flum[1,4]

[1]*Departments of Surgery, University of Washington, Seattle, USA*

[2]*Department of Cellular and Integrative Physiology, Indiana University School of Medicine, Indianapolis, USA*

[3]*Departments of Medicine, University of Washington, Seattle, USA*

[4]*Departments of Health Services, University of Washington, Seattle, USA*

ABSTRACT

Background: The impact of Roux-en-Y gastric bypass (RYGB) on type 2 diabetes mellitus is thought to result from upper and/or lower gut hormone alterations. Evidence supporting these mechanisms is incomplete, in part because of limitations in relevant bariatric-surgery animal models, specifically the lack of naturally insulin-resistant large animals. With overfeeding, Ossabaw swine develop a robust metabolic syndrome, and may be suitable for studying post-surgical physiology. Whe- ther bariatric surgery is feasible in these animals with acceptable survival is unknown. Methods: Thirty-two Ossabaws were fed a high-fat, high-cholesterol diet to induce obesity and insulin resistance. These animals were assigned to RYGB (n = 8), RYGB with vagotomy (RYGB-V, n = 5), gastrojejunostomy (GJ, n = 10), GJ with duodenal exclusion (GJD, n = 7), or sham operation (n = 2) and were euthanized 60 days post-operatively. Post-operative changes in weight and food intake are re- ported. Results: Survival to scheduled necropsy among surgical groups was 77%, living an average of 57 days post-operatively. Cardiac arrest under anesthesia occurred in 4 pigs. Greatest weight loss (18.0% ± 6%) and food intake decrease (57.0% ± 20%) occurred following RYGB while animals undergoing RYGB-V showed only 6.6% ± 3% weight loss despite 50.8% ± 25% food intake decrease. GJ (12.7% ± 4%) and GJD (1.2% ± 1%) pigs gained weight, but less than sham controls (13.4% ± 10%). Conclusions: A survival model of metabolic surgical procedures is feasible, leads to significant weight loss, and provides the opportunity to evaluate new interventions and subtle variations in surgical technique (e.g. vagus nerve sparing) that may provide new mechanistic insights.

Keywords: Ossabaw, Bariatric Surgery, Roux-en-Y Gastric Bypass, Weight Loss, Metabolic Syndrome

1. INTRODUCTION

Roux-en-Y gastric bypass surgery (RYGB) promotes substantial, sustained weight loss [1] [2] , and it is the most effective method to ameliorate obesity-related comorbidity including type 2 diabetes mellitus (T2DM) [3] - [5] . Because major improvements in T2DM typically occur prior and out of proportion to significant body weight loss [6] , the impact of RYGB on T2DM may not result from weight loss and reduced caloric intake alone. Important unanswered questions regarding the anti-diabetes effects of RYGB remain, including contributions of the proximal vs. distal intestines in T2DM remission, the reversibility of improved glycemic control with foregut exposure to food, and roles of the vagus nerve in these effects.

Over the past several years, our group has developed a porcine survival model for metabolic surgery [7] and more recently extended that work to address one of the major shortcomings of large animals as models for metabolic surgery. Over many generations, animal husbandry practices have limited the gene pool of large animals such that there are no natural models of "unhealthy" obesity. Farm animals rendered obese typically do not develop insulin resistance or heart disease, and consequently, large animal models of metabolic disease have been limited [8] . Classically, larger animals are given toxins (e.g. streptozotocin) to impair pancreatic function [9] , while small animal models depend on knockout or gene silencing techniques to mimic human insulin resistance.

Although a naturally occurring large animal model of diabetes does not exist, in the 1970s, on Ossabaw Island near Georgia, a colony of pigs was discovered that exhibited many features useful for the study of bariatric surgery. Abundant natural resources exist on the island only seasonally, followed by periods of food scarcity. Through generations of natural selection, surviving Ossabaw pigs gain large amounts of weight during times of plenty, allowing them to survive seasonal famines. Exhibiting "thrifty genome" characteristics when provided unrestricted access to high-calorie diets [10] -[12] , they develop obesity, insulin resistance and glucose intolerance [13] , dyslipidemia [14] [15] , and hypertension [14] [16] , which are widely accepted characteristics of metabolic syndrome (MetS) [16] -[18] .

The objective of this study was to determine the feasibility and appropriateness of bariatric surgical techniques in the Ossabaw survival model. The advantages of a large-animal model are its similarities with humans regarding techniques and anatomic features. More so than rodents, the upper GI tract is very similar in humans and pigs [19] , and allows application of nearly identical surgical techniques and instruments. In obese Ossabaw pigs, we describe our initial experience with long-term survival surgery including RYGB without vagotomy, RYGB with vagotomy (RYGB-V), gastrojejunostomy with duodenal exclusion (GJD, which creates a gastric- sparing bypass of the segment of proximal intestine excluded in RYGB), and gastrojejunostomy (GJ, which is identical to the latter operation with duodenal inclusion). These variations were selected to better elucidate what had been dubbed the upper and lower intestinal hypotheses for endocrine/metabolic pathways of diabetes development and resolution.

2. MATERIALS AND METHODS

All experimental procedures involving animals were approved by the Institutional Animal Care and Use Committee at the University of Washington (UW) with the recommendations outlined by the National Research Council and the American Veterinary Medical Association Panel on Euthanasia [20] [21] .

2.1. Animals and Environment

Since January 2010, 32 female Ossabaw swine have entered the study protocol. Pigs were obtained from the joint Indiana University School of Medicine (IUSM) and Purdue University Facility. This study excludes 1 pig which died at IUSM before shipment. To promote weight gain and insulin resistance, animals were maintained on excess calorie high-fat, high-cholesterol diet (vide infra) for ≥180 days before arriving at UW, at age 12 - 18 months. Pigs were acclimated to the UW research facility for at least 7 days before undergoing initial vascular access catheter placement. Animals lived in a 70°F temperature-controlled room on a 12:12-h light:dark cycle. They had free access to drinking water and were removed at least once daily for stall cleaning. Experimental pigs were fed twice daily at fixed times a customized obesogenic TestDiet® containing high levels of fructose, lipids, and cholesterol, with 16.1% proteins, 43.1% lipids, and 40.8% carbohydrates (3580 kcal/meal or 7160 kcal/day).

Twenty Ossabaw were fed standard, non-fat chow (~3000 kcal/day) and served as non-surgical, lean controls. Food intake was recorded daily. Change in food intake was derived from food volume left uneaten, measured at daily feeding times, and calculated as difference from preoperative baseline.

2.2. Intravascular Catheters

All animals were surgically implanted with vascular-access catheters to permit collection of serial blood samples and for medication administration (e.g., analgesics, antibiotics if necessary) postoperatively. Porcine animal models pose a special challenge for long-term vascular access given their size and lack of physical restraint [22] -[24] . Catheters were initially placed in the external jugular vein percutaneously, but early problems with this technique led us to switch to tunneled placement in the internal jugular vein under anesthesia. All tunneled placements used 12 Fr. Dual lumen Hickman Catheter (Bard Peripheral Vascular, Tempe, AZ) using a technique that has previously been described [25] [26] . Catheter replacements were required whenever catheters malfunctioned. Early in the study period, catheters were removed at 2 weeks post-operatively and replaced close to 60 days post-operatively, prior to necropsy. Following several anesthesia-related deaths, however, pigs in the later part of the study maintained catheters throughout the entire experimental period. Catheters were flushed at least twice daily with a heparin-saline solution then locked with a solution containing vancomycin (1.0 mg/ml).

2.3. Surgical Intervention

Pigs were randomly assigned to one of four GI operations (RYGB, RYGB-V, GJ, GJD) or a sham operation. All operations were standardized, and our pre-operative and anesthesia protocols have been previously reported [7] . For RYGB, a gastric pouch approximately 3 × 3 cm was created. The small bowel reconstruction approximated the human RYGB with approximately ~45 cm of biliary-pancreatic-duodenal (BPD) limb. However, given recognized variation in the intestinal length of Ossabawswine [19] , the entire small bowel length was measured in each animal and one third of the small bowel used for an antecolic, antegastric alimentary limb. The RYGB-V is identical, but includes division of anterior and posterior vagus nerves as they cross the gastro-esophageal junction. Resection of nervous tissue was confirmed by histologic assessment. The GJ operation creates a connection between the mid-stomach and jejunum (as with RYGB procedures approximately one third of the distance on the small bowel), with full preservation of the stomach and pylorus. The GJD operation is similar to GJ, with additional division and detachment of the pylorus from the proximal duodenum, bypassing the same length of small bowel as in RYGB and RYGB-V. Sham surgery involves a full midline incision and bowel manipulation for ~130 minutes, the time of an average RYGB.

Early in the study, gastro-enteric anastomoses were performed using 4.5-mm GIA (United States Surgical, Norwalk, CT) stapler around the anvil of a 25-mm circular stapler. However, the thickened esophageal and gastric walls resulted in early anastomotic dehiscence at the esophageal-gastric junction, resulting in mediastinitis and/or peritonitis, and ultimately in untimely death in 5 animals. After these complications, we revised our protocol to hand-sewing for the gastric-jejunostomy anastomoses, and we created better practices for monitoring post-surgical outcomes to identify dehiscence within the first few post-operative days, using endoscopic exploration when animals began to deteriorate clinically. We continued to perform the side-to-side jejuno-jejunosto- my using a GIA stapler and 3-0 Maxon (United States Surgical, Norwalk, CT) reinforcement suture as we have previously described [7] .

2.4. Statistical Analyses

Data analyses were descriptive rather than comparative, given the small number of animals involved. Pre-opera- tive weights are reported as means with standard deviations (SD). Weight change is reported as a percentage change (±SD) using difference between pre-operative and pre-necropsy weights. Food intake change is reported as a percentage change (±SD) using average daily intake before and serially after surgery. Excel (version 12.3.6, Microsoft) was used for statistical analysis.

3. RESULTS

Thirty-two Ossabaw swine have been studied to date, and survival and detailed perioperative complications are reported in Table 1. Six out of 8 (75%) pigs assigned to RYGB survived to scheduled necropsy. One that died post-operatively was euthanized on POD 35 after evisceration; necropsy revealed intra-abdominal adhesions but an intact anastomosis. The other died on POD 63, with necropsy showing severe necro-hemorrhagic enteritis

Table 1. Obesogenic-diet Ossabaw survival and complications.

	All	Sham	GJ	GJD	RYGB	RYGB-V
N*	32	2	10 (9 to OR)	7	8	5 (4 to OR)
N completing protocol/ N having GI surgery	23/30	1/2	8/9	6/7	6/8	1/2
(%)	77%	50%	89%	86%	75%	50%
Mean Postoperative survival, days	56.5	68.5	59.8	57.8	61.2	34
Death	9					
Intraoperative or immediate postoperative	5	1	2		0	2
During catheter placement	3	1 cardiac arrest during catheter replacement, POD 75	1 cardiac arrest during initial catheter placement 1 respiratory arrest during catheter replacement, POD 57			
During GI surgery	1					1 cardiac arrest during initial surgery
During re-exploration	1					1 explored for leak POD7, died of intra-operative hemorrhage
Death prior to completion of study period	4			1 died POD 5, pancreatitis	1 euthanized for evisceration on POD 35 1 died POD 63, presumed C. diff	1 euthanized for prolonged difficult recovery and decreased PO intake on POD11

*Pigs expiring of cardiac arrest during initial catheter placement (n = 2) not counted beyond first row as they did not undergo GI surgery.

and typhlocolitis, presumably from Clostridium difficile infection. Five pigs were assigned to RYGB-V. One had cardiac arrest during surgery and was not counted towards the overall survival. Two died before scheduled necropsy. One was re-explored on POD7 for anastomotic leak, and died during surgery. The other underwent revision of a laparotomy closure on POD2 and was subsequently euthanized on POD 11 from RYGB-V due to failure to thrive and PO intolerance. Necropsy revealed an intact but edematous gastro-jejunostomy.

Ten pigs were assigned to GJ surgery, with 8 of the 9 (89%) who underwent the operation surviving to scheduled necropsy. One pig died in the recovery phase of initial catheter placement and never underwent GJ, and was not counted towards the overall survival of the group. Another went into cardiac arrest during tunneled catheter replacement on POD 57; necropsy revealed pulmonary changes consistent with cardiac arrest. Seven pigs underwent GJD, and 6 (86%) survived to scheduled necropsy. One died on POD5; necropsy revealed pancreatitis. Two pigs underwent sham surgery, with only one surviving to scheduled necropsy. The other died on POD 75 from cardiac arrest during attempted tunneled catheter replacement.

The mean preoperative weight of high-fat-fed Ossabaws was 73.4 ± 8.6 kg compared to 63.2 ± 12.9 kg in chow-fed Ossabaws. Preoperative weights for each surgical procedure were: Sham 78.1 ± 15.2 kg; GJ 72.3 ± 8.0 kg; GJD 79.8 ± 8.1 kg; RYGB 68.5 ± 7.0 kg; and RYGB-V 70.7 ± 6.7 kg. Postoperative weight and food intake change, stratified by operation, are shown in Figure 1 and Figure 2, respectively. Ossabaws undergoing sham

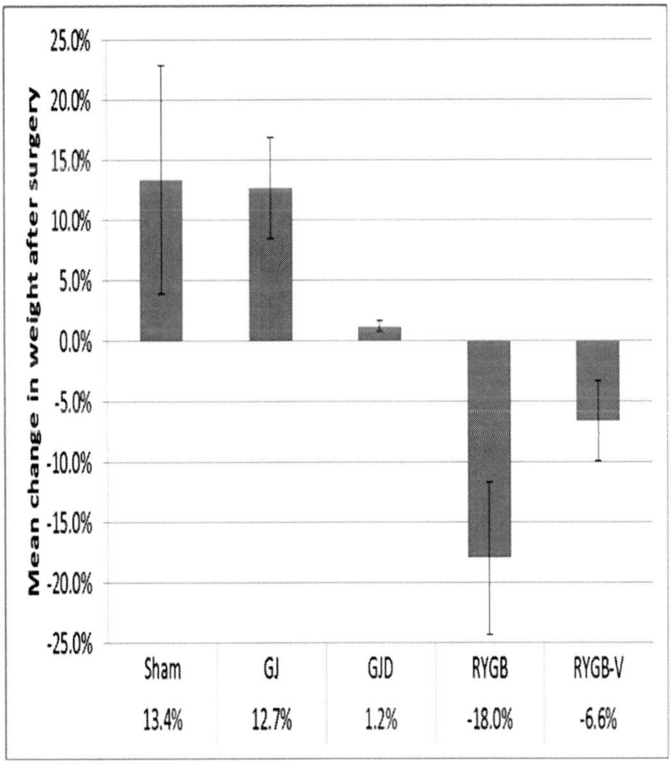

Figure 1. Weight change following surgery for Obesogenic-diet Ossabaws[a,b]. [a]Mean change in weight reported as percentage (±SD) change from preoperative weigh (Sham: 78.1 ± 15.2 kg; GJ: 72.3 ± 8.0 kg; GJD: 79.8 ±8.1 kg; RYGB: 68.5 ± 7.0 kg; RYGB-V: 70.7 ± 6.7 kg); [b]Only pigs surviving beyond immediate postoperative period (POD 35+) included in estimates ($n_{Sham} = 2$; $n_{GJ} = 9$; $n_{GJD} = 6$; $n_{RYGB} = 8$; $n_{RYGB\text{-}V} = 2$).

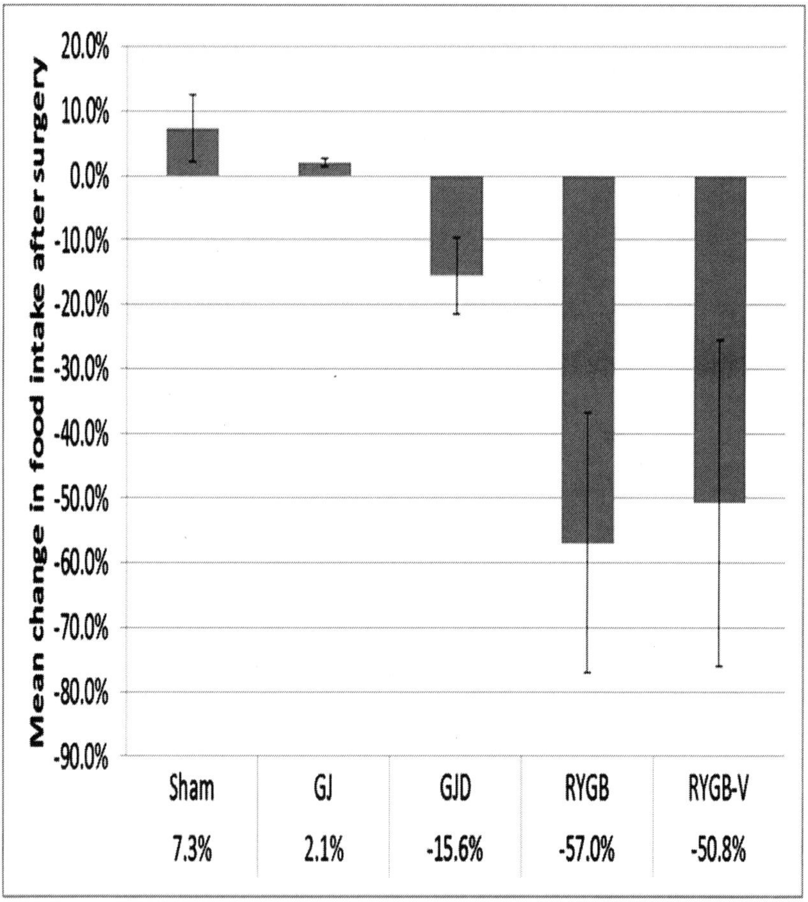

Figure 2. Change in food intake following surgery for Obese Ossabaws[a]. [a]Only pigs surviving beyond immediate postoperative period (POD 35+) included in estimates ($n_{Sham} = 2$; $n_{GJ} = 9$; $n_{GJD} = 6$; $n_{RYGB} = 8$; $n_{RYGB\text{-}V} = 2$).

operation gained 13.4% ± 9.5% weight and increased their food intake by 7.3% ± 5.2% over the experimental period. Over an equivalent period, GJ pigs gained 12.7% ± 4.2% weight and increased food intake by 2.1% ± 0.7%. Pigs in the GJD group gained 1.2% ± 0.5% weight but decreased their food intake by 15.6% ± 5.9%. Pigs in the RYGB and RYGB-V groups decreased their weight by 18.0% ± 6.4% and 6.6% ± 3.3% respectively, and decreased their food intake by 57.0% ± 20.2% and 50.8% ± 25.4%.

4. DISCUSSION

Surgical approaches to obesity consistently promote major, sustained weight loss, and improvement or remission of many, obesity-related comorbidities. This reduction in comorbidities, most notably T2DM, cannot be explained solely by the effects of weight loss and reduced caloric intake, and there are several hypotheses regarding how GI rearrangement ameliorates diabetes [6] [27] . The "upper intestinal hypothesis" postulates that exclusion of the proximal small bowel from ingested nutrients exerts direct anti-diabetes effects, potentially through incompletely identified, nutrient-regulated factors or processes that influence glucose homeostasis [28] [29] . The "lower intestinal hypothesis" postulates that enhanced delivery of nutrients to the distal bowel augments glucagon-like peptide-1 secretion, increasing insulin secretion [30] [31] . In addition, compromised secretion of the pro-diabetic hormone ghrelin after RYGB might help improve glucose tolerance [32] [33] . Unfortunately, surgical experimental GI manipulations to test these hypotheses are often not practical in humans [34] .

Rodents are the dominant model for metabolic surgical evaluations [19] , but concerns over anatomic differences and variation in physiologic responses compared with humans limit their utility in studying feasibility of surgical techniques [35] . Insulin resistance in obese, large animals amenable to routine biomedical study has heretofore been lacking. Although swine appear to be a more appropriate model to study bariatric surgical changes because of their anatomic and functional GI similarity to humans [36] [37] , as well as their equivalent hormonal responses to starvation and feeding [38] -[40] , the lack of a swine model with insulin resistance has been problematic [8] .

The Ossabaw miniature swine appears to have a form of naturally occurring, polygenic insulin resistance, along with a unique propensity to obesity [10] -[13] and dyslipidemia [14] [15] . Comparisons of Ossabaw swine to the well-established Yucatan swine model revealed that Ossabaws show greater propensity to obesity, especially visceral, than Yucatans on rigorously controlled experimental diets [14] -[17] . The miniature stature of Ossabaw pigs (50 - 70 kg at six months), coupled with metabolic-syndrome characteristics at this age, make them an ideal animal model for metabolic biomedical research. Having performed RYGB in other swine breeds [7] , we hypothesized that the Ossabaw would be a superior model for studies of metabolic surgery and glucose homeostasis. Our investigations confirm that high-fat-fed Ossabaws had higher pre-operative weights compared to lean Ossabaws. In addition, evidence of increased levels of fasting glucose, insulin, and insulin resistance in these animals and the impact of the GI surgeries on glucose homeostasis, have been separately reported by our group [41] .

The surgical survival model we developed is feasible and effective, but required trial and error manipulations, especially related to repeated anesthesia exposure. Although not all animals survived to necropsy, an important marker of a successful model is the ability to reproducibly perform "human-like" procedures and techniques and sustain enough animals to target survival dates. Three deaths occurred late in the study period from non-surgical causes, approximately at or beyond 60 days. We achieved 77% survival to scheduled necropsy, with intra-oper- ative cardiac arrest being the leading cause of premature death. Three of the animals died during catheter placements, with two deaths from catheter replacement prior to scheduled necropsy. Two animals required re-exploration, and one died of intra-operative hemorrhage. These anesthesia complications were early in the model development (first 10 cases) and prompted important changes in our protocol to minimize anesthesia exposure. Only 3 unplanned deaths occurred during the second 22 cases (86% survival).

Our GI operations were designed to clarify mechanisms involved in bariatric post-surgical changes. As expected, our sham operation served as a control and did not lead to a decrease in weight or food intake. The GJ operation creates a moderate shortcut for food to pass from the intact stomach into the proximal jejunum, equivalent to the intestinal shortcut in our porcine RYGB but without gastric restriction or proximal intestinal bypass. By expediting delivery of ingested nutrients to the distal bowel, the operation discretely engages physiologic phenomena described in the lower intestinal hypothesis, but only to a degree of RYGB. In our study, the GJ did not inhibit weight gain, and the pigs had only a slight increase in food intake. These animals could serve in future, more sophisticated studies of glucose homeostasis to explore the distal intestinal hypothesis without confounding from weight loss.

The GJD procedure creates the same shortcut from stomach to jejunum as does GJ and RYGB, but unlike GJ, it excludes proximal intestinal nutrient exposure similar to a traditional RYGB. Comparing the effects of GJD vs. GJ on glucose homeostasis will allow us to distinguish discrete consequences of phenomena described in the upper and lower intestinal hypotheses. Neither of these procedures is confounded by any added gastric alterations, nor do they cause weight loss in our pigs. We have previously reported how these operations should highlight the weight-independent effects of distal intestinal nutrient exposure vs. proximal intestinal nutrient exclusion on glucose homeostasis [41] , and studies describing hormonal alterations in these pigs are underway by our group.

Along with the upper and lower intestinal pathways influenced by the GJ and GJD operations, RYGB reduces gastric capacity. By comparing this operation with the other two, we can discriminate effects on glucose homeostasis that are related to gastric change and/or weight loss from those that result purely from the upper and lower intestinal hypotheses. The role of the vagus nerve in glucose homeostasis can be studied with the RYGB- V operation. The GI tract is heavily innervated by the vagus, which mediates many effects of gut peptides on satiety and glucose homeostasis [42] [43] . Hence, vagotomy might be expected to promote increased meal size and weight gain. This very preliminary data would hint that vagotomy attenuates the effects of RYGB on weight loss, without a major effect on food intake. These conclusions are limited by the small numbers and high mortality in our RYGB-V group and require further study.

There were two major learning curves with development of this model. First, we quickly learned that peripheral blood-draws were largely unsuccessful, and found placement of tunneled catheters superior. During our initial experience, we removed the tunneled catheters on POD 14 after glucose homeostasis testing and replaced them under anesthesia closer to final testing and scheduled necropsy. However, following two intra-operative arrests with catheter placements, including one in a sham surgery pig, we modified our protocol to allow tunneled catheters to remain until necropsy or catheter malfunction. Second, we learned that the thick stomach of the Ossabaw was not suitable for standard GIA staplers, and hand-sewn anastomoses became our practice.

Our model is limited by a relatively short, 60-day postoperative course, which restricts our ability to draw conclusions beyond the study time period or see complications that may develop beyond this time window. Furthermore, to date we have only studied a small number of Ossabaw in each subgroup, especially RYGB-V. Most of our deaths were early during the study period and part of our learning curve for operative technique. Lastly, while our preoperative weight, glucose, and insulin values demonstrate increased insulin resistance in high-fat- fed Ossabaws vs. lean Ossabaws, these values do not correlate perfectly with human values, and further work is needed to understand how the postoperative changes in GI physiology and glucose homeostasis relate to human counterparts.

5. CONCLUSION

In conclusion, the Ossabaw survival model for metabolic surgery appears to be appropriate, reproducible and should offer a novel way to evaluate the mechanism of effect of these procedures. This model is being used to evaluate the mechanism of T2DM resolution after bariatric surgery, the physiologic impact of novel surgical manipulations of the GI tract, and identification of novel targets for surgical and medical interventions.

FUNDING DISCLOSURES

Research reported in this publication was supported by the by the National Cancer Institute (NCI) under Award Number T32CA138312 and the National Institute of Diabetes and Digestive And Kidney Diseases (NIDDK) of the National Institutes of Health (NIH) under Award Numbers T32DK070555 and R01DK084324. The content is solely the responsibility of the authors and does not necessarily represent the official views of the NCI, NIDDK, or NIH.

AUTHOR CONTRIBUTIONS

VV Simianu, JG Sham: analysis and interpretation, data collection, writing the article, critical revision of the article SD Stewart: data collection, writing the article.

M Alloosh, M Sturek: conception and design, data collection, critical revision of the article, obtaining funding.

Wright: conception/design, data collection, writing the article, critical revision of the article.

DE Cummings, DR Flum: conception/design, analysis and interpretation, data collection, writing the article, critical revision of the article and obtaining funding.

ACKNOWLEDGEMENTS

The authors would like to acknowledge Yuki Aoki[1] and Kelly Morrisoe[1] for their contributions to and coordination of perioperative care of the experimental animals. [1]Departments of Surgery, University of Washington, Seattle, WA, USA.

CITE THIS PAPER

Vlad V. Simianu, Jonathan G. Sham, Andrew S. Wright, Skye D. Stewart, Mouhamad Alloosh, Michael Sturek, David E. Cummings, David R. Flum, (2015) A Large Animal Survival Model to Evaluate Bariatric Surgery Mechanisms. *Surgical Science*,**06**,337-345.

REFERENCES

1. Sjostrom, L., Lindroos, A.K., Peltonen, M., Torgerson, J., Bouchard, C., Carlsson, B., et al. (2004) Lifestyle, Diabetes, and Cardiovascular Risk Factors 10 Years after Bariatric Surgery. New England Journal of Medicine, 351, 2683-2693.
2. Sjostrom, L., Narbro, K., Sjostrom, C.D., Karason, K., Larsson, B., Wedel, H., et al. (2007) Effects of Bariatric Surgery on Mortality in Swedish Obese Subjects. New England Journal of Medicine, 357, 741-752.
3. Buchwald, H., Avidor, Y., Braunwald, E., Jensen, M.D., Pories, W., Fahrbach, K., et al. (2004) Bariatric Surgery: A Systematic Review and Meta-Analysis. JAMA, 292, 1724-1737.
4. Pories, W.J., Swanson, M.S., MacDonald, K.G., Long, S.B., Morris, P.G., Brown, D.M., et al. (1995) Who Would Have Thought It? An Operation Proves to Be the Most Effective Therapy for Adult-Onset Diabetes Mellitus. Annals of Surgery, 222, 339-350.
5. Schauer, P.R., Burguera, B., Ikramuddin, S., Cottam, D., Gourash, W., Hamad, G., et al. (2003) Effect of Laparoscopic Roux-en Y Gastric Bypass on Type 2 Diabetes Mellitus. Annals of Surgery, 238, 467-484.
6. Thaler, J.P. and Cummings, D.E. (2009) Minireview: Hormonal and Metabolic Mechanisms of Diabetes Remission after Gastrointestinal Surgery. Endocrinology, 150, 2518-2525.
7. Flum, D.R., Devlin, A., Wright, A.S., Figueredo, E., Alyea, E., Hanley, P.W., et al. (2007) Development of a Porcine Roux-en-Y Gastric Bypass Survival Model for the Study of Post-Surgical Physiology. Obesity Surgery, 17, 1332-1339.
8. Varga, O., Harangi, M., Olsson, I.A. and Hansen, A.K. (2010) Contribution of Animal Models to the Understanding of the Metabolic Syndrome: A Systematic Overview. Obesity Reviews, 11, 792-807.
9. Otis, C.R., Wamhoff, B.R. and Sturek, M. (2003) Hyperglycemia-Induced Insulin Resistance in Diabetic Dyslipidemicyucatan Swine. Comparative Medicine, 53, 53-64. [Citation Time(s):1]
10. Brisbin Jr., I.L. and Mayer, J.J. (2001) Problem Pigs in a Poke: A Good Pool of Data. Science, 294, 1280-1281.
11. Martin, R.J., Gobble, J.L., Hartsock, T.H., Graves, H.B. and Ziegler, J.H. (1973) Characterization of an Obese Syndrome in the Pig. Experimental Biology and Medicine, 143, 198-203. http://dx.doi.org/10.3181/00379727- 143-37285
12. Martin, R.J. and Herbein, J.H. (1976) A Comparison of the Enzyme Levels and the in Vitro Utilization of Various Substrates for Lipogenesis in Pair-Fed Lean and Obese Pigs. Experimental Biology and Medicine, 151, 231-235.
13. Wangsness, P.J., Martin, R.J. and Gahagan, J.H. (1977) Insulin and Growth Hormone in Lean and Obese Pigs. American Journal of Physiology, 233, E104-E108.
14. Edwards, J.M., Neeb, Z.P., Alloosh, M.A., Long, X., Bratz, I.N., Peller, C.R., et al. (2010) Exercise Training Decreases Store-Operated Ca2+ Entry Associated with Metabolic Syndrome and Coronary Atherosclerosis. Cardiovascular Research, 85, 631-640.

15. Etherton, T.D. and Kris-Etherton, P.M. (1980) Characterization of Plasma Lipoproteins in Swine with Different Propensities for Obesity. Lipids, 15, 823-829.
16. Neeb, Z.P., Edwards, J.M., Alloosh, M., Long, X., Mokelke, E.A. and Sturek, M. (2010) Metabolic Syndrome and Coronary Artery Disease in Ossabaw Compared with Yucatan Swine. Comparative Medicine, 60, 300-315.
17. Dyson, M.C., Alloosh, M., Vuchetich, J.P., Mokelke, E.A. and Sturek, M. (2006) Components of Metabolic Syndrome and Coronary Artery Disease in Female Ossabaw Swine Fed Excess Atherogenic Diet. Comparative Medicine, 56, 35-45.
18. Lee, L., Alloosh, M., Saxena, R., Van Alstine, W., Watkins, B.A., Klaunig, J.E., et al. (2009) Nutritional Model of Steatohepatitis and Metabolic Syndrome in the Ossabaw Miniature Swine. Hepatology, 50, 56-67.
19. Rao, R.S., Rao, V. and Kini, S. (2010) Animal Models in Bariatric Surgery—A Review of the Surgical Techniques and Postsurgical Physiology. Obesity Surgery, 20, 1293-1305.
20. National Research Council (US) Committee for the Update of the Guide for the Care and Use of Laboratory Animals (2011) Guide for the Care and Use of Laboratory Animals. Eighth Edition, The National Academies Press, Washington DC.
21. AVMA Panel on Euthanasia (2001) 2000 Report of the AVMA Panel on Euthanasia. Journal of the American Veterinary Medical Association, 218, 669-696.
22. Lombardo, C., Damiano, G., Cassata, G., Palumbo, V.D., Cacciabaudo, F., Spinelii, G., et al. (2010) Surgical Vascular Access in the Porcine Model for Long-Term Repeated Blood Sampling. Acta Bio-Medica, 81, 101-103.
23. Harris, W.H. (1974) A Technique for Chronic Venous Cannulation in Swine. Laboratory Animals, 8, 237-240.
24. Hu, C., Cheang, A., Retnam, L. and Yap, E.H. (1993) A Simple Technique for Blood Collection in the Pig. Laboratory Animals, 27, 364-367.
25. Swindle, M.M. (2004) Chronic Intravascular Catheterization in Swine: Peripheral Vessels. Sinclair Research Institute, Inc., Columbia.
26. Swindle, M.M. (2004) Chronic Intravascular Catheterization in Swine: General Principles. Sinclair Research Center, Inc., Columbia.
27. Rubino, F., Schauer, P.R., Kaplan, L.M. and Cummings, D.E. (2010) Metabolic Surgery to Treat Type 2 Diabetes: Clinical Outcomes and Mechanisms of Action. Annual Review of Medicine, 61, 393-411.
28. Cummings, D.E., Overduin, J., Foster-Schubert, K.E. and Carlson, M.J. (2007) Role of the Bypassed Proximal Intestine in the Anti-Diabetic Effects of Bariatric Surgery. Surgery for Obesity and Related Diseases, 3, 109-115.
29. Rubino, F., Forgione, A., Cummings, D.E., Vix, M., Gnuli, D., Mingrone, G., et al. (2006) The Mechanism of Diabetes Control after Gastrointestinal Bypass Surgery Reveals a Role of the Proximal Small Intestine in the Pathophysiology of Type 2 Diabetes. Annals of Surgery, 244, 741-749.
30. Cummings, D.E. (2005) Gastric Bypass and Nesidioblastosis—Too Much of a Good Thing for Islets? New England Journal of Medicine, 353, 300-302.
31. Thaler, J.P. and Cummings, D.E. (2008) Metabolism: Food Alert. Nature, 452, 941-942.
32. Cummings, D.E., Weigle, D.S., Frayo, R.S., Breen, P.A., Ma, M.K., Dellinger, E.P., et al. (2002) Plasma Ghrelin Levels after Diet-Induced Weight Loss or Gastric Bypass Surgery. New England Journal of Medicine, 346, 1623-1630.
33. Cummings, D.E. and Shannon, M.H. (2003) Ghrelin and Gastric Bypass: Is There a Hormonal Contribution to Surgical Weight Loss? The Journal of Clinical Endocrinology & Metabolism, 88, 2999-3002.
34. Courcoulas, A.P. and Flum, D.R. (2005) Filling the Gaps in Bariatric Surgical Research. Journal of the American Medical Association, 294, 1957-1960.
35. Arner, P. (2005) Resistin: Yet Another Adipokine Tells Us That Men Are Not Mice. Diabetologia, 48, 2203-2205.
36. Cefalu, W.T. (2006) Animal Models of Type 2 Diabetes: Clinical Presentation and Pathophysiological Relevance to the Human Condition. ILAR Journal, 47, 186-198.

37. Sheikh, S.P., Holst, J.J., Orskov, C., Ekman, R. and Schwartz, T.W. (1989) Release of PYY from Pig Intestinal Mucosa; Luminal and Neural Regulation. Regulatory Peptides, 26, 253-266.
38. Adrian, T.E., Bacarese-Hamilton, A.J., Smith, H.A., Chohan, P., Manolas, K.J. and Bloom, S.R. (1987) Distribution and Postprandial Release of Porcine Peptide YY. Journal of Endocrinology, 113, 11-14.
39. Hayashida, T., Murakami, K., Mogi, K., Nishihara, M., Nakazato, M., Mondal, M.S., et al. (2001) Ghrelin in Domestic Animals: Distribution in Stomach and Its Possible Role. Domestic Animal Endocrinology, 21, 17-24.
40. Salfen, B.E., Carroll, J.A. and Keisler, D.H. (2003) Endocrine Responses to Short-Term Feed Deprivation in Weanling Pigs. Journal of Endocrinology, 178, 541-551.
41. Sham, J.G., Simianu, V.V., Wright, A.S., Stewart, S.D., Alloosh, M., Sturek, M., Cummings, D.E. and Flum, D.R. (2014) Evaluating the Mechanisms of Improved Glucose Homeostasis after Bariatric Surgery in Ossabaw Miniature Swine. Journal of Diabetes Research, 2014, Article ID: 526972.
42. Berthoud, H.R. (2008) Vagal and Hormonal Gut-Brain Communication: From Satiation to Satisfaction. Neurogastroenterology & Motility, 20, 64-72.
43. Cummings, D.E. and Overduin, J. (2007) Gastrointestinal Regulation of Food Intake. Journal of Clinical Investigation, 117, 13-23.

CHAPTER 16

Clinical Use of a Multivariate Electroencephalogram (Narcotrend) for Assessment of Anesthetic Depth in Horses during Isoflurane–Xylazine Anesthesia

Julia Tünsmeyer[1], Klaus Hopster[2] and Sabine B. Kästner[1,2*†]

[1]*Small Animal Clinic, University of Veterinary Medicine Hannover Foundation, Hannover, Germany*

[2]*Equine Clinic, University of Veterinary Medicine Hannover Foundation, Hannover, Germany*

ABSTRACT

Objective: To investigate the use of the Narcotrend electroencephalogram (EEG) monitor for the assessment of anesthetic depth in horses undergoing xylazine balanced isoflurane anesthesia.

Study design: Blinded experimental study.

Animals: Seven healthy warm-blooded horses, aged 10.6 ± 5.9 years, weighing 535 ± 55 kg.

Methods: Horses were anesthetized for a terminal surgical trial with xylazine, thiopentone, and guaiphenesin for induction and isoflurane and xylazine continuous rate infusion for maintenance. After surgery, an EEG was recorded and processed by the Narcotrend monitor. It displays an index [Narcotrend index (NI)] between 0 and 100, which is supposed to indicate anesthetic depth. This index was recorded and correlated with eight different end tidal (ET) isoflurane concentrations between 0.8 and 2.2 vol%. In addition, anesthetic depth was numerically scored based on common clinical signs with a score of 1 (plane "too deep") to 4 (plane "too light"). After testing for normal distribution, both clinical scores and NI were correlated with different ET isoflurane concentrations using Spearman rank correlation.

Results: Correlation of NI with ET isoflurane concentrations was poor (r_s = 0.24). The NI ranged between maximal 48 and minimal 13 in the horses. The clinical scores decreased with increasing ET isoflurane concentrations (r_s = 0.80). They ranged from 1 to 4 in different horses at the concentrations investigated.

Conclusion: In this study, the NI did not seem to be useful for assessment of anesthetic depth in horses receiving isoflurane anesthesia balanced with a xylazine constant rate infusion.

Keywords: horses, EEG, Narcotrend, anesthesia, isoflurane

INTRODUCTION

The exact determination of anesthetic depth is still a challenge, and the electroencephalogram (EEG) has been extensively investigated as a tool to monitor anesthetic depth in human beings and also in horses. Precise assessment ofnervous system (CNS) activity in horses is of particular importance because too light anesthetic planes may be associated with movement, leading to injury of the horse or personnel. Too deep anesthetic planes increase the risk of complications, such as cardiorespiratory depression and postanesthetic myopathy (1, 2). This might be one reason for the high perioperative mortality rate in this species compared to small animals (3, 4).

Different studies investigated EEG power spectrum analysis in horses (5, 6). However, in human medicine, this technique has more and more been replaced by the use of multivariate monitors using algorithms, such as the bispectral index (BIS). In humans, BIS seems to be a reliable indicator of anesthetic depth with different anesthetic protocols (7). Similarly, in veterinary medicine, BIS was inversely correlated with inhalation anesthetic concentrations in dogs, cats, and pigs (8–10). However, in isoflurane-anesthetized horses, BIS did not precisely indicate the degree of CNS depression (11).

The Narcotrend is a different multivariate EEG monitor, which uses algorithms based on a visual classification of the human sleep EEG, classical spectral parameters, and algorithms for recognition of suppression lines (12). In the same way as BIS, it displays anesthetic depth by a dimensionless index between 0 (electrical silence) and 100 (awake). Multiple human clinical and validation studies investigating the Narcotrend are available, including comparison of Narcotrend to BIS, with good results (13, 14). Nevertheless, veterinary experience is limited (15, 16). In dogs anesthetized with inhalant agents, Narcotrend index (NI) differed between anesthetic planes assessed clinically as "too deep" and "adequate," but not between planes assessed as "too light" and "adequate" (15).

To the best of our knowledge, no published data are available about the Narcotrend assessing anesthetic depth in horses. Therefore, the aim of this study was to investigate if the NI correlates with different end tidal (ET) isoflurane concentrations in horses and to compare this correlation to common clinical signs that are routinely used to monitor anesthetic depth in horses.

We hypothesized that the NI would be able to detect different stages of anesthesia reflected by different ET isoflurane concentrations.

MATERIALS AND METHODS

Seven warm-blooded horses, aged 10.6 ± 5.9 years, weighing 535 ± 55 kg, were anesthetized for a terminal experimental abdominal surgery approved by the local animal welfare committee (number 33.14-42502-04-11/0572). The EEG part of the study was started after termination of the abdominal surgery, because it was not possible to vary anesthetic depth during the surgical study. The surgeries lasted approximately 2 h. After investigating the NI at different ET isoflurane concentrations, the horses were euthanized with 80 mg kg^{-1} pentobarbital, and the cadavers were used for anatomical studies.

Prior to anesthesia, all the horses were assessed to be healthy by physical examination and complete blood cell count (American Society of Anesthesiologists classification I/II). All the horses were premedicated intravenously (IV) with xylazine (Xylapan, Vetoquinol, Germany) until deeply sedated (0.8–1.1 mg kg^{-1}) and placed in stocks with a swinging door. Guaiphenesin (Myolaxin15%, Vetoquinol, Germany) was infused IV to effect (75–100 mg kg^{-1}) until horses became ataxic, and anesthesia was induced with 5 mg kg^{-1} thiopental IV (Trapanal, Nycomed Deutschland GmbH, Germany). Following orotracheal intubation (Cuffed endotracheal silicone tube, Smiths Medical ASD Inc., Germany; ID 25–30 mm), horses were hoisted on the surgical table, placed in dorsal recumbency, and connected to a large animal circle system. Anesthesia was maintained with isoflurane (Isofluran CP, CP-Pharma, Germany) delivered in 4–8 l $min^{-1}$of 100% oxygen (O_2), and a xylazine infusion at a continuous rate of 0.75 mg kg^{-1} h^{-1} delivered by a syringe driver (Braun Perfusor Compact S, B. Braun Melsungen AG, Germany), which was started immediately after induction.

Lactated Ringer's solution (Ringer-Laktat-Lösung, B. Braun Melsungen AG, Germany) was administered at a rate of 10 ml kg^{-1} h^{-1}, and dobutamine (Dobutamin-ratiopharm 250 mg, ratiopharm GmbH, Germany) (0.5 µg kg^{-1} min^{-1}) was administered *via* syringe driver (Braun Perfusor Compact S, B. Braun Melsungen AG, Germany) to maintain mean arterial blood pressure (MAP) above 60 mmHg. A MAP below 60 mmHg was defined as hypotension. For the evaluation of anesthetic depth, a MAP >90 mmHg was presumed as "elevated," a MAP <90 and >70 mmHg as "normal," a MAP <70 and >65 mmHg as "minimally depressed," a MAP <65 and >60 mmHg as "moderately depressed," and a MAP <60 mmHg as "markedly depressed" (Table 1).

TABLE 1. Clinical scoring system of anesthesia depth in horses [modified from hubbell and muir (**17**)].

Anesthetic plane	Clinical score	Pupil position/size	Eye reflex (P, palpebral; C, corneal)	Heart rate/blood pressure/(spontaneous respiration)
Too light	4	Central/large or small	P: active; C: active	Normal or elevated/breathing against ventilator; occasional swallowing
Medium	3	Ventromedial/small or medium	P: depressed; C: mildly depressed	Normal or minimally depressed
Medium–deep	2	Central/medium	P: depressed; C: depressed	Minimally to moderately depressed
Too deep	1	Central/large	P: absent; C: markedly depressed	Markedly depressed

A mean arterial pressure (MAP) >90 mmHg was presumed as "elevated," a MAP ≤90 and ≥70 mmHg as "normal," a MAP <70 and ≥65 mmHg as "minimally depressed," a MAP <65 and >60 mmHg as "moderately depressed," and a MAP ≤60 mmHg as "markedly depressed".
A heart rate (HR) >45 beats per minute (bpm) was supposed to be elevated, a HR between 30 and 45 bpm was thought to be normal, and a HR <30 bpm was thought to be depressed.

Pressure-cycled intermittent positive pressure ventilation (Bird Mark 7, Vet.-Tec. Model JAVC 2000 J.D. Medical Distributing Company, Phoenix, AZ, USA) was initiated immediately after connection to the anesthetic machine in all horses and adjusted to maintain end tidal carbon dioxide (ET CO_2) between 4.7 and 6 kPa (35–45 mmHg). The transverse facial artery was cannulated with a 21-G catheter (Introcan-W, B. Braun Melsungen AG, Germany) for invasive blood pressure monitoring and arterial blood sampling. The catheter was connected to a pre-calibrated pressure transducer (Gould Statham Transducer, PD 23 ID, USA) *via* fluid-filled, non-compliant extension lines, and the pressure transducer was positioned at the level of the shoulder joint and zeroed to atmospheric pressure. Respiratory gases were sampled continuously from the *y*-piece. Electrocardiogram (ECG), respiratory rate, inspiratory and ET CO_2, O_2 and isoflurane concentrations, O_2 saturation of hemoglobin *via* pulse oximetry

(SpO_2), and arterial blood pressure were monitored continuously with a multiparametric anesthesia monitor (Kardiocap 5 monitor, Datex-Ohmeda GmbH, Germany) and recorded every 10 min. Arterial blood gases were sampled every 30 min and immediately analyzed (AVL995, AVL Medizintechnik, Germany).

The Narcotrend monitor (Monitor Technik, Bad Bramstedt, Germany) was used for recording and processing of the raw EEG. The raw EEG signal was recorded by standard silver ECG needle electrodes using a single-channel registration. The recording electrodes were subcutaneously placed modified from Mayhew and Washbourne (18), with the active electrodes bilaterally placed over the zygomatic processes and the reference placed caudal to the poll. Correct needle placement was checked by automatic impedance measurements of the monitor. The positions of needles were changed when the impedance was 6 kΩ or higher. The EEG was continuously recorded and processed to the Narcotrend stages and indices, and the data were stored by the monitor for off-line analysis.

The Narcotrend monitor analyzes the EEG signal in epochs of 20 s, and the display is updated every 5 s. The monitor displays an anesthetic stage between A (awake) and F (very deep hypnosis), which is further subdivided in the NI ranging between 100 (awake) and 0 (isoelectricity). Due to greater precision and better comparability to other EEG monitors, such as the BIS, only the NI was analyzed in this study.

Randomly assigned by drawing labeled papers from a box to either an ascending or a descending order, vaporizer settings were adjusted to achieve eight different ET isoflurane concentrations (0.8, 1.0, 1.2, 1.4, 1.6, 1.8, 2.0, and 2.2 vol%). The observer assessing anesthetic depth by clinical signs was blinded to the vaporizer settings and the EEG monitor readings. ET concentrations were kept constant for an equilibration phase of 15 min. After 15 min, the NI was recorded, and anesthetic depth was assessed clinically using a score between 1 ("too deep" plane) and 4 ("too light" plane), with a score of 2 reflecting a "medium–deep" anesthetic plane and a score of three a "medium" plane (Table 1). If horses were between two planes based on their clinical parameters, half scores were recorded. For the clinical score adapted from Hubbell and Muir (17), spontaneous movement led automatically to the evaluation as "too light." For possible major movement, a thiopentone bolus of 5 mg kg^{-1} was prepared for IV injection.

Statistical Analysis

The Kolmogorov–Smirnov test and histograms were used for evaluation of normal distribution. Correlation was analyzed by Spearman rank correlation coefficient. A $p < 0.05$ was considered significant.

RESULTS

Vital parameters ranged in clinical acceptable limits for anesthetized horses in all individuals during anesthesia. None of the horses experienced severe hypotension (MAP < 60 mmHg), hypoxemia (PaO_2 < 10.7 kPa) (<80 mmHg), severe hypercapnia ($PaCO_2$ > 8 kPa) (>60 mmHg), hypocapnia ($PaCO_2$ < 4 kPa) (<30 mmHg), or hypothermia (rectal temperature <34°C). Therefore, it was not necessary to increase the dobutamine infusion rate in any of the horses. During the study, none of the horses showed movement requiring thiopentone injection.

Over a wide range of lower ET isoflurane concentrations, the NI did not change considerably. Only at high ET concentrations of 2.0 and 2.25 vol%, a sudden decrease of the NI was observed in four of seven horses (Figure 1), and the correlation between NI and isoflurane concentration was poor ($r_s = 0.24$; $p = 0.078$). However, the clinical score decreased continuously with increasing isoflurane concentrations (Figure 2), and correlation was good ($r_s = 0.80$; $p < 0.001$).

DISCUSSION

In the present study, NI did not change over a wide range of isoflurane concentrations in all of the seven horses. In contrast to the clinical evaluation, NI did not detect a decrease in anesthetic depth at the lower ET isoflurane concentrations. Only at very high isoflurane concentrations (2.0–2.2 vol%), a decrease in NI was observed.

The results presented here are in accordance with an investigation about a different multivariate EEG monitor, the BIS, in horses (11). In this study, BIS was not able to distinguish between the sedated and the anesthetized state and paradoxically increased at an ET isoflurane concentration of 1.9 vol% when compared to 1.4 vol%. In contrast, in a study in horses receiving halothane or sevoflurane anesthesia, BIS detected a difference in CNS depression between anesthetized and awake horses (19).

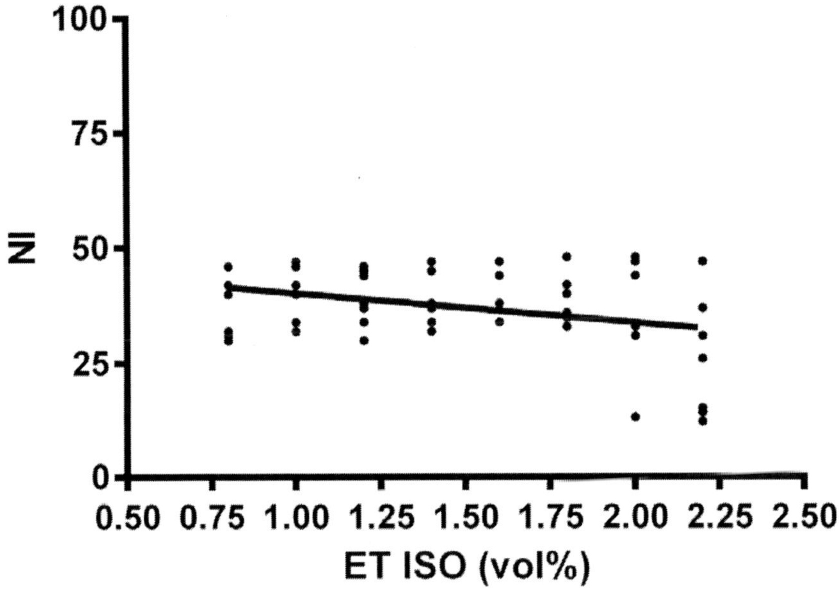

FIGURE 1. RELATIONSHIP OF THE DIMENSIONLESS NARCOTREND INDEX (NI) WITH EIGHT DIFFERENT END TIDAL ISOFLURANE CONCENTRATIONS (ET ISO) IN VOLUME PERCENT (VOL%) AFTER AN EQUILIBRATION PHASE OF 15 MIN IN SEVEN HORSES. The bold line included in the scatter plot is the regression line.

One explanation for these conflicting results could be a specific effect of isoflurane on the equine EEG. In a study investigating the effects of volatile anesthetics on EEG spectral parameters, an increase in ET isoflurane concentration from 1.25 to 1.5 MAC paradoxically increased spectral edge frequency of the equine EEG (19). In a different study, isoflurane similarly increased spectral edge frequency, whereas halothane and methoxyflurane reduced it with increasing concentrations (20). In a recent experimental study about the effects of monoanesthetics with different volatile agents on the equine EEG, none of the quantitative EEG data provided useful information to monitor anesthesia and offered little advantage over the use of changes in MAP for this purpose (21). The BIS reversely increased with increasing isoflurane MAC levels in that study (21). However, when propofol was used instead of volatile agents for anesthetic maintenance in horses, BIS was also not different between multiples of the propofol minimum infusion rate (MIR) in anesthesia and correlation with MIR was only moderate (22).

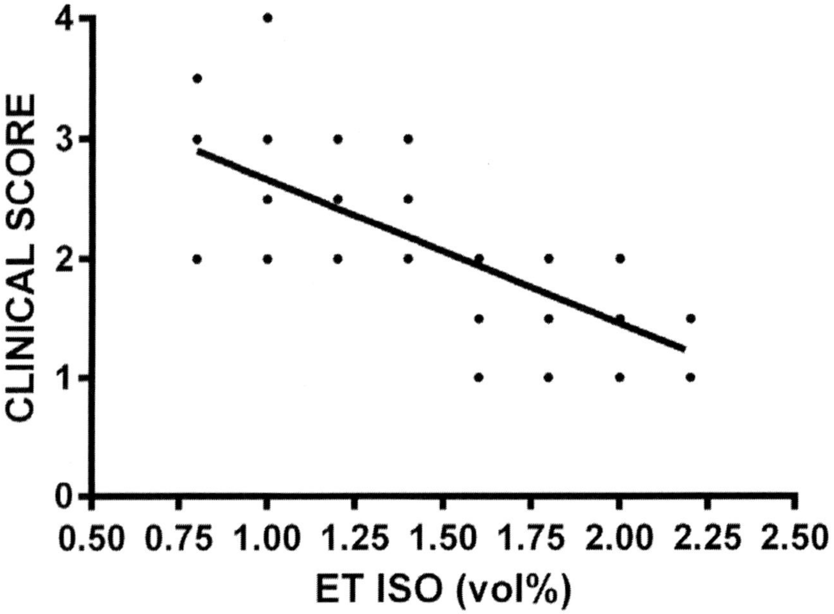

FIGURE 2. RELATIONSHIP OF A CLINICAL SCORE BETWEEN 1 AND 4 EVALUATING ANESTHETIC DEPTH WITH EIGHT DIFFERENT END TIDAL ISOFLURANE CONCENTRATIONS (ET ISO) IN VOLUME PERCENT (VOL%) AFTER AN EQUILIBRATION PHASE OF 15 MIN IN SEVEN HORSES. The clinical score is based on common physical signs of anesthetic depth, such as eye reflexes, pupil size and position, respiration, heart rate, and arterial blood pressure [modified from Hubbell and Muir (17)], with a score of 4 reflecting a "too light" anesthetic plane, score of 3 reflecting a "medium" anesthetic plane, score of 2 reflecting a "medium–deep" anesthetic plane, and score of 1 reflecting a "too deep" anesthetic plane. The bold line included in the scatter plot is the regression line.

Another possible explanation for the poor correlation in our study is that multivariate monitors, such as Narcotrend, have been developed for use in humans. For example, the Narcotrend incorporates algorithms based on the human sleep EEG, which may not completely be transferable to other species. In dogs anesthetized with isoflurane or sevoflurane, similar to the horses reported here, NI was not different between anesthetic planes assessed clinically as "too light" and as "adequate" but differentiated those planes from a clinical evaluation as "too deep" (15). The difficulty of evaluating the use of such EEG monitors is the lack of some "gold standard" reference method, which determines the hypnotic component of anesthesia precisely. In veterinary anesthesia, commonly used variables include reflex response, muscle tone, respiratory, and hemodynamic parameters, which only indirectly reflect the hypnotic state of a patient.

Clinical evaluation of anesthetic depth seemed to be superior to NI in the examined horses due to better correlation with isoflurane concentrations and differentiation between a "too light" and an "adequate" plane of anesthesia. However, the loss of reflex movement is usually due to inhibition of sensory or motor pathways at the segmental level of spinal cord or brainstem (23, 24), and isoflurane has been shown to inhibit spinal motor neurons and produce immobility primarily by its effect at the spinal cord level (25, 26). Furthermore, it has been shown that the inhalant agent concentration, required to induce unconsciousness and amnesia, is only 20–45% of that required to suppress purposeful movement (27). Consequently, it is possible that the horses reported here did not experience "awareness," while increase of reflex response and muscle tone were leading to the clinical assumption of a "too light" anesthesia.

Autonomic nervous responses, such as hemodynamic responses, are also commonly used as indicators of anesthetic depth and have been included in the clinical score used in the horses presently reported. However, hemodynamic responses are primarily subcortical in origin and are not depending on corticocerebral input. Therefore, they may not indicate the conscious perception of noxious stimuli (28). Additionally, they are affected by several factors, such as the pharmacodynamic properties of the different anesthetics, the duration of anesthesia, and physiologic factors. Anesthetic concentrations preventing autonomic reflexes and responses are even higher than anesthetic concentrations preventing movement (29). This corroborates the assumption that horses may still have been centrally depressed, while clinical scoring already evaluated anesthetic plane as being "too light."

The Narcotrend is primarily a monitor of the hypnotic component of anesthesia (30). Therefore, its use in horses might be questionable, because immobility is particularly mandatory for the equine surgeon, and the use of neuromuscular blocking agents is less common than in human anesthesia. However, if reliable monitoring ofdepression would be available, the use of drugs with muscle relaxing properties in a multimodal anesthesia approach could reduce doses of general anesthetics and thereby their risk for side effects.

Xylazine slows the EEG in horses (31–33); therefore, the xylazine constant rate infusion used in our horses might have contributed to the fact that the NI did not increase at lower isoflurane concentrations. However, because xylazine was infused at a constant dose over a long infusion period throughout the study, we assume that it causes a continuous background dampening of the EEG. This should not completely mask a dose-dependent effect of isoflurane on the EEG. When investigating EEG effects of centrally acting agents, drug administration is usually restricted to the investigated drug (34). Due to animal welfare, this was not possible in the horses reported here, because they were part of a surgical trial. Furthermore, we wanted to mimic clinical conditions to test the use of the NI to monitor clinical patients. Other drugs included here, as in other EEG studies, were guaiphenesin and thiopentone (35, 36). Guaiphenesin has been shown to have minimal influence on the equine EEG (37), and thiopentone significantly reduced the spectral edge frequency in halothane-anesthetized horses only for 10 min (38).

In four of the seven horses, NI began to decrease at the high ET isoflurane concentrations of 2.0 and 2.2 vol%, respectively, which implicates that NI can distinguish a "too deep" anesthetic plane from an "adequate" plane. Isoflurane is known to suppress the EEG dose dependently and to cause burst suppression even at clinical relevant doses (39, 40). A special algorithm for the recognition of burst suppressions is incorporated in the Narcotrend, which leads to calculation of low indices. Therefore, the decrease of NI could be explained by an increasing amount of burst suppressions or suppression lines at high isoflurane concentrations. The clinical score assessed anesthesia as "too deep" at lower ET isoflurane concentrations, beginning from 1.6 vol%. Therefore, the clinical advantage of NI for recognition of a deep anesthetic plane is questionable.

However, our study is not free of limitations. First, ET isoflurane concentrations used for the correlation were only indicators of effective concentrations, because equilibration phases of 15 min for the different end expiratory isoflurane concentrations were probably too short for equilibration of alveolar anesthetic partial pressure with arterial partial pressure and brain partial pressure. However, differences between the inspiratory and ET isoflurane concentrations were within 10% at the end of the equilibration phases, as recommended by Eger and Bahlman (41). Second, we used a very general clinical score modified from Hubbell and Muir to assess anesthetic depth in horses. This score might not have been ideal for a xylazine balanced inhalation anesthesia, because horses may appear more superficial than they really are, and the palpebral reflex is less dampened in an alpha$_2$-agonist balanced inhalation anesthesia (42). Third, clinical usefulness of NI was investigated in a limited number of horses, and the lack of a reference method to measure unconsciousness precisely aggravates the evaluation of EEG monitors.

CONCLUSION

Correlation of NI with different ET isoflurane concentrations in horses in balanced anesthesia was poor. Common clinical parameters seemed to be more useful to ensure appropriate surgical conditions.

AUTHOR CONTRIBUTIONS

JT contributed to the design of the study, collection and analysis of the data, and prepared the manuscript. KH contributed to the design and collection of data. SK designed the study, analyzed the data, reviewed and finalized the manuscript.

CONFLICT OF INTEREST STATEMENT

The authors declare that the research was conducted in the absence of any commercial or financial relationships that could be construed as a potential conflict of interest.

FUNDING

The study was University funded, and no external funds were used.

REFERENCES

1. Grandy JL, Steffey EP, Hodgson DS, Woliner MJ. Arterial hypotension and the development of postanaesthetic myopathy in halothane anesthetized horses. *Am J Vet Res* (1987) **48**:192–7.
2. Raisis AL. Skeletal muscle blood flow in anaesthetized horses. Part II: effects of anaesthetics and vasoactive agents. *Vet Anaesth Analg* (2005) **32**:331–7.
3. Johnston GM, Eastment JK, Wood JLN, Taylor PM. The confidential enquiry into perioperative equine fatalities (CEPEF): mortality results of phases 1 and 2. *Vet Anaesth Analg* (2002) **29**:159–70.
4. Brodbelt DC, Blissitt KJ, Hammond RA, Neath PJ, Young LE, Pfeiffer DU, et al. The risk of death: the confidential enquiry into perioperative small animal fatalities. *Vet Anaesth Analg* (2008) **35**:365–73.
5. Otto K, Short CE. Electroencephalographic power spectrum analysis as a monitor of anesthetic depth in horses. *Vet Surg* (1991) **20**:362–71.

6. Johnson CB, Young SS, Taylor PM. Analysis of the frequency spectrum of the equine electroencephalogram during halothane anaesthesia. *Res Vet Sci* (1994) **62**:159–63.
7. Sebel PS, Lang E, Rampil IJ, White PF, Cork R, Jopling M, et al. A multicenter study of bispectral electroencephalogram analysis for monitoring anesthetic effect. *Anesth Analg* (1997) **84**:891–9.
8. Greene SA, Tranquilli WJ, Benson GJ, Grimm KA. Effect of medetomidine administration on bispectral index measurements in dogs during anesthesia with isoflurane. *Am J Vet Res* (2003) **64**:316–20.
9. Lamont LA, Greene SA, Grimm KA, Tranquilli WJ. Relationship of feline bispectral index to multiples of isoflurane minimum alveolar concentration. *Comp Med* (2005) **55**:269–74.
10. Martín-Cancho MF, Lima JR, Luis L, Crisóstomo V, Ezquerra LJ, Carrasco MS, et al. Bispectral index, spectral edge frequency 95%, and median frequency recorded for various concentrations of isoflurane and sevoflurane in pigs. *Am J Vet Res* (2003) **64**:866–73.
11. Haga HA, Dolvik NI. Evaluation of the bispectral index as an indicator of degree ofnervous system depression in isoflurane-anesthetized horses. *Am J Vet Res* (2002) **63**:438–42.
12. Schultz B, Kreuer S, Wilhelm W, Grouven U, Schultz A. The Narcotrend monitor. Development and interpretation algorithms. *Anaesthesist* (2003) **52**:1143–8.
13. Kreuer S, Bruhn J, Larsen R, Bialas P, Wilhelm W. Comparability of Narcotrend index and bispectral index during propofol anaesthesia. *Br J Anaesth* (2004) **93**:235–40.
14. Kreuer S, Bruhn J, Larsen R, Grundmann U, Shafer SL, Wilhelm W. Application of bispectral index and Narcotrend index to the measurement of the electroencephalographic effects of isoflurane with and without burst suppression. *Anesthesiology* (2004) **101**:847–54.
15. Tünsmeyer J, Kramer S. Processed electroencephalogram monitoring of anaesthetic depth in dogs under inhalational anaesthesia. *Vet Anaesth Analg* (2008) **35**:17.
16. Kulka AM, Otto KA, Bergfeld C, Beyerbach M, Kästner SB. Effects of isoflurane anesthesia with and without dexmedetomidine or remifentanil on quantitative electroencephalographic variables before and after nociceptive stimulation in dogs. *Am J Vet Res*(2012) **73**:602–9.
17. Hubbell JA, Muir WW. Monitoring anesthesia. In: Muir WW, Hubbell JA, editors. *Equine Anesthesia*. St. Louis: Saunders Elsevier (2009). p. 149–70.
18. Mayhew IG, Washbourne JR. A method of assessing auditory and brainstem function in horses. *Br Vet J* (1990) **146**:509–18.
19. Belda E, Blissitt KJ, Duncan JC, Laredo FG, Escobar Gil de Montes M, Clutton RE. The bispectral index during recovery from halothane and sevoflurane anaesthesia in horses. *Vet Anaesth Analg* (2010) **37**:25–34.
20. Johnson CB, Taylor PM. Comparison of the effects of halothane, isoflurane and methoxyflurane on the electroencephalogram of the horse. *Br J Anaesth* (1998) **81**:748–53.
21. Williams DC, Aleman MR, Brosnan RJ, Fletcher DJ, Holliday TA, Tharp B, et al. Electroencephalogram of healthy horses during inhaled anesthesia. *J Vet Intern Med* (2016) **30**:304–8.
22. Yamashita K, Akashi N, Katayama Y, Uchida Y, Umar MA, Itami T, et al. Evaluation of bispectral index (BIS) as an indicator ofnervous system depression in horses anesthetized with propofol. *J Vet Med Sci* (2009) **71**:1465–71.
23. Rampil IJ, Laster MJ. No correlation between quantitative electroencephalographic measurements and movement response to noxious stimuli during isoflurane anesthesia in rats. *Anesthesiology* (1992) **77**:920–5.
24. Rampil IJ, King BS. Volatile anesthetics depress spinal motor neurons. *Anesthesiology* (1996) **85**:129–34.
25. Antognini JF, Carstens E, Buzin V. Isoflurane depresses motoneuron excitability by a direct spinal action: an F-wave study. *Anesth Analg* (1999) **88**:681–5.

26. Antognini JF, Carstens E. In vivo characterization of clinical anaesthesia and its components. *Br J Anaesth* (2002) **89**:156–66.
27. Dwyer R, Bennett HL, Eger EI II, Heilbron D. Effects of isoflurane and nitrous oxide in subanesthetic concentrations on memory and responsiveness in volunteers. *Anesthesiology* (1992) **77**:888–98.
28. Zbinden AM, Petersen-Felix S, Thomson DA. Anesthetic depth defined using multiple noxious stimuli during isoflurane/oxygen anesthesia. II. Hemodynamic responses. *Anesthesiology* (1994) **80**:261–7.
29. Schmeling WT, Ganjoo P, Staunton M, Drexler C, Farber NE. Pretreatment with dexmedetomidine: altered indices of anesthetic depth for halothane in the neuraxis of cats. *Anesth Analg* (1999) **88**:625–32.
30. Schmidt GN, Bischoff P, Standl T, Voigt M, Papavero L, Schulte am Esch J. Narcotrend, bispectral index, and classical electroencephalogram variables during emergence from propofol/remifentanil anesthesia. *Anesth Analg* (2002) **95**:1324–30.
31. Purohit RC, Mysinger PW, Redding RW. Effects of xylazine and ketamine hydrochloride on the electroencephalogram and the electrocardiogram in the horse. *Am J Vet Res* (1981) **42**:615–9.
32. Mysinger PW, Redding RW, Vaughan JT, Purohit RC, Holladay JA. Electroencephalographic patterns of clinically normal, sedated, and tranquilized newborn foals and adult horses. *Am J Vet Res* (1985) **46**:36–41.
33. Williams DC, Aleman M, Tharp B, Fletcher DJ, Kass PH, Steffey EP, et al. Qualitative and quantitative characteristics of the electroencephalogram in normal horses after sedation. *J Vet Intern Med* (2012) **26**:645–53.
34. Smith NT, Westover CJ Jr, Quinn M, Benthuysen JL, Dec Silver H, Sanford TJ Jr. An electroencephalographic comparison of alfentanil with other narcotics and with thiopental. *J Clin Monit* (1985) **1**:236–44.
35. Otto KA, Voight S, Piepenbrock S, Deegen E, Short CE. Differences in quantitated electroencephalographic variables during surgical stimulation of horses anesthetized with isoflurane. *Vet Surg* (1996) **25**:249–55.
36. Murrell JC, Johnson CB, White KL, Taylor PM, Haberham ZL, Waterman-Pearson AE. Changes in the EEG during castration in horses and ponies anaesthetized with halothane. *Vet Anaesth Analg* (2003) **30**:138–46.
37. Johnson CB, Bloomfield M, Taylor PM. Effects of guaiphenesin on the equine electroencephalogram during anaesthesia with halothane in oxygen. *Vet Anaesth Analg* (2000) **27**:6–12.
38. Johnson CB, Bloomfield M, Taylor PM. Effects of thiopentone on the equine electroencephalogram during anaesthesia with halothane in oxygen. *Vet Anaesth Analg* (2000) **27**:82–8.
39. Eger EI II, Stevens WC, Cromwell TH. The electroencephalogram in man anesthetized with forane. *Anesthesiology* (1971) **35**:504–8.
40. Murrell JC, Waters D, Johnson CB. Comparative effects of halothane, isoflurane, sevoflurane and desflurane on the electroencephalogram of the rat. *Lab Anim* (2008) **42**:161–70.
41. Eger EI II, Bahlman SH. Is the end-tidal anesthetic partial pressure an accurate measure of the arterial anesthetic partial pressure? *Anesthesiology* (1971) **35**:301–3.
42. Kalchofner KS, Ringer SK, Boller J, Kästner SBR, Lischer C, Bettschart-Wolfensberger R. Clinical assessment of anesthesia with isoflurane and medetomidine in 300 equidae. *Pferdeheilkunde* (2006) **22**:301–8.

CHAPTER 17

Comparison of Hemodynamic and Inflammatory Changes between Transoral and Transthoracic Thoracoscopic Surgery

Yen Chu[1]., Chien-Ying Liu[2]., Yi-Cheng Wu[1] , Ming-Ju Hsieh[1] , Tzu-Ping Chen[1] , Ying-Kai Chao[1] , ChingYang Wu[1] , Hsu-Chia Yuan[3] , Po-Jen Ko[1] , Yun-Hen Liu[1] *, Hui-Ping Liu[1] *

[1] *Department of Thoracic Surgery, Chang Gung Memorial Hospital, Chang Gung University, Taoyuan, Taiwan,*

[2] *Department of Thoracic Medicine, Chang Gung Memorial Hospital, Chang Gung University, Taoyuan, Taiwan,*

[3] *Laboratory Animal Center, Chang Gung Memorial Hospital, Chang Gung University, Taoyuan, Taiwan*

ABSTRACT

Background

Natural orifice transluminal endoscopy has been developed for abdominal surgical procedures. The aim of this study was to compare the surgical outcome between a novel transoral approach and a standard transthoracic approach for the thoracic cavity in a canine model.

Methods

Twenty-eight dogs were assigned to transoral (n=14) or standard thoracoscopy (n=14). Each group underwent thoracic exploration, pre-determined surgical lung biopsy, and pericardial window creation. Blood draws were obtained before surgery and at postoperative days 1, 3, 7, and 14. Operative time, complications, laboratory parameters, hemodynamic parameters, and inflammatory parameters were compared between the two procedures. The animals were monitored for two weeks and necropsy were performed for surgical outcome evaluation.

Results

The thoracic procedures were successfully performed in all of the dogs, with the exception of one animal in the transoral group. There were no serious acute or delayed complications related to surgery. There was no difference between the two surgical groups for each of the hemodynamic parameters that were evaluated. Regarding the immunological impact of the surgeries, transoral thoracoscopy was associated with significant elevations in interleukin 6 and c-reactive protein levels on postoperative days 1 and 3, respectively, when compared with the standard thoracoscopy. All dogs recovered well, without signs of mediastinitis or thoracic infection. Necropsy revealed absence of infection, no injury to vital organs, and confirmed the success of the novel procedure.

Conclusions

This study suggests that both techniques were comparable with respect to procedure success rate, hemodynamic impact, and inflammatory changes. Furthermore, there was no difference in the incidence of postoperative discomfort between groups.

INTRODUCTION

Natural orifice transluminal endoscopic surgery (NOTES) has been used as a diagnostic and therapeutic platform to perform a wide variety of abdominal surgical procedures, including peritoneoscopy, cancer staging, appendectomy, cholecystectomy, and nephrectomy [1]–[4]. By eliminating conventional abdominal incisions, several potential advantages have been proposed, including faster recovery, shorter hospital stay, less postoperative pain, better post-surgery pulmonary function, and improved cosmetic outcome. Recent reports have also confirmed the safety and benefits of NOTES in abdominal surgery in human studies [5]–[7].

The feasibility of NOTES thoracoscopy has been demonstrated in various access methods (transvesical, transesophageal, transtracheal, and transoral) in porcine and canine models [8]–[10]. However, there have been no large, randomized, controlled trials of NOTES in thoracic surgery. We previously reported the use of a transoral technique to perform thoracoscopy, pericardial window creation, and surgical lung biopsy without complications in a survival canine model [11]–[13]. To further evaluate the potential benefits of transoral thoracoscopy, we aimed to compare the surgical outcome of minor thoracic intervention using the recently developed transoral method compared with the standard transthoracic thoracoscopy in a randomized study. The experiment focused on surgical outcomes and evaluated relevant safety issues, including operative complications, cardiopulmonary impact, and inflammatory changes. Oxidative activity of neutrophils, inducible nitric oxide synthase (iNOS) expression of monocytes, and cytokine interleukin 6 (IL-6) are good guides in the assessment of the activation of the systemic inflammation to tissue injury. The expression of these mediators as well as c-reactive protein (CRP) play a major role in the acute-phase response against surgical stress [14]–[15]. They are objective and quantitative parameters in the assessment of inflammatory system activation in response to operative trauma. Therefore, these mediators were measured to clarify the surgical stress of thoracoscopy via different approaches.

METHODS

Ethics Statement

The study was conducted in accordance with the Guide for the Care and Use of Laboratory Animals as promulgated by Institutional of Laboratory Animal Resources, National Research Council, U.S.A. The protocol of the animal experiment was approved by the Institutional Animal Care and Use Committee (IACUC) of Chang Gung Memorial Hospital in Taiwan (No. 2009121003).

Dogs and surgical procedures

Twenty-eight dogs with a mean body weight of 8.5 kg (7.2–10.6 kg) were obtained from BioVet Beagle Farm, Pingtung, Taiwan. They were housed under standard lighting (12 hours light/dark cycle) and temperature conditions, and received humane care for the animal research at laboratory animal center of Chang Gung Memorial Hospital in Taouyan, Taiwan. All animals were fed with regular diet, and then fast overnight with unrestricted to water prior to surgery. The procedures were performed with the animals under general anesthesia using intramuscular injection of ketamine (5 mg/kg, Pfizer Inc, Taipei, Taiwan) and xylazine hydrochloric acid (HCL, 10 mg/kg). The animals were then placed in the supine position and intubated with a homemade endotracheal tube into a mainstem bronchus to provide single lung ventilation during surgery. The general anesthesia was maintained with 2% isoflurane. Before performing the thoracoscopy, an antibiotic (cefazolin, 20 mg/kg, Standard Chem & Pharm, Tainan, Taiwan) was administered for prevention of postoperative infection.

A 9 mm homemade metal tube was introduced through the vestibular incision and advanced down to the mandibular periosteum. The metal tube was advanced carefully from the premandibular space, through the pretracheal space, and into the submanubrial space. The mediastinal incision was carefully performed according to a previously reported procedure [11]–[13]. Briefly, the blunt tip of the metal tube was used to make a linear incision over the upper anterior mediastinal pleura by blunt dilation. Once the incision was complete, the metal tube was advanced into the thoracic cavity. The bronchoscope was then advanced through the metal tube for thoracic cavity exploration. For pericardial window creation, a needle knife (KD-10Q-1, Olympus Optical Co. Ltd., Tokyo, Japan) was introduced through the working channel of the flexible bronchoscope. The pericardium was incised anterior to the phrenic nerve with the tip of needle knife. The endoscope was then advanced through this window into the pericardial cavity using the needle knife as a guide, and was used to explore the pericardial cavity. For surgical lung biopsy, the predetermined lung region was resected with the electrocautery loop (Olympus Optical Co. Ltd., Tokyo, Japan) and endoscopic grasper through a 1.9 mm bronchoscopic working channel. The resected lung margin was reinforced with a homemade (Ethicon, Somerville, NJ, USA) endoloop ligature to prevent air leakage. The vestibular incision was then closed with a running 3-0 Vicryl suture (Ethicon, San Angelo, TX, USA) as previously described [11]–[13]. Soon after completion of wound closure, postoperative local anesthesia with xylocaine 20 mg was administrated over the incision. The animal was then recovered from anesthesia and returned to their cages. The arterial blood gases were analyzed immediately before and after surgery and on the 14th day.

Care after surgery

Regular oral feeding was restarted after the animals were recovered from anesthesia. The animals were closely monitored daily for general status, eating status, social behavior, and signs of mediastinitis, sepsis, clinical signs of postoperative discomfort (decreased food intake, weight reduction, or inappropriate social behavior), and survival for 2 weeks after experiment. The animals were euthanized with xylocaine 200 mg intravenously after sedation with the ketamine and xylazine hydrochloric acid combination described above. The necropsy was performed via medial sternotomy incision and evaluated for evidence of infection, bleeding, or signs of injury to adjacent organs.

Hemodynamic study

Immediately following intubation and then at 5 min intervals until the animal completely recovered from anesthesia, in addition to respiratory rate (breaths per minute, bpm), the values of mean arterial pressure (MAP, mmHg), heart rate (HR, beats per minute, bpm), cardiac index (CI, L/[min·m^2]), systemic vascular resistance (SVRI, [dyn·s]/[cm^{-5}·m^2]), and global end-diastolic volume index (GEDVI, mL/m^2) were recorded by transpulmonary thermodilution technique operated by the pulse contour cardiac output (PiCCO®) system (Pulsion Medical Systems, Munich, Germany)

Leukocyte subset analysis and isolation of neutrophils and mononuclear cells

Total leukocyte count was measured preoperatively and postoperatively at day 1, day 3, day 7, and day 14. Specific mouse anti-canine CD3, CD4, CD8, CD14, or CD45 monoclonal antibodies (BD PharMingen™, BD Biosciences, Franklin Lakes, NJ, USA) were used to define the subpopulations of leukocytes (CD4$^+$ and CD8$^+$ lymphocytes, monocytes and neutrophils) by flow cytometry (Becton Dickinson, Mountain View, CA, USA).

Intracellular oxidative activity of neutrophils

The neutrophils (2×10^5 cells in 200 μL of supplemented RPMI medium) were supplemented with cell-permeable, fluorogenic 2′,7′-dichlorohydrofluorescein-diacetate (DCFH-DA, 5 μM; Sigma, St. Louis, MO, USA) for 30 min. Upon intracellular hydrolysis and subsequent oxidization, fluorescent DCFH was generated and analyzed quantitatively by flow cytometry.

iNOS expression of monocytes

After cytospin preparation of peripheral blood mononuclear cells and fixation in methanol, the slides were prepared as previously described [16]. A rabbit polyclonal antibody to canine iNOS (Abcam Biochemicals, Cambridge, UK) was used for immunological staining. Antibody labeling was visualized by an avidin-biotin complex method (LSAB 2 kit, Dako, Carpenteria and Vector Laboratories, Burlingame, CA, USA). An image analyzer was used to determine the intensity of iNOS expression of monocytes.

Enzyme-linked immunosorbent assay for IL- 6 and CRP

Serum for IL-6 and CRP analysis was stored at −80°C, and the levels of IL-6 and CRP were determined using a commercially available enzyme-linked immunosorbent assay (ELISA) kit with monoclonal antibodies specific for IL-6 (Quantikine, R & D Systems Inc., Minneapolis, Minnesota, USA) and CRP (PharMingen, BD Biosciences, San Diego, California, USA), according to the manufacturers' instructions.

Statistical analysis

Data are presented as mean and 95% confidence interval (C.I.) except otherwise mentioned. Since the data did not approximate a Gaussian distribution (for example, the mean value did not approximate the median value), nonparametric statistical analyses were used. The Mann–Whitney *U* test was performed to assess the significance of differences between 2 groups. The frequency distributions between 2 groups were tested by using a Fisher exact probability test. GraphPad Prism (version 4.0, GraphPad Software, San Diego, CA, USA) was used for all statistical analyses. Statistical significance was defined as $P<0.05$.

RESULTS

All animals survived the follow-up period of 14 days. Thoracoscopy with surgical biopsy and pericardial window creation was successfully performed in 27 animals, with the exception of 1 animal in the transoral group. There was no different in the mean procedure time and mean weight gain between groups. None of the animals in either group had intraoperative or perioperative complications related to surgery. No postoperative complications such as infection or mediastinitis were found during the 2-week postoperative period (Table 1).

Table 1. Characteristics of the Experimental Animals.

Characteristics	Total	Thoracic Surgery		P value
		Transoral	Transthoracic	
Dogs (No.)	28	14	14	
Body weight (kg, mean ± SD)	8.5±1.2	8.8±1.1	8.3±1.3	0.2229
Surgical time (min, mean ± SD)	55.8±14.4	55.4±13.1	56.3±16.1	0.9632
Successful procedure	27	13	14	
Weight gain (after 14 days, kg, mean ± SD)	0.33±0.53	0.20±0.59	0.46±0.13	0.1956
Postoperative complications (No.)				
Fever	6	2	4	0.6483
Infection or mediastinitis	0	0	0	
Pain	4	2 (1 mild, 1 severe)	2 (severe)	1.0
Necropsy after 14 days (No.)				
Infection	0	0	0	
Organ injury	0	0	0	
Correct lung biopsy	27	13	14	1.0
Good healing of pericardial window	17	8	9	1.0
Local adhesion	25	12	13	1.0

Temperature

Six animals showed rectal temperature elevation after surgery. In 2 dogs from the transoral group, the temperature rose above normal on day 1–3 and returned to normal 7 days after surgery. In the transthoracic group, 4 animals had elevated temperatures at 3–7 days after surgery, which returned to baseline at 2 weeks in 3 of the animals (Table 1).

Hemodynamic study

Comprehensive hemodynamic changes were recorded in 24 animals. We observed a significant decline in HR and a moderate variation in CI during surgery. We also observed a minor decrease in GEDI and MAP compared to baseline. However, the change in the cardiopulmonary parameters between the groups was not statistically significant, and no clinically relevant immediate or late complications were identified after surgery (Fig. 1).

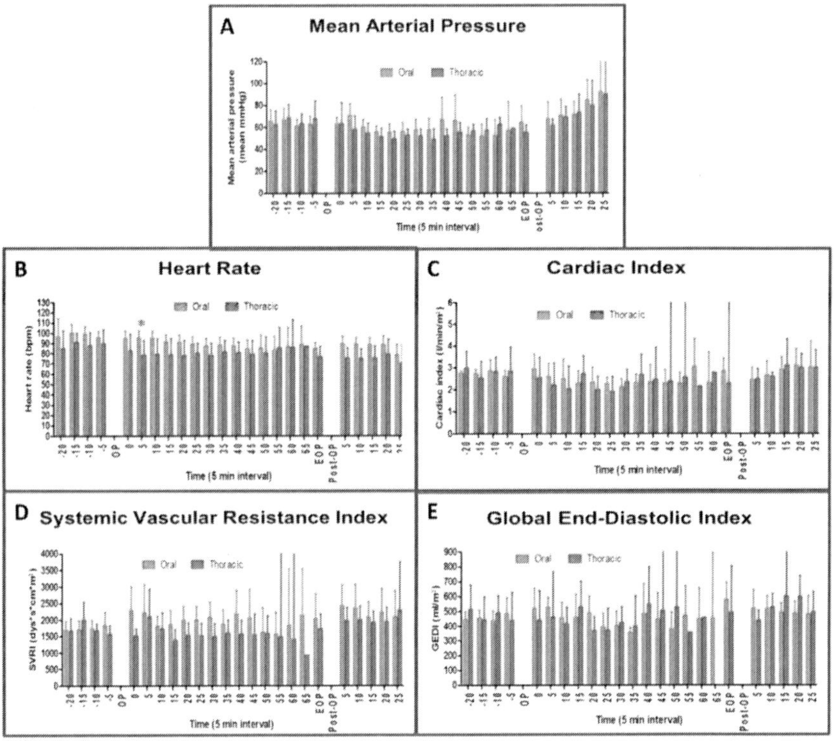

Figure 1. Hemodynamic analyses, including mean arterial pressure (A), heart rate (B), cardiac index (C), systemic vascular resistance index (D), and global end-diastolic index (E).

N=12 for the transoral approach and n=12 for the transthoracic approach. *P<0.05, transoral vs. transthoracic approach for the same time interval, by Mann–Whitney *U* test.

White blood cells

Among the 28 animals, 25 animals showed an increased white blood cell count above baseline 1 day after surgery, which included 15 animals from the transthoracic group and 10 from the transoral group. The white blood cell (WBC) count returned to baseline in 24 of the 28 animals at 2 weeks. The WBC count of transthoracic group was significantly higher on day 1 after surgery relative to the transoral group, but there was no significant difference in WBC count between the groups thereafter (Fig. 2).

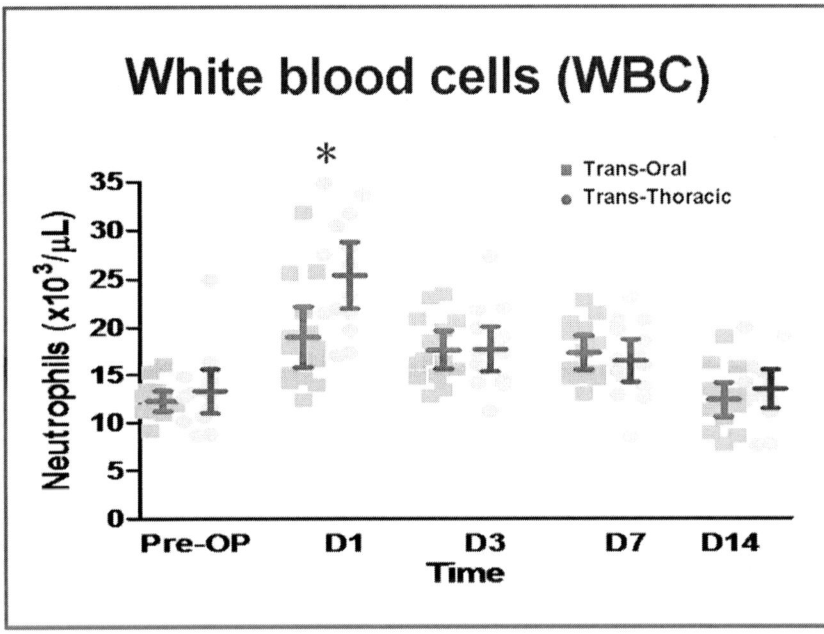

Figure 2. White blood cell counts.

N=14 for the transoral approach and n=14 for the transthoracic approach. *P<0.05, transoral vs. transthoracic approach for the same time point, by Mann–Whitney *U* test.

Neutrophils and oxidative activity

The number of neutrophils was elevated on postoperative day 1, 3, and 7 in both groups. This variation was significantly increased in the transthoracic group relative to the transoral group on the first postoperative day. Neutrophil activation was revealed by expression of DCFH, which increased significantly in both groups on postoperative day 1 and returned to preoperative levels at 1 week. However, DCFH expression was not significantly different between the groups (Fig. 3A and 3B).

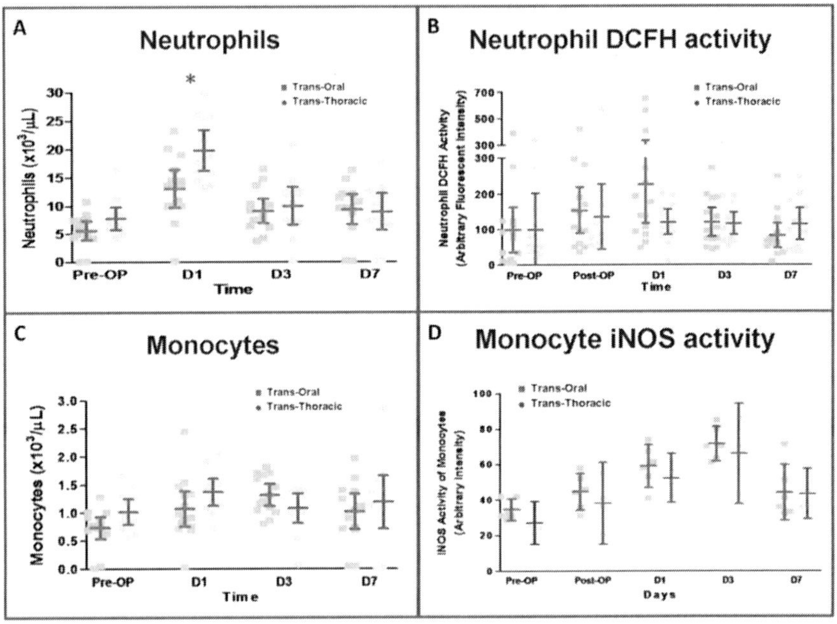

Figure 3. Neutrophil cell counts (A) and dichlorodihydrofluorescein (DCFH) activity (B) as well as monocyte cell counts (C) and iNOS activity (D).

N=14 for the transoral approach and n=14 for the transthoracic approach. *$P<0.05$, transoral vs. transthoracic approach for the same time point, by Mann–Whitney *U* test.

Monocytes and the intensity of iNOS expression

During the postoperative period, the number of monocytes increased significantly in both groups following surgery, and remained elevated above the baseline until 7 days post-surgery. Significantly increased iNOS levels were observed in both groups during the postoperative period; however, there were no statistically significant differences between the transoral and transthoracic groups with regard to monocyte cell counts and iNOS expression (Fig. 3C and 3D).

Lymphocytes and CD4/CD8 T lymphocyte subset analysis

There was a significant decrease in the total lymphocyte count on postoperative day 1 in both groups. Thereafter, the total lymphocyte count increased significantly above the baseline on postoperative day 3. With regard to T lymphocyte subsets, the number of CD4+ T cells increased on postoperative days 3 and 7, the number of CD8+ T cells did not differ significantly from preoperative levels at any of the time points throughout the study, and a significant elevation in the CD4:CD8 ratio was also noted on postoperative day 3 in both groups. However, there were no significant differences in total lymphocyte count, T lymphocyte subsets, or CD4:CD8 ratio between the groups at any of the study time points (Fig. 4).

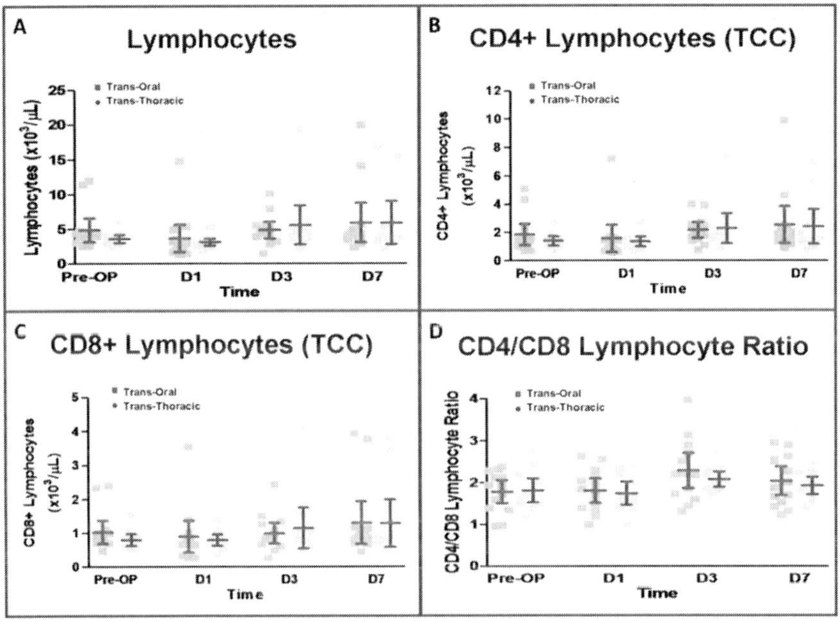

Figure 4. Lymphocyte cell counts (A), the CD4 (B) and CD8 (B) subsets, and CD4/CD8 lymphocyte ratio (D).

N=14 for the transoral approach and n=12 for the transthoracic approach. *$P<0.05$, transoral vs. transthoracic approach for the same time point, by Mann–Whitney *U* test.

Interleukin 6 and C-reactive protein levels

Plasma levels of CRP increased significantly 1 day after surgery in both groups. The CRP levels were significantly higher in the transoral group compared with the transthoracic group at day 3. However, there was no group difference in the plasma levels of CRP on day 7 and day 14. The serum concentration of IL-6 was increased over preoperative levels in early postoperative period in both groups. This elevation in IL-6 levels was greater in the transoral group than in the transthoracic group at day 1 after surgery. However, no differences were observed between the groups 3 days after surgery (Fig. 5).

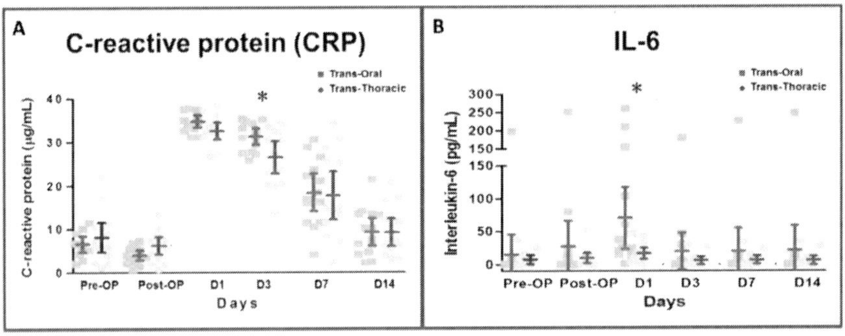

Figure 5. Serum levels of c-reactive protein (A) and interleukin 6 (B).

N=14 for the transoral approach and n=14 for the transthoracic approach. *P<0.05, transoral vs. transthoracic approach for the same time point, by Mann–Whitney *U* test.

Arterial blood gas

There was no significant change in PaO2 values before and after the procedure. The pH value was decreased and PaCO2 level was elevated immediately after the procedure. Both values returned to the pre-operative levels before day 14 post-operatively. There was no difference in the values of pH, PaO2, and PaCO2 between the two approaches at the same time point (Fig. 6).

Postoperative pain

In the study animals, postoperative discomfort manifested as decreased food intake, weight reduction, or inappropriate social behavior after surgery; 4 of 28 dogs presented with moderate to severe postoperative wound pain. There was no significant difference in the severity of pain between the groups (1 animal with mild pain and 1 animal with severe pain in the transoral group vs. 2 animals with severe pain in the transthoracic group; Table 1).

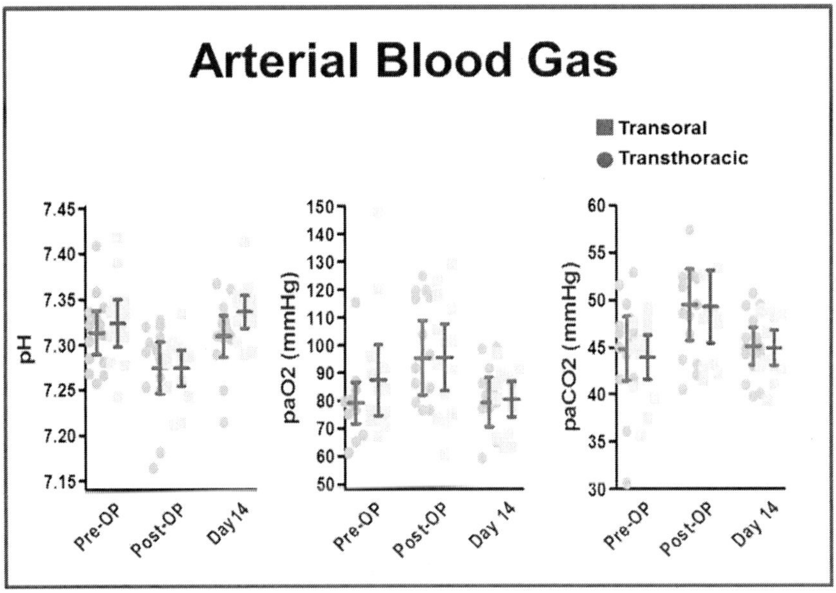

Figure 6. Arterial blood gas analysis.

N=14 for the transoral approach and n=14 for the transthoracic approach. *$P<0.05$, transoral vs. transthoracic approach for the same time point, by Mann–Whitney *U* test.

Necropsy

At the time of necropsy, there were no signs of infection or evidence of organ injury in either the thoracic cavity or the mediastinum. The lung biopsy was correctly performed in 27 out of the 28 animals (n=13 in the transoral group, n=14 in the transthoracic group, P=1.0). The pericardial window had obliterated and the surgical wounds were difficult to identify in 17 out of the 28 animals (n=8 in the transoral group, n=9 in the transthoracic group, P=1.0). Twenty-five animals revealed adhesions either on the pleural or mediastinum at the surgical lung biopsy region (n=12 in the transoral group, n=13 in the transthoracic group, P=1.0; Table 1).

DISCUSSION

We recently published a study demonstrating the successful performance of a minor thoracic procedure via a transoral approach in a canine model [11]–[13]. In line with these findings, our present study further demonstrated the safety and efficacy of the transoral approach in hemodynamic, inflammatory conditions. These findings are clinically relevant, as our study included thoracic exploration, surgical lung biopsy, and pericardial window creation in a survival canine model, and each of these procedures is commonly performed in daily practice.

Since Kallo et al. reported the first transgastric peritoneoscopy in 2004, NOTES has been increasingly used in the diagnosis and therapy of both benign and malignant disease of the abdomen [17] Natural orifice surgery (NOS) has potential advantages over conventional laparoscopic surgery; however, the development of NOTES in thoracic surgery was delayed. The critical problem that hindered the implementation of NOTES in thoracic disease was the lack of a reliable platform for approaching the thoracic cavity via a natural orifice. Furthermore, potential complications during the procedure, including incision leakage, mediastinal organ injury, and infection were important concerns.

Several studies have investigated the feasibility of using the natural orifice as a novel platform for thoracic intervention and emphasized that an optimal approach was key to successful natural orifice thoracic surgery. Lima et al. reported their experience with 6 pigs that underwent transvesical lung biopsy. Their overall success rate was 100%. The authors found that a major disadvantage of this intervention is difficulty in performing the procedure over the upper mediastinum and pleural region [8]. Palma et al. reported another study of 4 pigs that underwent transgastric lung biopsy. Their success rate was 100%; however, results from a few of the experimental animals failed to effectively assure the safety of a transgastric approach [18]. Sumiyama and colleagues performed transesophageal thoracoscopy in 9 pigs and demonstrated an increased risk of tension pneumothorax and massive bleeding [19]. Our group previously described the utilization of transtracheal thoracoscopy in 12 dogs; we encountered 3 cases of tension pneumothorax due to inadvertent injury to the lung during creation of a tracheal access site [20]. Recently, we have made significant progress in using a transoral route for surgical lung biopsy and pericardial window creation, without major intraoperative or postoperative complications [10]–[13]. Our findings suggest that a transoral approach can be used as a novel platform for studies of NOTES for intrathoracic surgery. However, its clinical application has yet to be implemented, as there is uncertainty regarding the impact of hemodynamic and inflammatory status during the procedure.

In 2011, our research group showed that identification of the anterior medial aspect of the thoracic cavity is possible with the transoral approach, and creation of a pericardial window and surgical lung biopsy is technically feasible in a canine model [11]–[13]. In this comparative study, 93% of the target lung regions were successfully accessed by the present approach. We expect that the present transoral method will play arole in the development of NOTES in intrathoracic surgery. The safety of novel surgical procedures is one of the most critical barriers to gaining acceptance in clinical practice. Our previous study demonstrated that 90% to 95% of surgical lung biopsies could be completed by using the transoral route without major complications. The present study also demonstrates that using the transoral route for thoracic intervention is safe and practical in a canine model. This method did not cause infection or injury to the vital organs in all 14 animals, suggesting that the use of transoral thoracoscopy is a potential alternative to the current transthoracic endoscopic method, especially in young adults that are concerned about postoperative cosmesis.

Many recent reports in the literature regarding the hemodynamic impact of NOTES during surgery have been studied in a porcine model. In 2011, Suzuki et al. evaluated the cardiopulmonary effect of transvaginal NOTES cholecystectomy in 10 pigs [21]. This study demonstrated that there was similar cardiopulmonary stability in both the transvaginal NOTES and the laparoscopic cholecystectomy groups. With regard to cardiopulmonary safety of NOTES in intrathoracic surgery, Von Delius et al. studied hemodynamic impact in 8 pigs that underwent transesophageal mediastinostomy [22]. They observed hemodynamic instability in 3 animals, due to an injury in the parietal pleura, which resulted in tension pneumothorax and mortality in 1 pig. These results highlight the importance of using a safe and effective surgical platform and monitoring the hemodynamic status during NOTES. Our current study revealed a similar impact of hemodynamic status in the transoral and standard transthoracic approaches in a comparative animal study, which suggests that transoral access to the thoracic cavity is a safe platform compared with the standard transthoracic approach.

When the clinical application of the novel platform is considered, the invasiveness of the procedure is another major concern that must be addressed. Several researchers have evaluated inflammatory and immunologic responses in both animal and human studies following endoscopic and open surgery, and suggest that endoscopic surgery is associated with less surgical stress than the open surgical approach [14]–[15]. In the current study, the transthoracic group showed higher IL-6 and CRP levels than the transoral group. However, only 1 of 6 IL-6 and CRP measurements was found to be significantly different. Furthermore, there were no clinically apparent complications in both groups. Regarding the cellular responses to the surgical procedure, we found a similar change in iNOS, DCFH, and CD4/CD8 indicators in the transoral group compared with those that underwent conventional thoracoscopy. Based on these results, our study suggests that transoral thoracoscopy results in similar surgical stress as the current transthoracic approach. Additionally, the arterial blood gas analysis revealed no difference between groups, suggesting that transoral thoracoscopy did not increase pulmonary dysfunction, compared to the transthoracic approach.

The present transoral method offers several potential advantages over the transthoracic technique that is currently in use. First, the technique demonstrated good cosmetic outcomes, as the surgery was completed without any external thoracic incision. Second, the transoral group experienced similar acute postoperative pain compared with the animals undergoing transthoracic surgery. However, many studies reported that thoracic incisions may result in post-thoracotomy discomfort and intercostal neuralgia, which can be avoided by using the transoral approach.

The present study has some limitations. First, our transoral approach for thoracic cavity techniques in a canine model may face difficulty in clinical practice, particularly in obese patients with a thick or short neck. Second, a relatively small sample size was obtained for the cellular immunity study, which may contribute to the non-significant difference observed in the cellular immune reactions between the groups. Third, we were unable to measure the complete cardiopulmonary data that were influenced by arterial line dysfunction. Fourth, since the IL-6 and CRP assays were performed 1 day after surgery in our study, we could not explore the change in inflammatory cytokine secretion immediately after surgery. Finally, we only performed the surgical lung biopsy over the anterior aspect of thorax. Thus, we cannot determine whether approaching the posterior thorax under the decubital position would be successful. Nevertheless, we designed the present experiment to be as relevant to surgical practice as possible, and in clinical daily practice surgical lung biopsy is one of the most commonly performed endoscopic procedures in thoracic surgery.

In summary, this study revealed that utilizing a transoral approach for thoracic cavity procedures was feasible, efficacious, and safe in a canine survival study. We also found that the immunologic, inflammatory, and hemodynamic responses to the transoral approach were similar to those of the conventional thoracoscopic approach, with the exception of minimal differences in inflammatory reactions. We conclude that clinical trials are needed to further clarify the clinical role of NOTES in thoracic disease.

AUTHOR CONTRIBUTIONS

Conceived and designed the experiments: YC CYL YCW YHL HPL. Performed the experiments: YC HCY CYW MJH TPC PJK YHL. Analyzed the data: CYL MJH YKC. Contributed reagents/materials/analysis tools: CYL MJH YKC. Wrote the paper: YC CYL YHL. Supervision and coordination of project: YC YHL HPL.

REFERENCES

1. Hazey JW, Narula VK, Renton DB, Renton DB, Reavis KM, et al. (2008) Natural orifice transgastric endoscopic peritoneoscopy in humans: initial clinical trial. Surg Endosc 22: 16–20.
2. Dallemagne B, Perretta S, Allemann P, Asakuma M, Marescaux J (2009) Transgastric hybrid cholecystectomy. Br J Surg 96: 1162–1166.
3. Palanivelu C, Rajan PS, Rangarajan M, Parthasarathi R, Senthilnathan P, et al. (2008) Transvaginal endoscopic appendectomy in humans: a unique approach to NOTES - world's first report. Surg Endosc 22: 1343–1347.
4. Alcaraz A, Musquera M, Peri L, Izquierdo L, García-Cruz E, et al. (2011) Feasibility of transvaginal natural orifice transluminal endoscopic surgery-assisted living donor nephrectomy: is kidney vaginal delivery the approach of the future? Eur Urol 59: 1019–1025.
5. Chukwumah C, Zorron R, Marks JM, Ponsky JL (2010) Current status of natural orifice translumenal endoscopic surgery (NOTES). Curr Probl Surg 47: 630–668.
6. Lehmann KS, Ritz JP, Wibmer A, Gellert K, Zornig C, et al. (2010) The German registry for natural orifice translumenal endoscopic surgery: report of the first 551 patients. Ann Surg 252: 263–270.
7. Coomber RS, Sodergren MH, Clark J, Teare J, Yang GZ, et al. (2012) Natural orifice translumenal endoscopic surgery applications in clinical practice. World J Gastrointest Endosc 16: 65–74.
8. Lima E, Henriques-Coelho T, Rolanda C, Pêgo JM, Silva D, et al. (2007) Transvesical thoracoscopy: a natural orifice translumenal endoscopic approach for thoracic surgery. Surg Endosc 21: 854–858.
9. Fritscher-Ravens A, Cuming T, Eisenberger CF, Ghadimi M, Nilges A, et al. (2010) Randomized comparative long-term survival study of endoscopic and thoracoscopic esophageal wall repair after NOTES mediastinoscopy in healthy and compromised animals. Endoscopy 42: 468–474.
10. Liu YH, Chu Y, Liu CY, Liu HP, Wu YC, et al. (2011) Feasibility of the transtracheal approach for the thoracic cavity in a large animal model. Surg Endosc 25: 1652–1658.
11. Liu YH, Chu Y, Wu YC, Chen TP, Lu MS, et al. (2012) Feasibility of endoscopic transoral surgical lung biopsy in a live canine model. Surg Innov 19: 162–170.
12. Ko PJ, Chu Y, Wu YC, Liu CY, Hsieh MJ, et al. (2012) Feasibility of endoscopic transoral thoracic surgical lung biopsy and pericardial window creation. J Surg Res 175: 207–214.
13. Liu YH, Chu Y, Wu YC, Yeh CJ, Liu CY, et al. (2011) Single-dose antimicrobial prophylaxis in transoral surgical lung biopsy: a preliminary experience. Surg Endosc 25: 3912–3917.

14. Wichmann MW, Hüttl TP, Winter H, Spelsberg F, Angele MK, et al. (2005) Immunological effects of laparoscopic vs open colorectal surgery. Arch Surg 140: 692–697.
15. Romeo C, Cruccetti A, Turiaco A, Impellizzeri P, Turiaco N, et al. (2002) Monocytes and neutrophils activity after minor surgical stress. J Pediatr Surg 37: 741–744.
16. Liu CY, Wang CH, Chen TC, Lin HC, Yu CT, et al. (1998) Increased level of exhaled nitric oxide and upregulation of nitric oxide synthase in alveolar macrophages in patients with primary lung cancer. Br J Cancer 78: 534–541.
17. Kalloo AN, Singh VK, Jagannath SB, Niiyama H, Hill SL, et al. (2004) Flexible transgastric peritoneoscopy: a novel approach to diagnostic and therapeutic interventions in the peritoneal cavity. Gastrointest Endosc 60: 114–117.
18. De Palma GD, Siciliano S, Addeo P, Salvatori F, Persico M, et al. (2010) A NOTES approach for thoracic surgery: transgastric thoracoscopy via a diaphragmatic incision in a survival porcine model. Minerva Chir 65: 11–15.
19. Sumiyama K, Gostout CJ, Rajan E, Bakken TA, Knipschield MA (2007) Transesophageal mediastinoscopy by submucosal endoscopy with mucosal flap safety valve technique. Gastrointest Endos 65: 679–683.
20. Sumiyama K, Gostout CJ, Rajan E, Bakken TA, Knipschield MA, et al. (2008) Pilot study of transesophageal endoscopic epicardial coagulation by submucosal endoscopy with the mucosal flap safety valve technique (with videos). Gastrointest Endosc 67: 497–501.
21. Suzuki K, Yasuda K, Kawaguchi K, Yoshizumi F, Inomata M, et al. (2010) Cardiopulmonary and immunologic effects of transvaginal natural-orifice transluminal endoscopic surgery cholecystectomy compared with laparoscopic cholecystectomy in a porcine survival model. Gastrointest Endosc 72: 1241–1248.
22. von Delius S, Wilhelm D, Feussner H, Sager J, Becker V, et al. (2010) Natural orifice transluminal endoscopic surgery: cardiopulmonary safety of transesophageal mediastinoscopy. Endoscopy 42: 405–412.

CHAPTER 18

Prospective Study of The Primary Evaluation of 1016 Horses with Clinical Signs of Abdominal Pain by Veterinary Practitioners, and The Differentiation of Critical and Non-Critical Cases

Laila Curtis, John Harold Burford, Jennifer Sara Marian Thomas, Marise Linda Curran, Tom Curtis Bayes, Gary Crane William England and Sarah Louise Freeman

School of Veterinary Medicine and Science, University of Nottingham, College Road, Sutton Bonington, Loughborough, LE LE12 5RD, UK

ABSTRACT

Background

The majority of research on the evaluation of horses with colic is focused on referral hospital populations. Early identification of critical cases is important to optimise outcome and welfare. The aim of this prospective study was to survey the primary evaluation of horses with clinical signs of abdominal pain by veterinary practitioners, and compare the initial presentation of critical and non-critical cases.

Results

Data from 1016 primary evaluations of horses presenting with clinical signs of colic were submitted by 167 veterinary practitioners across the United Kingdom over a 13 month period. The mean age of the study population was 13.5 years (median 12.0, range 0–42). Mean heart rate on primary presentation was 47 beats/min (median 44, range 18–125), mean respiratory rate was 20 breaths/min (median 16, range 6–100), and median gastrointestinal auscultation score (0–12, minimum–maximum) was 5 (range 0–12). Clinical signs assessed using a behavioural severity score (0–17, minimum–maximum), were between 0 and 6 in 70.4 % of cases, and 7–12 for 29.6 % of cases. Rectal examination was performed in 73.8 % of cases. Cases that responded positively to simple medical treatment were categorised retrospectively as 'non-critical'; cases that required intensive medical treatment, surgical intervention, died or were euthanased were categorised as 'critical'. Eight-hundred-and-twenty-two cases met these criteria; 76.4 % were 'non-critical' and 23.6 % were 'critical'. Multivariable logistic regression was used to identify features of the clinical presentation associated with critical cases. Five variables were retained in the final multivariable model: combined pain score: (OR 1.19, $P < 0.001$, 95 % CI 1.09–1.30), heart rate (OR 1.06, $P < 0.001$, 95 % CI 1.04–1.08), capillary refill time >2.5 s (OR 3.21, $P = 0.046$, 95 % CI 1.023–10.09), weak pulse character (OR 2.90, $P = 0.004$, 95 % CI 1.39–5.99) and absence of gut sounds in ≥1 quadrant (OR 3.65, $P < 0.001$, 95 % CI 2.08–6.41).

Conclusions

This is the first study comparing the primary presentation of critical and non-critical cases of abdominal pain. Pain, heart rate, gastrointestinal borborygmi and simple indicators of hypovolaemia were significant indicators of critical cases, even at the primary veterinary examination, and should be considered essential components of the initial assessment and triage of horses presenting with colic.

Keywords: Horse Colic Evaluation Critical Outcome

BACKGROUND

The term 'colic' refers to clinical signs of abdominal pain. Abdominal pain in the horse can be caused by a plethora of pathological processes, and manifest itself in many forms. Colic remains a major cause of morbidity and mortality in the horse [1, 2]. The majority of horses presenting with colic resolve with simple medical treatment [1], but a significant proportion may be critical, requiring intensive medical or surgical treatment for a successful outcome. Many of the critical cases show a rapid deterioration due to intestinal compromise, hypovolaemia and systemic inflammatory response syndrome. Any delay in diagnosis can affect prognosis [3], and prolong pain and suffering in cases where intensive care or surgical intervention are not an option. In human emergency medicine, the process of triage whereby patients are rapidly assessed on primary presentation to determine the priority of their treatment, is well established [4, 5]. It is often used in conjunction with "red flag" protocols; these assist in the early identification of symptoms associated with pathologies which require rapid diagnosis and treatment. In equine veterinary medicine, decision-making can be more difficult, both due to the many different types of colic in the horse, but also because there are significant gaps in the evidence. The majority of evidence on decision-making in the horse with abdominal pain is based on studies of referral hospital populations. There are two main limitations of this evidence. Firstly, referral hospital cases represent only a small subset of horses suffering from abdominal pain which is subject to bias; referral cases are predominantly those that are considered to be critical or surgical and require referral treatment. In addition the cost of referral treatment means that it is limited to cases where the owner can afford the higher cost, and has decided that it is appropriate in their horse based on the financial and/or emotional value of the horse. Secondly, the data from referral hospital studies is based on the clinical presentation of the horse when it arrived in the referral hospital, rather than the first assessment performed by the practitioner who performed the primary examination. There are very few studies which have reported on the primary presentation and evaluation of horses presenting with clinical signs of abdominal pain [6, 7, 8]. The current evidence includes two studies of incidence and/or causes in the UK [6, 9], one study on clinical parameters of horses presenting in primary practice in France [10], and one study in the UK reporting on the clinical parameters of horses with recurrent colic [11]. This highlights a significant gap in the evidence from the primary population of horses, which is surprising, considering the prevalence of abdominal pain in the horse and its importance to veterinary practitioners [12]. Research on the primary evaluation of horses with clinical signs of abdominal pain by veterinary practitioners is required to fill gaps in the current evidence and aid decision-making.

The aim of this study was to survey the primary evaluation of horses with clinical signs of abdominal pain by veterinary practitioners, and compare the primary presentation of critical and non-critical cases.

The objectives of the study were: (1) To describe the clinical presentation of horses with signs of abdominal pain on the primary evaluation by a veterinary surgeon; (2) To document the diagnostic approaches used by veterinary surgeons on the primary evaluation of horses with clinical signs of abdominal pain and (3) To identify clinical features which differ between non-critical and critical cases at the primary presentation to a veterinary practitioner.

METHODS

The study was reviewed and approved by the School of Veterinary Medicine and Science, University of Nottingham. The study design and data collection forms were initially discussed and piloted with four veterinary practices, and modified on the basis of feedback, prior to the survey launch.

Recruitment of participating veterinary surgeons

A prospective 13 month study (September 2012–October 2013) was designed to collect data from veterinarians involved in the primary assessment of horses with abdominal pain [13, 14]. The sampling frame consisted of all veterinary practices that were published on the RCVS (Royal College of Veterinary Surgeons) Directory of Veterinary Practices in the United Kingdom in 2010 (n = 3640). This was systematically searched, and practices were excluded if they did not treat horses, or were branch practices, producing a final list of principal practices (n = 850). These were contacted by post and asked to participate, and the survey was also publicised through two veterinary journals (Veterinary Record and Equine Veterinary Education), and websites. Participating veterinary surgeons completed a registration form which included information on their practice and experience, and were given a participant code to provide confidentiality when submitting data. Weekly emails were used to remind participants to report cases, and quarterly newsletters were sent to participating practices.

Case selection

Colic was defined as 'Any incidence of abdominal pain as assessed by the veterinary surgeon in attendance, and seven days free of colic was required for a case to be considered unrelated to a previous episode.

Data collection

Participating veterinary surgeons were asked to complete data collection forms detailing the information that they collected on the primary presentation of any cases of abdominal pain. The case assessment data collection forms were available either as an online form (Adobe Forms Central, Adobe Systems Inc., San Jose, CA, USA) through the project website, or paper-based versions. The case assessment form was divided into five sections using a mix of open and closed format questions relating to dependent and independent variables, requiring continuous and/or discrete information [15]. The specific sections were: (1) case presentation and history; (2) physical examination on presentation; (3) diagnostic tests; (4) treatment and diagnosis; (5) additional case information (Additional file 1).

Section 1 included information on the history and case description, and behavioural severity score. The behavioural severity score was developed based on previously described systems [16, 17, 18]. It consisted of six assessments of pain and behaviour: kicking, pawing, sweating, flank-watching, and attempts to lie down (which were scored from 0: none, 1: slight, infrequent, 2: moderate, occasional and 3: severe/continuous), and demeanour (scored 0: normal, 1: lower head, no response to auditory stimulus, and 2: twitching, agitations and continuous movement), giving a total maximum score of 17. These were analysed individually for each behavioural parameter, and summed to give a maximum total behavioural severity score of 17.

Section 2 included cardiovascular and respiratory indices, rectal temperature and assessment of gastrointestinal borborygmi. Gastrointestinal sounds were assessed by auscultation, using descriptors which were converted to a numerical value (0 = absent, 1 = reduced, 2 = normal and 3 = hypermotile) for each flank quadrant [19]. These were analysed for each quadrant of the abdomen and summed to calculate an overall score for analysis (maximum of 12).

Section 3 consisted of closed questions were used to determine whether rectal examination, nasogastric intubation, blood samples or abdominocentesis were performed, with open questions for findings of each test, and details of any additional tests used. Practitioners were also asked to identify whether any factors affected their choice of diagnostic tests, using specific options and a free text option.

Section 4 recorded the treatments and outcomes of each case. Treatments used by practitioners were recorded using closed questions according to category (non-steroidal anti-inflammatory drugs, opioids, oral fluids, sedatives, spasmolytics, anthelmintics, laxatives or others). Open questions were used to record the names and dosages of treatments, and the detail of other treatments.

The case diagnosis was recorded as an open text comment. Case diagnoses from all cases were reviewed at the conclusion of the study and categorised into four main categories: (1) no definitive diagnosis (subcategorised into spasmodic, gas/tympanitic, and unknown) (2) SCOD (simple colonic obstruction and distension [20]), subcategorised into large colon impaction and large colon displacement, (3) surgical/strangulating lesion (subcategorised according to lesion location), (4) other conditions (subcategorised according to lesion location/type) (Table 1). The diagnosis category was determined by reviewing the veterinary surgeon's diagnosis, presenting signs, physical examination findings, diagnostic test findings, further information provided by the veterinary surgeon and final outcomes recorded by the veterinary surgeons. If final outcomes were pending when the form was submitted, participating veterinary surgeons could provide details if they wished to be contacted subsequently for further information. The category definitions and inclusions were generated and discussed by three researchers (LC, JB and SF), and the data were reviewed and categorised by one researcher (SF).

Table 1. Disease categories for 1016 horses from a survey of veterinary practitioners' primary assessment of colic

Disease category	Definition/inclusion criteria for disease category and sub-category	Number of cases
No definitive diagnosis	Defined as cases in which a definitive diagnosis was not determined at the primary or any subsequent assessments	580
Sub category: spasmodic	Inclusion criteria: described by veterinary surgeon as spasmodic, no abnormalities on rectal examination, and resolved with medical treatment Exclusion criteria: cases subsequently found to have other lesions were excluded and categorised according to the final diagnosis/outcome	254
Sub category: gas	Inclusion criteria: with gas distension of intestines on rectal examination, with no underlying cause of distension identified Exclusion criteria: cases with gas distension associated with another lesion were excluded and categorised according to the primary lesion	68
Sub category: unknown	Inclusion criteria: diagnosis undetermined by veterinary surgeon, or where proposed diagnosis could not be confirmed from diagnostic work up (including cases where there were no clinical or diagnostic findings to support the veterinary surgeon's proposed diagnosis (e.g. cases reported as 'impactions' where no rectal examination was performed, recurrent or geriatric cases euthanased for colic with mild signs of pain, no rectal findings and no post mortem results)	258
SCOD (impaction or simple displacement)	Defined as simple obstruction with subsequent distension of the large colon [2] diagnosed on the basis of positive findings on rectal examination either at the primary or any subsequent assessments, and resolved with medical treatment	155
Sub category: large colon impaction	Inclusion criteria: positive rectal finding of a primary large colon impaction Exclusion criteria: impactions with a positive sand test were categorised as a separate category under category 4 Cases which required surgical intervention, euthanasia or died were excluded and categorised as category 3. Surgical/strangulating lesion	121
Sub category: large colon displacement	Inclusion criteria: positive rectal finding of a large colon displacement on rectal examination (including palpation of a LDD displacement, RDD displacement, Pelvic flexure retroflexion or abnormal taenial bands) Exclusion criteria: cases which required surgical intervention, euthanasia or died were excluded and categorised as category 3. Surgical/strangulating lesion	34
Surgical/strangulating lesion	Defined as cases that required surgical treatment, were euthanased or died due to surgical or strangulating lesions either at the primary or any subsequent assessments	178
Sub category of SI lesion	Inclusion criteria: identification of small intestinal lesion at surgery or post mortem, or where these were not available, positive rectal findings of small intestinal distension	72

Sub category of LI lesion	Inclusion criteria: identification of large intestinal lesion at surgery or post mortem, or where these were not available, positive rectal findings of large intestinal distension. This sub category includes large colon displacements which had surgical treatment or were euthanased	36
Sub category of other location	Inclusion criteria: identification of other intestinal lesion (gastric obstruction n = 1 and small colon strangulation n = 1) at surgery or post mortem	2
Sub category of no lesion site identified	Inclusion criteria: cases where the site of the surgical lesion was not determined, including surgical cases where the data could not be obtained, and horses that were euthanased or died with no rectal examination, or no findings on rectal examination and no post mortem	68
Other	Defined as cases where a definitive diagnosis was obtained either at the primary assessment or subsequent investigations, and which did not have either SCOD or a surgical/strangulating lesion	103
Gastric diseases (EGUS)	Inclusion criteria: EGUS diagnosed by endoscopy	2
Simple SI obstruction	Inclusion criteria: distended small intestine on rectal or ultrasound examination of thickened small intestine, and resolved with medical treatment	7
Caecal disease	Inclusion criteria: abnormalities of the caecum identified on rectal examination, including caecal tympany (n = 4), caecal impaction (n = 2), and typhlitis (n = 1) which resolved with medical treatment	7
Small colon obstruction	Inclusion criteria: positive rectal finding of impaction of the small colon, which resolved with medical treatment	6
Rectal impaction	Inclusion criteria: positive rectal finding of impaction of the rectum (n = 5), or meconium impaction (n = 1), which resolved with medical treatment	6
Grass sickness	Inclusion criteria: euthanased with a diagnosis of grass sickness confirmed by biopsy, post mortem or clinical signs (ptosis, dysphagia, sweating)	13
Neoplasia	Inclusion criteria: neoplasia confirmed on surgery or post mortem	2
Parasitic	Inclusion criteria: worms seen in faeces, high faecal egg count (>800 eggs per gram), positive surgical or post mortem findings, clinical history and laboratory results consistent with cyathastomiasis	9
Peritonitis/PUO	Inclusion criteria: pyrexic, with no underlying cause identified (PUO) (n=3) or peritonitis confirmed on abdominocentesis (n=4)	7
Enteritis, colitis or enterocolitis	Inclusion criteria: presence of diarrhoea, or ultrasound or surgical findings consistent with colitis or enteritis	13
Sand colic	Inclusion criteria: positive sand test (faecal sand test or on radiography)	14
Rupture of the gastro-intestinal (GI) tract	Inclusion criteria: rupture or tear of GI tract identified at surgery or post mortem, regardless of location Lesion location: 1 large colon, 1 small colon, 1 rectal, 1 unrecorded	4
Non-GI causes	Inclusion criteria: non-gastrointestinal problem confirmed by other clinical findings or diagnostic tests. This included cardiac disease diagnosed on auscultation and clinical signs of cardiac failure (n = 1), choke (n = 2), haematuria (n = 1), hepatic disease diagnosed based on blood biochemistry (n = 5) and 'maggots in sheath', diagnosed on physical examination (n = 2), urticarial/allergic reaction (n = 1), muscle abscess (n = 1)	13

Disease categories were determined retrospectively by reviewing the veterinary surgeon's diagnosis, presenting signs, physical examination findings, diagnostic test findings, further information provided by the veterinary surgeon and final outcomes recorded

Disease categories were determined retrospectively by reviewing the veterinary surgeon's diagnosis, presenting signs, physical examination findings, diagnostic test findings, further information provided by the veterinary surgeon and final outcomes recorded

Section 5 recorded additional case information, including current management and use of the horse, any recent changes, previous health problems and preventative health regimes (dental and anthelmintic). This information was recorded separately to the presenting history, in response to feedback from veterinary surgeons during the pilot phase.

Case outcome

Case outcome was also recorded as closed questions on category and free text sections for further detail. Cases in which outcome was not completed or not known at the time of submission were followed up by contacting the veterinary practice if consent had been given by the veterinary practitioner.

The case outcome was recorded as closed questions from options of 'resolved before visit', 'resolved with treatment at visit', 'referred', 'euthanased' or 'other'. For cases that were referred and for those with outcomes pending, the veterinary practitioners were able to indicate if they consented to be contacted for case follow up to be obtained; cases with consent were followed up to determine final outcome and diagnosis.

For the purposes of this study, two sub-groups (non-critical and critical) were extracted from the overall case population by retrospective classification of cases after they had reached outcome. Non-critical cases were defined as cases exhibiting signs of gastro-intestinal pain at the time of the clinical examination which responded positively to simple medical treatment. This group therefore did not include cases which had resolved prior to examination. Critical cases were defined as all instances where the animal was hospitalised to receive critical care (either intensive medical treatment and/or surgical intervention) at any point during the single episode of abdominal pain, or where the animal died or was euthanased on humane grounds as a result of the condition. All other scenarios where individuals were hospitalised for treatment deemed non-critical, or if euthanasia was performed due to factors not directly associated with the current disease were excluded. Cases were excluded from the study if information regarding the nature of the disease or outcome did not allow the case to be identified in one of these two categories.

An a priori power calculation based on a binary response variable (Y) suggested that 719 observations would achieve an 80 % power with 95 % confidence intervals to detect an odds ratio of 1.3 assuming a baseline probability that $Y = 1$ of 0.2 (PASS 11, NCSS, UT, USA). These calculations were based on data from a similar study (Mair, unpublished data).

Data analysis

Descriptive statistics were used for preliminary exploration of the data. The mean, median, mode, range and standard deviation was calculated for each of the continuous variables and percentage frequencies were calculated for all categorical data. Free text responses were analysed and categorised, and descriptive analysis performed to generate frequencies for each category.

Recruitment of diagnostic tests overall and between non-critical and critical groups was assessed using Chi squared tests. Indices relating to the signalment, history and clinical presentation of cases were compared between critical and non-critical cases using logistic regression. Screening was performed to determine the degree of association of independent continuous and categorical variables with the dependent outcome variable (non-critical/critical) using univariable logistic regression. Variables with a likelihood ratio test statistic (LRTS) of <0.2 were considered for inclusion in the multivariable model. Linearity of the continuous variables was assessed using generalised additive models (GAM) (Additional file 2). Pearson correlation coefficients were used for continuous data to investigate the association of these variables. The importance of biologically plausible interactions was assessed by including these terms within the model. Variables with >20 % missing data were initially excluded before being retested. Terms were added to the model in a forward, stepwise manner, with each included if they significantly improved the fit using the LRTS ($P < 0.05$). Analysis of residuals was performed to assess outlying data which were tested by exclusion from the data to ensure they did not apply excessive leverage to the model. Data analyses were performed using SPSS V21.0 (IBM Corporation) and R ×64 3.0.2 (http://www.r-project.org).

RESULTS

Participating veterinary surgeons

Data were submitted on cases from 167 veterinary surgeons working at 108 different practices (12.3 % of the practices that were contacted). Participants categorised their type of practice as: equine first and second opinion (30 %), equine first-opinion (24 %), mixed practice—mainly small animal (21 %), mixed practice—mainly large animal (19 %) and mixed practice—mainly equine (6 %).

Development of case forms and case submission

1064 case assessment forms were submitted. Cases which had been submitted in duplicate were excluded (n = 17). Forms containing information on subsequent visits (n = 31) were used to inform the categorisation of outcome and final diagnosis but were not used for any other analyses. 1016 case forms related to the primary assessment of individual cases and were subject to further analysis.

Participating veterinary surgeons did not complete all data fields for some cases, and therefore the number of cases where data were recorded is given for each parameter.

Study population

The study population consisted of 55.5 % (559/1008) geldings, 41.2 % (415/1008) mares and 3.4 % (34/1008) stallions, with a mean age of 13.5 (median 12.0, range 0–42) years. Estimated body condition of the cases was 69.8 % (692/992) moderate, 15.5 % (154/992) overweight and 14.7 % (146/992) thin. Fifty different breeds/types of equid were described.

History and management

Management history was recorded in 759 (74.7 %) case forms. A recent management change was reported in 47.0 % (357/759) cases; alterations in diet/bedding [21.5 % (77/357)] and in turnout [20.7 % (74/357)] were the most frequently reported changes (Additional file 3). Data on frequency of dental care was recorded in 588 cases, and of these, 17.5 % (103/588) were reported as having received no routine dental care, 10.2 % (60/588) as receiving dental care every 0–6 months, 47.8 % (281/588) every 6–12 months, and 24.5 % (144/588) every 1–2 years. Dental care was carried out by the veterinary surgeon in 57.4 % (271/472) and by an equine dental technician in 42.6 % (201/472) of cases. Horses were not ridden in 39.5 % (341/864) of cases, ridden 1–2 times per week in 21.5 % (186/864), ridden 3–6 times per week in 33.0 % (285/864) and ridden 7 times per week in 6 % (52/864) of cases. The mean (±SD) duration of colic signs (time since horse was last seen 'normal') was 8.7 ± 18.64 h for non-critical cases, and 10.64 ± 19.43 h for critical cases; the duration was not significant within the univariate statistical model ($P = 0.453$).

Clinical findings

Using an abdominal pain behavioural severity scale of 0–17 (minimum-maximum), 70.4 % of cases (716/1017) were scored 0–6 and 29.6 % (301/1017) scored 7–12; scores and frequencies for individual behaviours are shown in Table 2.

Table 2. Behavioural severity score for signs of colic in 1016 horses in a prospective study of the primary assessment of colic

Behavioural severity score	Score 0 (none/normal)	Score 1 (mild/occasional)	Score 2 (moderate/frequent)	Score 3 (severe/continuous)
Kicking	57.5 % (558/971)	30.3 % (294/971)	10.9 % (106/971)	1.3 % (13/971)
Pawing	40.1 % (388/968)	38.4 % (372/968)	19.4 % (188/968)	2.1 % (20/968)
Sweating	55.8 % (542/972)	24.1 % (234/972)	14.3 % (139/972)	5.9 % (57/972)
Flank watching	31.8 % (306/961)	46.4 % (446/961)	20.2 % (194/961)	1.6 % (15/961)
Attempts to lie down	22.8 % (226/991)	29.7 % (294/991)	32.0 % (317/991)	15.5 % (154/991)
Demeanour	50.3 % (486/967)	18.0 % (174/967)	31.7 % (307/967)	

Mean heart rate was 47 beats/min (median 44, range 18–125; SD 15.4), mean respiratory rate was 20 breaths/min (median 16, range 6–100; SD 12.4) and mean rectal temperature was 37.6 °C (range 33.0–40.3). Mucous membranes were pink in 91.7 % (911/993) of cases, red in 5 % (50/993) and cyanotic in 3.2 % (32/993) of cases. Capillary refill time was <2.5 s in 92 % (905/984) of cases and >2.5 s in 8 % (79/984) of cases. The median total gastrointestinal auscultation score was 5 (range 0–12).

Diagnostic approach

At the primary examination, veterinary practitioners assessed pain and behaviour in 100 % of cases (1016/1016), heart rate in 98.9 % (1005/1016) of cases, respiratory rate in 89.4 % (908/1016) and rectal temperature in 81.4 % of cases (827/1016). Gastrointestinal sounds were recorded in 98.7 % (1003/1016) of cases.

A rectal examination was performed in 73.8 % (743/1007) of cases, 35.6 % (348/978) underwent nasogastric intubation, 18.1 % (175/969) had a blood sample taken for various haematological and biochemical measurements, and abdominocentesis was carried out in 7.3 % (70/964) of cases. Additional diagnostic tests performed in some cases included ultrasound, faecal sedimentation test and faecal worm egg count which were performed in 3.4 % (35/1016), 2.5 % (25/1016) and 2.0 % (20/1016) of cases, respectively.

In 52.1 % (529/1016) cases, veterinary surgeons recorded that there were factors that affected their decision making. The three most commonly given identified factors that affected the choice of diagnostic tests were, "Mild nature of colic, diagnostic tests unnecessary (21 % frequency)", "Co-operation of the horse" (19 % frequency) and, "Financial situation of the owner (16 % frequency)" (Additional file 4).

Treatment

Treatment information was provided for 97.0 % (985/1016) of cases, and the frequency and categories of treatments are described in Additional file 5.

Case diagnosis

57.1 % of cases (580/1016) met the criteria for disease category 1: no definitive diagnosis, subcategorised as 'spasmodic', 'gas', and 'unknown' (Table 1).

15.3 % of cases (155/1016) met the criteria for disease category 2: SCOD, and were subcategorised as 'large colon impaction', and 'large colon displacement' (Table 1). 17.5 % of cases (178/1016) met the criteria for disease category 3: surgical/strangulating lesions. These were subcategorised as 'small intestinal lesion', 'large intestinal lesion', and 'other location' or 'no site identified'.

10.1 % of cases (103/1016) met the criteria for disease category 4: other diagnosis, which included a range of other disease types, including non-gastrointestinal causes of colic.

Categorisation and outcomes of critical and non-critical cases

One hundred and ninety-four cases were removed using the exclusion criteria (19.1 %, 194/1016) for critical/non-critical cases. These consisted of 85 cases which resolved prior to the initial visit (43.8 %, 85/194), 49 cases which were referred with outcome unknown (25.3 %, 49/194), 26 which were described as ongoing with insufficient follow-up detail to allow them to be categorised (13.4 %, 26/194), 20 cases which had no data, diagnosis or outcome (10.3 %, 20/194), and 14 cases with cause of colic/death identified as a non-gastro-intestinal associated diagnosis (7.2 %, 14/194). The latter 14 cases consisted of five with hepatic disease, two euthanased due to old age, two horses with maggots in their sheath, two with choke, one case of cystitis, one muscle abscess and one with urticaria.

822/1016 (80.9 %) cases met the inclusion criteria and were subject to further analysis; 628 were categorised as non-critical cases (76.4 %, 628/822), and 194 were categorised as critical cases (23.6 %, 194/822).

Of the 628 non-critical cases, 497 resolved with treatment at initial visit (79.1 %, 497/628), 93 resolved after further visits (14.8 %, 93/628), 37 resolved medically after referral (5.9 %, 37/628), and one case was euthanased for other reasons (a history of other ongoing medical problems which influenced the owners decision) although the symptoms of abdominal pain were resolving (0.2 %, 1/628).

Of the 194 critical cases, 135 were euthanased (69.6 %, 135/194) following the initial visit, 1 was euthanased after further visits (0.5 %, 1/194), 23 had surgery after referral from the first visit (11.9 %, 23/194), 16 were euthanased after referral (8.2 %, 16/194), 12 died (none of which had been referred) (6.2 %, 12/194), 6 resolved medically after referral for critical care (3.1 %, 6/194) and one case was referred for surgery after further visits (0.5 %, 1/194). Reasons for euthanasia were reported in 65 cases, and were categorised based on the veterinary surgeon's description as 'owner elected euthanasia' n = 15, 'owner unable to afford referral/surgery' (n = 11), travel/surgery not an option due to pain/age (n = 7), no response to pain relief (n = 30), and 'ileal biopsy confirmed grass sickness' (n = 2). Post mortem outcomes were recorded in five of the 135 cases that were euthanased.

Clinical presentation and diagnostic approach of critical and non-critical cases

There were significant differences in the utilisation of diagnostic tests between cases with critical and non-critical outcomes. Rectal examination, nasogastric intubation, blood sampling and abdominal paracentesis were performed significantly more frequently in cases with critical outcomes compared to those with non-critical outcomes (Table 3). Other diagnostic tests were performed in 48 non-critical cases and 22 critical cases. These were listed as faecal sand tests (performed in 15 non-critical and four critical cases), ultrasound examination (performed in 18 non-critical and 12 critical cases), faecal worm egg count or tapeworm ELISA (12 non-critical, four critical cases), dental examination (one non-critical case), *Streptococcus equi* antibody ELISA (one non-critical case), percussion of the abdominal wall (one non-critical case), histopathology (one critical case) and administration of phenylephrine eye drops (one critical case).

Table 3. Comparison of use of diagnostic tests in 194 critical cases of colic and 624 non-critical cases

	Non-critical	Critical	Significance
Rectal examination	72.7 % (454/624)	83.9 % (162/193)	$P \leq 0.002$
Nasogastric intubation	34.0 % (207/608)	47.9 % (91/190)	$P \leq 0.001$
Blood sample	16.3 % (99/606)	22.8 % (43/189)	$P < 0.05$
Abdominal paracentesis	5.7 % (34/597)	15.2 % (29/191)	$P < 0.0001$

Univariable analysis of critical and non-critical cases

The functional forms of the continuous variables were initially evaluated using GAM plots (Additional file 2) which demonstrated a linear functional relationship for each of the variables apart from weight, for which a quadratic function was tested but this did not result in a LRTS sufficient to merit consideration in the multivariable model. All variables with a LRTS of less than 0.2 were put forward into the multivariable model, including recoding using dummy variables of categorical groups appearing to show some evidence of association (Table 4).

Multivariable model of critical and non-critical cases

The final multivariable model indicated five variables to be significantly associated with the likelihood of a case being classified as critical (Table 5). Although individual pain behavioural indices showed a high degree of association in the univariable analysis, the combined pain score resulted in the best model fit with an odds ratio (OR) of 1.19 for each unit increase in pain score ($P < 0.001$). Increasing heart rate was also associated with the likelihood of being critical ($P < 0.001$) (Table 5).

DISCUSSION

The study aimed to survey the different presentations of cases of abdominal pain encountered by veterinary surgeons, their diagnostic approach, and to identify how severe cases may be differentiated on the primary examination. Previous studies of first opinion colic included a prospective study of 200 cases from a single veterinary practice in the UK [6], a cross-sectional questionnaire based study of 509 cases of abdominal pain across thoroughbred training premises in the UK [9], a stratified randomised prospective survey of the national incidence and risk factors for colic in 916 horses across the US [21], and a survey by the British Equine Veterinary Association in 2003–2004 [22]. The current study analysed data recorded by different veterinary surgeons, across the general population of horses, which provides information across a range of areas and different practices. It does not describe risk factors or the incidence of different types of abdominal pain within the populations which would require comparisons within the population as a whole and data from every case seen by each veterinary surgeon. The study design may be subject to selection bias and observer bias. Observer bias is a particular problem with subjective assessments, or where decisions, such as treatments for example, will depend on individual opinion. The study therefore used a standardised

Table 4. Univariable analysis for clinical variables from 822 horses on the primary presentation to veterinary practitioners for clinical signs of colic, categorised as critical or non-critical on the basis of case outcome

Variable	Coefficient	Odds ratio	95 % CI	*P* value
Continuous				
Age (years)	0.06	1.06	1.04–1.08	<0.001
Estimated weight (kg)	−0.00	1.0	1.00–1.00	0.247
Combined pain score (max. 17)	0.30	1.35	1.28–1.43	<0.001
Heart rate (beats per minute)	0.08	1.08	1.07–1.10	<0.001
Respiratory rate (breaths per minute)	0.06	1.06	1.05–1.08	<0.001
Rectal temperature	−0.20	0.82	0.64–1.05	0.114
Total gut sounds	−0.38	0.68	0.64–0.73	<0.001
Categorical				
Breed				
Pony	Ref.			
Arab/Arab ×	0.41	1.51	0.97–2.38	0.072
Cob	0.87	2.38	0.90–6.31	0.081
Cross breed	0.81	2.24	1.36–3.69	0.002
Warm blood/sports horse	−0.07	0.93	0.30–2.87	0.901
Heavy type	0.23	1.25	0.77–2.05	0.368
Thoroughbred (TB)/ TB ×	−0.71	0.49	0.11–2.19	0.352
Sex				
Mare	Ref.			
Gelding	−0.09	0.91	0.66–1.27	0.588
Stallion	−1.01	0.37	0.11–1.23	0.105
Body condition score				
Thin	Ref.			
Moderate	0.41	1.51	0.85–2.66	0.158
Overweight	−0.12	0.89	0.56–1.41	0.604
Kicking				
None	Ref.			
Slight/occasional	0.11	1.11	0.76–1.63	0.583
Moderate/frequent	0.98	2.65	1.65–4.27	<0.001
Severe/continuous	2.39	10.89	2.83–41.91	0.001
Pawing				
None	Ref.			
Slight/occasional	−0.39	0.68	045–1.02	0.064
Moderate/frequent	0.82	2.27	1.49–3.47	<0.001
Severe/continuous	1.29	3.65	1.32–10.10	0.013
Sweating				
None	Ref.			
Slight/occasional	0.59	1.80	1.14–2.83	0.011
Moderate/frequent	1.98	7.28	4.59–11.52	<0.001
Severe/continuous	3.03	20.70	10.32–41.5	<0.001
Flank watching				
None	Ref.			

Slight/occasional	−0.28	0.76	0.50–1.15	0.194
Moderate/frequent	0.82	2.26	1.45–3.53	<0.001
Severe/continuous	2.13	8.44	2.49–28.58	0.001
Attempts to lie down				
None	Ref.			
Slight/occasional	0.10	1.10	0.58–2.09	0.773
Moderate/frequent	1.05	2.85	1.61–5.05	<0.001
Severe/continuous	2.49	12.02	6.53–22.13	<0.001
Demeanour				
Standing normally, BAR	Ref.			
Lowered head	1.75	5.73	3.48–9.43	<0.001
Twitching/agitated	1.99	7.32	4.70–11.40	<0.001
Absence of gut sounds				
No	Ref.			
Yes	2.12	8.33	5.77–12.03	<0.001
Capillary refill time				
<2.5 s	Ref.			
>2.5 s	2.78	16.14	8.97–29.04	<0.001
Mucous membrane colour				
Pink	Ref.			
Red	2.06	7.86	4.13–14.93	<0.001
Cyanotic	4.87	130.1	17.6–964.1	<0.001
Pulse character				
Strong	Ref.			
Weak	1.20	7.39	4.72–11.56	<0.001
Recent management change				
None	Ref.			
Box rest	−1.72	0.180	0.04–.077	0.020
Increased stabling	−0.26	0.77	0.34–1.76	0.538
Decreased exercise	−1.65	0.19	0.03–1.50	0.115
Moved yards	−0.25	0.78	0.37–1.66	0.522
Field change/turnout	−0.82	0.44	0.21–0.93	0.032
Diet/bedding change	−0.33	0.72	0.38–1.36	0.306
Weather	−0.76	0.47	0.13–1.65	0.238
Other	−0.18	0.84	0.32–2.17	0.711
Recent faeces				
Faeces passed in past 6 h	Ref.			
No faeces within last 6 h	1.04	2.82	1.90–4.19	<0.001

Table 5. Multivariable model for clinical variables from 822 horses on the primary presentation to veterinary practitioners for clinical signs of colic, categorised as critical or non-critical on the basis of case outcome

Variable	Coefficient	SE	Odds ratio	95 % CI	*P* value
Combined pain score	0.171	0.045	1.187	1.086–1.297	<0.001
Heart rate (beats per minute)	0.057	0.010	1.058	1.037–1.080	<0.001
Capillary refill time					
<2.5 s	Ref.				
>2.5 s	1.167	0.372	3.213	1.023–10.09	0.046
Pulse character					
Strong	Ref.				
Weak	1.061	0.372	2.889	1.394–5.987	0.004
Absence of gut sounds					
No	Ref.				
Yes	1.295	0.287	3.652	2.082–6.406	<0.001

Three categorical variables were retained in the final model: capillary refill time >2.5 s ($P = 0.046$), weak pulse character ($p = 0.004$) and an absence of gut sounds in at least one quadrant ($P < 0.001$) (Table 5).

recording form with descriptors and categorisation where possible, and differentiation of critical cases was focused on differences on clinical parameters. To minimise selection bias, the data collection form was developed with input from practitioners, and participants were asked to submit data from all cases, irrespective of severity. The researchers also used weekly emails and quarterly newsletters to encourage participation, and participants were requested and reminded throughout the study to submit data on any cases seen, regardless of severity. There was still a relatively high proportion of surgical/strangulating cases (17.5 %) compared to previous studies of incidence and risk, although interestingly, a similar proportion of surgical/strangulating cases was also reported in the BEVA survey in 2003–2004 [23]. The BEVA Evidence Based Medicine (EBM) Colic Study reported on data from 1015 cases of colic in the UK collected over a 12 month period. The outcomes of their cases were 82.9 % survival, 15.5 % euthanased and 1.6 % died. This is very similar to the current study, where 13.5 % (138/1016) of cases were euthanased (135 critical cases, one non-critical case and two non-gastrointestinal colic cases) and 1.2 % (12/1016) horses died. There is likely to be a reporting bias towards severe cases with this type of multi-participant survey study design, and this study design is not appropriate for estimating incidence of different types of abdominal pain. Both studies however include data from a large range of cases illustrating the variation and challenges facing the veterinary practitioner. The present study highlighted that practitioners are assessing cases that vary from those in which clinical signs had resolved prior to examination, to those that died or were euthanased at or following the primary examination.

The information recorded by veterinary surgeons on history and signalment varied, with limited data collected in some cases. The section on management and preventative healthcare was designed based on research evidence of risk factors for abdominal pain, but this study showed that this information was not used by veterinary surgeons in decision-making in 25.4 % of cases of colic (258/1016), with areas such as dental care, current feeding and ridden management having the lowest completion rates. Pain and demeanour were assessed in all cases, highlighting the perceived importance associated with this aspect of the examination. Basic assessments of cardiovascular indices and gastrointestinal sounds were performed in 98 % of cases; exceptions were usually where the temperament of the horse or severity of signs precluded these assessments. Respiratory rate and rectal temperature were assessed less frequently (89.4 and 81.4 %, respectively), which probably reflects the greater significance of pain and cardiovascular indices in decision making in equine colic.

Rectal examination was the most commonly used diagnostic test (73.8 % of cases). A number of reasons were given in cases where it was not used, including 'lack of co-operation of the horse', 'recumbency of the horse', 'lack of facilities' and 'not required for diagnosis'. Most primary assessments are made in conditions with limited facilities and therefore there may be reasonable considerations against its use in some situations. The second most common diagnostic test was nasogastric intubation (35.6 % of cases); this was often associated with administration of oral fluids, and therefore its use may have been diagnostic, therapeutic or both [24]. Other diagnostic tests were used infrequently during the first assessment of cases, but these often resulted in positive findings, which may reflect their selection in cases with strongly indicative clinical features. Based on veterinary surgeons' rating of factors that affected their choice of diagnostic tests, most considered that clinical examination and rectal palpation were the key tests required for diagnosis (98 % stated that other tests were unnecessary). Lack of experience in techniques was the least frequently identified reason (<10 %) that affected choice of diagnostic tests and therefore is not currently a major limiting factor affecting veterinary surgeons' decision-making. There was marked variation in veterinary practitioner's approaches and the tests used in both critical and non-critical cases. Although out of scope for the present study, more evidence is required on how different tests contribute to decision making, why veterinarians preferentially use tests in some cases and not in others, and the influence of factors such as cost, condition of the horse, facilities, and owner wishes.

There are many potentially confounding factors which will influence decision-making including the veterinary surgeon's opinion, client preference, finance constraints, and facilities. Therefore analysis of the non-critical and critical cases focused on clinical features of the horse at presentation rather than diagnostic approach and treatments used. Cases were categorised non-critical vs critical as this was considered most appropriate for a primary care setting. Most previous studies are based on referral populations, and have categorised cases as medical/surgical or on the basis of survival/death outcomes [25, 26, 27]. This does not allow comparison with mild cases seen in a primary care setting, and will exclude a proportion of the population for whom referral or surgery is not an option. In this study, the majority of the critical cases were euthanased in the primary practice setting (69.6 %, 135/194); only 23.7 % (46/194) of the critical cases were referred for surgery or medical treatment, and only 12.4 % (24/194) horses had surgery. Categorisation of cases as non-critical and critical reflect the key decision-making by practitioners; it encompasses all the severe cases, and captures data from other diseases such as grass sickness and enterocolitis which require also critical decision-making but would not be considered in a surgical category. The parameters that remained in the final multivariate model as significantly associated with critical cases (higher heart rate, increased pain/behaviour scores, reduced gastrointestinal sounds and simple indicators of hypovolaemia or shock) are similar to those identified as prognostic indicators in previous studies from referral hospital population. This is however the first study to assess and highlight the importance of these parameters at the primary assessment.

Scoring systems that have been previously described for behavioural assessments of pain associated with abdominal pain, have included assessment of individual behaviours such as rolling, flank gestures or kicking [28]. This study found that all six of the pain/behaviour assessments (kicking, pawing, sweating, flank-watching, attempts to lie down and demeanour) were significant in the univariable model, but the total sum of all six scores was most predictive in the final model. The findings of this study are consistent with previous research on the importance of pain/behaviour in identifying critical cases, but again highlights that this is also significant in the early presentation of cases.

Heart rate has been consistently associated with abdominal pain severity and mortality [25], and identified as a useful prognostic indicator of survival [26, 27, 29, 30, 31, 32]. The use of heart rate as a prognostic indicator for surgery is less well documented; it was omitted from a multivariable logistic regression model by Reeves et al. [33] despite being used in the death/survival model in the same study. Other indices found to be significant in predicting outcome of equine colic include packed cell volume (PCV) [31], total protein (TP) albumin [34] and peritoneal lactate concentration [35] all of which were rarely utilised at the primary examinations in this study. This present study shows that changes in cardiovascular indices may predict critical cases even at the primary presentation of the case.

Gastrointestinal sounds were also retained in the final model. Absence or decrease in intestinal sounds was significantly associated with survival [36] or the need for surgery in studies by Parry et al. and White et al. [37, 38]. The present study looked at changes in gastrointestinal sounds within individual quadrants of the abdomen, as well as overall gastrointestinal sounds (summed score from all four quadrants). The final model had absence of gut sounds within one quadrant and this was more predictive than an overall decrease in the total score gut sounds across all four quadrants, which suggests a localised absence may be an earlier indicator of severe gastrointestinal disease.

There were a number of cases in the present study which had severe clinical parameters comparable with levels found on admission to hospital (data not shown), consistent with relatively advanced pathology. The advanced nature of these cases highlights some of the challenges faced by primary care practitioners, and identifies a potential requirement for improved owner education on recognising and seeking assistance for horses with abdominal pain. A significant number of cases died (n = 12) or were euthanased at the primary evaluation (n = 136) in this study. This presents a potential welfare concern, and more research is needed to investigate the possible reasons behind this, including how owners are recognising colic and deciding to seek veterinary assistance, and the impact of duration of abdominal pain on clinical signs and outcome.

CONCLUSIONS

There are still substantial gaps in the evidence relating to primary veterinary care of horses with abdominal pain. This study is an initial step in gathering evidence and identifying areas for future research. It illustrates the ranges of severity and outcome, and some of the challenges in diagnosis that practitioners face. It is the first study comparing non-critical and critical cases of abdominal pain on primary presentation to the veterinary practitioner. It highlights the importance of behavioural manifestations of pain, heart rate, pulse character, mucous membrane colour and gastrointestinal borborygmi in the primary triage of critical cases, and should be considered essential aspects of first veterinary examination.

AUTHORS' CONTRIBUTIONS

LC was the main researcher, undertaking this work as part of her Ph.D., with a major role in data collection, data analysis and manuscript preparation. JSMT, MC and TB undertook data collection. GE, JB and SF contributed to study design, data collection and study execution, and preparation of the final manuscript. All authors contributed to manuscript preparation. All authors read and approved the final manuscript.

ACKNOWLEDGEMENTS

The authors would like to thank the participating practices and veterinary practitioners.

COMPLIANCE WITH ETHICAL GUIDELINES

COMPETING INTERESTS The authors declare that they have no competing interests.

ETHICAL ANIMAL RESEARCH The study was reviewed and approved by the Ethics Committee, School of Veterinary Medicine and Science, University of Nottingham.

SOURCES OF FUNDING L. Curtis was funded by the School of Veterinary Medicine and Science, University of Nottingham. The survey was supported by a grant from Petplan Charitable Trust.

REFERENCES

1. Tinker MK, White NA, Lessard P, Thatcher CD, Pelzer KD, Davis B, et al. Prospective study of equine colic incidence and mortality. Equine Vet J. 1997;29(6):448–53.
2. Ireland JL, Clegg PD, McGowan CM, Platt L, Pinchbeck GL. Factors associated with mortality of geriatric horses in the United Kingdom. Prev Vet Med. 2011;101(3):204–18.
3. Fischer AT. Advances in diagnostic techniques for horses with colic. Vet Clin N Am Eq Prac. 1997;13(2):203–19.
4. Canonico M, Plu-Bureau G, Lowe GDO, Scarabin P-Y. Hormone replacement therapy and risk of venous thromboembolism in postmenopausal women: systematic review and meta-analysis. BMJ. 2008;336(7655):1227–31.
5. Guyatt GH, Oxman AD, Vist GE, Kunz R, Falck-Ytter Y, Alonso-Coello P, et al. GRADE: an emerging consensus on rating quality of evidence and strength of recommendations. BMJ. 2008;336(7650):924–6.
6. Proudman CJ. A two year, prospective survey of equine colic in general practice. Equine Vet J. 1992;24(2):90–3.
7. Hillyer MH, Mair TS. Recurrent colic in the mature horse: a retrospective review of 58 cases. Equine Vet J. 1997;29(6):421–4.
8. Hillyer MH, Taylor FGR, Proudman CJ, Edwards GB, Smith JE, French NP. Case control study to identify risk factors for simple colonic obstruction and distension colic in horses. Equine Vet J. 2002;34(5):455–63.
9. Hillyer MH, Taylor FGR, French NP. A cross-sectional study of colic in horses on Thoroughbred training premises in the British Isles in 1997. Equine Vet J. 2001;33(4):380–5.
10. Goncalves S, Leblond A, Drogoul C, Julliand V. Using feces characteristics as a criterion for the diagnosis of colic in the horse: a clinical review of 207 cases. Revue Med Vet. 2006;157(1):3–10.
11. Scantlebury CE, Archer DC, Proudman CJ, Pinchbeck GL. Recurrent colic in the horse: incidence and risk factors for recurrence in the general practice population. Equine Vet J. 2011;43:81–8.
12. Traub-Dargatz J, Salman M, Voss J. Medical problems of adult horses, as ranked by equine practitioners. JAVMA. 1991;198(10):1745–7.
13. Issaoui L. Survey of colic cases seen in first-opinion practice. Vet Rec. 2013;172(24):643–4.
14. Issaoui L. Equine practitioner participation required for colic survey. Vet Rec. 2012;171(17):430.
15. Iarossi G. The power of survey design: a user's guide for managing surveys, interpreting results, and influencing respondents. Washington DC: The World Bank.

2006. https://openknowledge.worldbank.org/handle/10986/6975. Accessed 31 May 2015.

16. Price J, Catriona S, Welsh EM, Waran NK. Preliminary evaluation of a behaviour-based system for assessment of post-operative pain in horses following arthroscopic surgery. Vet Anaesth Analg. 2003;30(3):124–37.
17. Mair TS, Smith LJ. Survival and complication rates in 300 horses undergoing surgical treatment of colic. Part 1: short-term survival following a single laparotomy. Equine Vet J. 2005;37(4):296–302.
18. Sutton GA, Dahan R, Turner D, Paltiel O. A behaviour-based pain scale for horses with acute colic: scale construction. Vet J. 2013;196(3):394–401.
19. Sasaki N, Murata A, Lee I, Yamada H. Evaluation of equine cecal motility by ausculation, ultrasonography and electrointestinography after jejunocecostomy. Res Vet Sci. 2008;84(2):305–10.
20. Hillyer MH, Taylor FG, Proudman CJ, Edwards GB, Smith JE, French NP. Case control study to identify risk factors for simple colonic obstruction and distension colic in horses. Equine Vet J. 2002;34(5):455–63.
21. Traub-Dargatz JL, Kopral CA, Seitzinger AH, Garber LP, Forde K, White NA. Estimate of the national incidence of and operation-level risk factors for colic among horses in the United States, spring 1998 to spring 1999. JAVMA. 2001;219(1):67–71.
22. Concato J, Shah N, Horwitz RI. Randomized, controlled trials, observational studies, and the hierarchy of research designs. N Engl J Med. 2000;342(25):1887–92.
23. Mair TS, Mellor D. BEVA evidence based medicine study of analgesia for colic: preliminary findings. 8th International Colic Research Symposium, Quebec; 2005.
24. Hallowell GD. Retrospective study assessing efficacy of treatment of large colonic impactions. Equine Vet J. 2008;40(4):411–3.
25. Furr MO, Lessard P, White NA. Development of a colic severity score for predicting the outcome of equine colic. Vet Surg. 1995;24(2):97–101.
26. Thoefner MB, Ersboll AK, Jensen AL, Hesselholt M. Factor analysis of the interrelationships between clinical variables in horses with colic. Prev Vet Med. 2001;48(3):201–14.
27. Van der Linden MA, Laffont CM, van Oldruitenborgh-Oosterbaan MMS. Prognosis in equine medical and surgical colic. J Vet Intern Med. 2003;17(3):343–8.
28. Pritchett LC, Ulibarri C, Roberts MC, Schneider RK, Sellon DC. Identification of potential physiological and behavioral indicators of postoperative pain in horses after exploratory celiotomy for colic. Appl Anim Behav Sci. 2003;80(1):31–43.
29. Puotunen-Reinert A. Study of variables commonly used in examination of equine colic cases to assess prognostic value. Equine Vet J. 1986;18(4):275–7.
30. Reeves MJ, Curtis CR, Salman MD, Hilbert BJ. Prognosis in equine colic patients using multivariable analysis. Can J Vet Res. 1989;53(1):87–94.
31. Ihler CF, Venger JL, Skjerve E. Evaluation of clinical and laboratory variables as prognostic indicators in hospitalised gastrointestinal colic horses. Acta Vet Scand. 2004;45(1):109–18
32. Proudman CJ, Dugdale AHA, Senior JM, Edwards GB, Smith JE, Leuwer ML, et al. Pre-operative and anaesthesia-related risk factors for mortality in equine colic cases. Vet J. 2006;171(1):89–97.
33. Reeves MJ, Curtis CR, Salman MD, Stashak TS, Reif JS. Validation of logistic-regression models used in the assessment of prognosis and the need for surgery in equine colic patients. Prev Vet Med. 1992;13(3):155–72.
34. Grosche A, Schrodl W, Schusser GF. Specific parameters of blood and peritoneal fluid to indicate the severity of intestinal ischemia in colic horses. Tierarztl Prax G N. 2006;34(6):387–96.
35. Van den Boom R, Butler CM, Van Oldruitenborgh-Oosterbaan MMS. The usability of peritoneal lactate concentration as a prognostic marker in horses with severe colic admitted to a veterinary teaching hospital. Equine Vet Ed. 2010;22(8):420–5.

36. Grulke S, Olle E, Detilleux J, Gangl N, Caudron I, Serteyn D. Determination of a gravity and shock score for prognosis in equine surgical colic. J Vet Med. 2001;48(8):465–73. doi
37. Parry BW, Gay CC, Anderson GA. Assessment of the necessity for surgical intervention in cases of equine colic: a retrospective study. Equine Vet J. 1983;15(3):216–21.
38. White NA, Elward A, Moga KS, Ward DL, Sampson DM. Use of web-based data collection to evaluate analgesic administration and the decision for surgery in horses with colic. Equine Vet J. 2005;37(4):347–50.

CHAPTER 19

Geographic Trends in Research Output and Citations in Veterinary Medicine: Insight into Global Research Capacity, Species Specialization, and Interdisciplinary Relationships

Mary M Christopher[1]* and Ana Marusic[2]

[1]*School of Veterinary Medicine, 4206 VM3A, University of California–Davis, One Shields Ave, Davis, CA 95616, USA*

ABSTRACT

Background

Bibliographic data can be used to map the research quality and productivity of a discipline. We hypothesized that bibliographic data would identify geographic differences in research capacity, species specialization, and interdisciplinary relationships within the veterinary profession that corresponded with demographic and economic indices.

Results

Using the SCImago portal, we retrieved veterinary journal, article, and citation data in the Scopus database by year (1996–2011), region, country, and publication in species-specific journals (food animal, small animal, equine, miscellaneous), as designated by Scopus. In 2011, Scopus indexed 165 journals in the veterinary subject area, an increase from 111 in 1996. As a percentage of veterinary research output between 1996 and 2010, Western Europe and North America (US and Canada) together accounted for 60.9% of articles and 73.0% of citations. The number of veterinary articles increased from 8815 in 1996 to 19,077 in 2010 (net increase 66.6%). During this time, publications increased by 21.0% in Asia, 17.2% in Western Europe, and 17.0% in Latin America, led by Brazil, China, India, and Turkey. The United States had the highest number of articles in species-specific journals. As a percentage of regional output, the proportion of articles in small animal and equine journals was highest in North America and the proportion of articles in food animal journals was highest in Africa. Based on principal component analysis, total articles were highly correlated with gross domestic product (based on World Bank data). The proportion of articles in small animal and equine journals was associated with gross national income, research and development, and % urban population, as opposed to the proportion of food animal articles, agricultural output, and % rural population. Co-citations linked veterinary medicine with medicine in the United States, with basic sciences in Eastern Europe and the Far East, and with agriculture in most other regions and countries.

Conclusions

Bibliographic data reflect the demographic changes affecting veterinary medicine worldwide and provide insight into current and changing global research capacity, specialization, and interdisciplinary affiliations. A more detailed analysis of species-specific trends is warranted and could contribute to a better understanding of educational and workforce needs in veterinary medicine.

Keywords: Agriculture Bibliometrics Economics Education Journal Medicine Research publication

BACKGROUND

Veterinary medicine is a diverse profession with strong historic roots in war (horses) and agriculture (livestock for food and fiber) [1, 2]. During the past few decades, the veterinary profession has undergone a profound shift in focus from agricultural animals to companion animals in the United States (US) and to a lesser extent in Western Europe [1, 2, 3, 4]. Several demographic and socioeconomic factors have contributed to this shift, including consolidation of the livestock and food production industries [5, 6], urbanization, changing social attitudes [4, 7], a gender shift towards women in veterinary medicine [7, 8], species-based tracking in veterinary curricula [9], and expanding clinical specialization [10]. The number of veterinary graduates entering food animal practice in the US has declined significantly since 1989 [11] and more than 75% of new graduates now enter small animal practice or specialized training in fields such as oncology and orthopedic surgery [12]. Today, 76.9% of veterinarians in the US work primarily with companion animals and 7.8% work primarily with food animals [13]. This shift has had major implications on educational and research programs; created shortages of veterinarians in laboratory animal medicine, research, food animal practice, regulatory veterinary medicine, and public health; and led to concerns about the profession's ability to meet global animal health and food security needs [1, 2, 3, 4, 7]. Despite rapid globalization of the veterinary profession, our understanding of the scope and quality of these demographic changes in other regions of the world, their potential effects on veterinary education and research, and their relationship to quantitative demographic indicators is limited.

Veterinary journals reflect the research and clinical practice trends and priorities of the veterinary profession around the world. Bibliometric data for scientific publications can be used to assess journal quality, map worldwide trends in research productivity and quality, and evaluate interdisciplinary alliances [14, 15]. Bibliometric analyses also can be used to mirror geographic trends within a discipline and provide insight into current and changing research priorities in different regions and countries. We hypothesized that bibliographic data would identify geographic differences and changes over time in veterinary research capacity, trends in species specialization, and interdisciplinary relationships in the veterinary profession that corresponded with demographic and economic indices. To assess this we evaluated veterinary journals indexed in the Scopus citation database. Scopus indexes a broader geographic range of journals and more core veterinary journals than other indexes and is the only database that designates journal subgroups based on species [16, 17]. We compared overall and species-specific research productivity to demographic and economic variables for each country. We also assessed the frequency of co-citations between veterinary journals and journals in other disciplines. The results of this study could help identify the leading edge of global shifts in veterinary specialization and research and indicate possible underlying cultural contexts and evolving directions of the veterinary profession.

METHODS

Data from the Scopus database were retrieved between Dec 2011 and Sept 2012 by using the SCImago portal [18]. Countries were organized into eight geographic regions: Western Europe, Eastern Europe, Africa (Central and Southern Africa), North America (US and Canada), Latin America (including the Caribbean), Middle East (Middle East and North Africa), Asia, and the Pacific region. These geographic regions were the same as those designated in SCImago, with the exception of North Africa and the Middle East, which were grouped together (as grouped by the World Health Organization), because of the low number of articles from African regions and because North African countries are closely aligned with Middle Eastern countries with regard to agriculture and cultural attitudes about animals [19].

The most current journal data (2011) obtained for the subject area "Veterinary" included the SCImago Journal Rank (SJR) indicator, H-index, total and citable articles (last 3 years), total citations (last 3 years), self-citations (last 3 years), citations/article (last 2 years), number of cited and uncited articles, years of coverage, country of publication, publisher, and subject category (food animal [FA], small animal [SA], equine [EQ], and miscellaneous [MISC]). The SJR is a prestige metric of journal impact based on the Scopus database that considers both the quantity of citations (average number of citations received in a specific year by documents published in the journal in the previous three years) and the quality of citations (based on the scientific influence of the journals that cite them) [20]. The SJR has been shown to be a good alternative to the impact factor for journal evaluation, although journal ranks differ, in part because the SJR restricts the inflationary effect of self-citations [20, 21]. The journal rankings used in this study were based on the SJR. The number and country of publication of veterinary journals indexed in 1999 (for 1996 citation data), 2003, and 2007 were recorded to document changes in the database over time. The 2011 impact factors for veterinary journals indexed in Scopus were obtained from Journal Citation Reports [22]. The Directory of Open Access Journals was queried to ascertain which of the journals were open-access [23] Language of publication was determined by examining journal websites and instructions for authors and through indexers.

Veterinary article data for regions and countries were obtained for the 14-year period 1996–2010 combined and for 1996 and 2010 (the most recent year for which data were available) separately. Data included the H index (for countries), total and citable articles, citations, citations/article, self-citations, number of cited and uncited articles, and % of articles with international collaboration. Percent international collaboration was defined as the proportion of articles whose author affiliations included more than one country. Regional data were also obtained for each individual year between 1996 and 2010. The number and percentage of articles in species-specific journals was recorded for each country and region from 1996 to 2010 and for the entire period combined. World data were calculated manually from regional data.

The following demographic and economic indicators were retrieved from the database of the World Bank [24]: gross domestic product (GDP; in current US$), gross national income (GNI; per capita, Atlas method, in current US$), agriculture value added (as a % of GDP), research and development (R&D; % of GDP), total population, % urban population, and % rural population. "Agriculture, value added" corresponds to the net output of agricultural sectors (forestry, hunting, fishing, crop cultivation, and livestock production). "Research and development" includes current and capital expenditures on basic and applied research and experimental development. Data were obtained for 2010 for all variables except R&D, for which 2009 data were used (too few countries had data reported for 2010). Where data were not available in 2010 (or 2009 for R&D), data from the most recent year available was used (but not earlier than 2000). The number of veterinary faculties in each country was obtained from the American Veterinary Medical Association [25] and the World Association of Veterinarians [26]. Where discrepancies occurred between sources, the highest number was used. The number of veterinary faculty per population and articles per population, GDP, and veterinary faculty were calculated.

Co-citation maps were visualized in SCImago for all countries for which data were available between 1996–97 and 2009–10. The discipline to which veterinary medicine was directly linked by co-citations in each year was recorded. In addition, co-citation links between veterinary subject categories (FA, SA, EQ, MISC) and other subject categories were also visualized and recorded. Where subject categories were linked to more than one discipline, both disciplines were recorded. The strength of co-citation links was not determined.

Data were collated and computations were done in Excel (Microsoft, Redmond, WA, USA). Data were analyzed statistically using JMP (v. 10.0.0, SAS Institute, Cary, NC, USA). Because most data were not normally distributed, median and range of values were reported. Principal component analysis was used to evaluate correlations among total articles, number and proportion of species-specific articles (using data from 1996–

2010), and demographic and economic indicators. Statistical significance was set at $P < 0.05$.

RESULTS

Veterinary journals in the Scopus database

The number of veterinary journals indexed by Scopus was 109 in 1996, 111in 1999, 116 in 2005, 128 in 2007, and 165 in 2011. The net increase (n = 56) in veterinary journals in 2011 as compared with 1996 came from Western Europe (n = 17), Asia (n = 14), Latin America (n = 9), Eastern Europe (n = 6), North America (n = 5), and the Middle East (n = 5). Veterinary journal citation data, impact factor, accessibility, language, and other data were tabulated for each region (Table 1). Citation data were incomplete for 14 journals (2 FA, 12 MISC). Of the 52,322 articles published in veterinary journals in the last three years, 46,297 (88.5%) were classified by Scopus as citable articles.

Table 1. Veterinary journals indexed by scopus in 2011, by region of publication*

Variable	Western Europe	Eastern Europe	Africa	North America	Latin America	Middle East	Asia	Pacific	World
No. journals (% of world)	82 (49.7)	14 (8.4)	2 (1.2)	25 (15.2)	13 (7.9)	6 (3.6)	20 (12.1)	3 (1.8)	165 (100)
No. open-access (% of world)	6 (17.6)	4 (11.8)	2 (5.9)	1 (2.9)	11 (32.4)	4 (11.8)	6 (17.6)	0 (0)	34 (100)
Impact factor, median (range)†	1.26 (0.03-4.25)	0.49 (0.17-0.67)	0.51, 0.52	0.94 (0.29-3.32)	0.31 (0.18-0.61)	0.29 (0.19-0.48)	0.32 (0.02-1.21)	0.95 (0.05-1.16)	0.92 (0.02-4.25)
SJR, median (range)	0.05 (0.03-0.25)	0.03 (0.03-0.05)	0.04	0.05 (0.03-0.15)	0.03 (0.03-04)	0.03 (0.03)	0.03 (0.03-0.15)	0.05 (0.03-0.06)	0.04 (0.03-0.25)
Journal H-index, median (range)	19 (1-73)	9 (2-20)	20 (18, 22)	21 (1-62)	6 (2-16)	3 (1-14)	6 (1-22)	22 (10-31)	14 (1-73)
No. articles last 3 years (% of world)	27,906 (53.3)	3428 (6.6)	245 (0.5)	7375 (14.1)	4463 (8.5)	1112 (2.1)	7147 (13.7)	646 (1.2)	52,322 (100)
No. citations last 3 years (% of world)	20,077 (70.4)	680 (2.4)	64 (0.2)	4857 (17.0)	950 (3.3)	270 (0.9)	1379 (4.8)	236 (0.8)	28,513 (100)
Uncited articles last 3 years (% of articles)	17,723 (63.5)	2907 (84.8)	196 (80.0)	4970 (67.4)	3704 (83.0)	905 (81.4)	6170 (86.3)	504 (78.0)	37,079 (70.8)
Self-citations last 3 years (% of citations)§	3739 (18.6)	146 (21.5)	12 (18.8)	912 (18.8)	226 (23.8)	119 (44.1)	141 (10.2)	35 (14.8)	5358 (18.8)
Citations/article last 2 years, median (range)	0.52 (0-1.93)	0.17 (0.08-0.43)	0.23 (0.17, 0.29)	0.42 (0-2.04)	0.08 (0-0.32)	0.13 (0.07-0.36)	0.11 (0-0.79)	0.51 (0-0.57)	0.28 (0-2.04)
Language of publication (No. journals)	English (55), German (7), German-English (5), French (4), French-English (2), English-German-French (1), Spanish (2), Spanish-English (1), Dutch (1), Dutch-English (1), Italian-English (2), Japanese-English (1)	English (11), Hungarian (1), Lithuanian-English (1), Polish-English (1)	English (1), English-Afrikaans (1)	English (25)	Portuguese-English (6), Portuguese-Spanish-English (3), Spanish-English (3), Spanish (1)	English (1), Turkish-English (3), Arabic-English (1), Hebrew-English (1)	English (19), Korean-English (1)	English (3)	English only (115), non-English only (16), mixed (34)
Species-specific journals	FA (11), SA (10), EQ (7)	FA (1), SA (1)	—	FA (3), SA (6)	—	—	FA (3), EQ (1)	SA (1)	FA (18), SA (18), EQ (8)

*Data for number of journals and species-specific journals, SJR and H-index, and articles (total documents) and citations are from Scopus, obtained via SCImago [18]. Regional location of journals was based on the publisher location indicated in Scopus in 2011. The number of open access journals was obtained from the Directory of Open Access Journals [23]. Impact factor data were obtained from Journal Citation Reports [22]. Language of publications was obtained from journal indexers, websites, and guidelines for authors.

†The number of journals in the Scopus database that had 2011 impact factors were as follows: Western Europe (65), Eastern Europe (6), Africa (2), North America (19), Latin America (4), Middle East (4), Asia (7), and Pacific (3).

§Refers to within-journal self-citations.

See Appendix for complete list of species-specific journals. *FA* Food Animal, *SA* Small Animals, *EQ* Equine. Other journals were categorized as Miscellaneous.

Veterinary article output and citations by country and region

Veterinary article and citation data were reported for 193 countries in the 8 regions for the combined 14-year period of 1996–2010 (Table 2). All but 10 countries had ≥90% citable articles. As a percentage of veterinary research output, Western Europe and North America together accounted for 60.9% of total articles and 73.0% of citations; countries in these regions also had higher H indexes than those in other regions. Eastern Europe had the fewest citations per article, whereas the Pacific region (Australia and New Zealand) had the most. The Pacific region also had the lowest proportions of uncited articles and self-citations, whereas Asia (largely India) had the highest. Africa had the highest percentage of articles with international collaboration, whereas the Middle East had the lowest.

Table 2. Veterinary Article And Citation Data For 193 Countries In Eight Geographic Regions During 1996-2010*

Variable	Western Europe	Eastern Europe	Africa	North America	Latin America	Middle East	Asia	Pacific	World
No. countries	25	22	46	2	42	20	26	10	193
Country H index, median (range)	34 (0-83)	8 (0-32)	5 (1-37)	84 (67, 98)	4 (0-41)	10 (1-67)	13 (1-47)	5 (0-59)	7 (0-98)
No. articles (% of world)	56,745 (34.8)	11,096 (6.8)	5229 (3.2)	42,566 (26.1)	13,173 (8.1)	10,567 (6.5)	23,922 (14.7)	6580 (4.0)	163,250 (100)
No. citations (% of world)	430,298 (40.0)	37,122 (3.4)	30,859 (2.8)	359,000 (33.0)	52,174 (4.8)	33,453 (3.0)	87,608 (8.0)	57,259 (5.3)	1,087,769 (100)
Uncited articles (% of articles in region)	13,847 (24.4)	3069 (27.8)	1179 (22.5)	9067 (21.3)	5016 (38.1)	4118 (39.0)	10,366 (43.3)	975 (14.8)	47,637 (29.2)
Self-citations (% of citations in region)†	36,270 (8.4)	2646 (7.1)	1500 (4.9)	28,650 (8.0)	3653 (7.0)	2331 (7.0)	10,625 (12.1)	2774 (4.8)	88,449 (8.1)
Citations/article, median (range)	8.69 (0.65-12.00)	3.86 (0.37-5.52)	7.26 (0.63-9.83)	9.72 (0.79-13.77)	7.26 (0.39-9.20)	4.33 (0.46-7.33)	5.01 (0.54-6.91)	9.90 (1.22-14.29)	7.26 (0.37-14.29)
Average % articles with international collaboration	31.2	20.4	52.9	23.0	32.2	16.7	24.3	39.0	30.2

*Data are from Scopus, obtained via SCImago [18]. Articles are total documents.
†Refers to within-country self-citations.

Regional trends were led by specific countries, and did not necessarily apply to individual trends for other countries in the region. In 2010, the US alone published 17.7% of all veterinary articles, followed by the United Kingdom (UK), India, Germany, Brazil, Canada, France, Turkey, and Japan (Table 3). Of countries with >100 articles during the 14-year period, the number of citations/article was highest for Iceland (12.9 citations/article), Denmark (12.6), Norway (11.6), Sweden (11.5), and Hong Kong (11.1). India and Iran had the highest proportion of uncited articles; Taiwan and Denmark had the least. India and Turkey had the highest proportion of self-citations.

Table 3. Article and citation data (1996-2010) for top countries in each region, based on h index*

Region	Country and Rank	H Index	Articles	Citations	Citations per Article	% Self-Citations	% Uncited Articles	% Int'l Collaboration	% World (1996)	% World (2010)	% Change
Western Europe	1. UK	83	12,888	134,410	9.2	30.1	27.8	34.3	8.5	7.4	-1.1
	2. France	65	6084	47,501	7.4	24.2	36.1	32.6	3.7	3.1	-0.6
	3. Netherland	63	4084	42,705	10.4	21.1	17.4	41.7	2.7	2.1	-0.6
	4. Germany	57	10,024	55,387	5.5	30.2	28.5	42.2	5.2	5.0	-0.2
	5. Denmark	57	2303	27,512	12.6	22.2	11.4	44.4	1.2	1.1	-0.1
Eastern Europe	1. Czech Rep	32	1636	9780	6.3	40.3	19.6	24.0	0.8	0.7	-0.1
	2. Hungary	31	1768	7236	4.0	26.9	38.1	36.6	1.4	0.8	-0.6
	3. Poland	30	4934	12,316	2.7	51.8	35.0	11.6	2.4	3.3	+0.9
	4. Slovakia	23	856	3657	4.2	31.4	28.4	32.0	0.6	0.3	-0.3
	5. Croatia	19	659	2071	3.7	28.2	35.7	28.3	0.4	0.4	0
Africa	1. So. Africa	37	1767	11,688	6.8	30.8	18.1	36.0	1.1	0.8	-0.3
	2. Kenya	27	584	4819	7.8	24.4	14.6	67.8	0.6	0.3	-0.3
	3. Ethiopia	23	476	2702	7.4	34.2	14.3	62.4	0.2	0.4	+0.2
	4. Tanzania	21	270	2118	8.0	23.6	15.4	72.0	0.2	0.1	-0.1
	5. Zimbabwe	21	267	2006	6.7	17.7	17.0	71.7	0.3	0.1	-0.2
North America	1. U.S.	98	35,208	316,515	9.0	47.0	21.2	22.8	24.4	17.7	-6.7
	2. Canada	67	6138	56,027	9.2	22.4	19.8	38.9	4.3	3.0	-1.3
Latin America	1. Brazil	41	8066	27,452	5.8	49.1	26.4	25.3	1.7	8.1	+6.4
	2. Argentina	34	1204	8216	7.9	22.0	20.5	44.0	0.6	0.8	+0.2
	3. Mexico	29	1442	7090	7.5	30.5	20.8	42.2	0.5	1.3	+0.8
	4. Uruguay	21	194	1659	9.4	22.3	17.3	60.2	0.1	0.2	+0.1
	5. Chile	20	688	2617	4.1	27.8	28.2	42.4	0.5	0.4	-0.1
Middle East	1. Israel	34	804	7287	9.3	18.3	14.8	32.3	0.5	0.5	0
	2. Turkey	24	5954	12,661	2.6	54.5	36.1	8.1	1.0	4.8	+3.8
	3. Iran	20	1663	3559	3.7	37.6	40.1	22.3	0.3	2.0	+1.7
	4. Egypt	23	663	3131	6.3	14.6	19.6	51.0	0.3	0.6	+0.3
	5. UAE	18	198	1284	6.6	17.8	15.5	55.4	0.2	0.1	-0.1
Asia	1. Japan	47	5113	35,062	7.2	32.2	16.5	24.0	3.1	2.3	-0.8
	2. China	34	2344	12,231	9.5	37.6	13.6	40.9	0.2	3.4	+3.2
	3. So. Korea	32	1719	7990	9.6	24.9	22.2	33.0	0.2	2.0	+1.8
	4. India	30	12,147	17,676	1.6	55.5	57.5	22.5	5.2	6.9	+1.7
	5. Taiwan	29	623	5493	10.9	23.7	11.0	22.6	0.2	0.6	+0.4
Pacific	1. Australia	59	4662	42,185	9.4	24.9	13.9	39.9	3.5	2.7	-0.7
	2. N Zealand	45	1794	17,037	9.6	28.5	13.4	43.7	1.1	0.9	-0.2

*Data are from Scopus, obtained via SCImago [18]. Articles are total documents. These countries together account for 84% of veterinary article output.

Worldwide, the number of veterinary articles published was 8815 in 1996 and 19,077 in 2010, for a net increase of 66.6% (Figure 1). The net increase in articles between 1996 and 2010 was highest in Asia (21.0% increase) followed by Western Europe (17.2%), Latin America (17.0%), the Middle East (12.3%), Eastern Europe (6.7%), North America (5.6%), Africa (1.5%), and the Pacific (1.4%). Between 1996 and 2010, China had the highest relative increase in article output (2767%, from 21 to 602 articles), followed by South Korea (1867%, from 18 to 354 articles), Iran (1207%, from 28 to 365 articles), Brazil (800%, from 162 to 1458 articles), Turkey (758%, from 100 to 858 articles), Italy (349%, 119 to 534 articles), Poland (157%, 233 to 599 articles), India (152%, 493 to 1241 articles), Germany

(80%, 498 to 895 articles), and the US (36%, 2337 to 3179 articles). High relative increases also were seen in Mexico and Thailand. In Africa, Nigeria had the largest increase, from 49 to 117 articles. A net decrease in articles between 1996 and 2010 was found only in Zimbabwe (from 27 to 9 articles) and Kenya (from 60 to 45 articles). Changes in total article output from 1996 to 2010 coincided with concomitant changes in percentage worldwide article contribution in all countries except the US and Germany, which had increased numbers of articles, but decreased relative contributions to world output (Table 3).

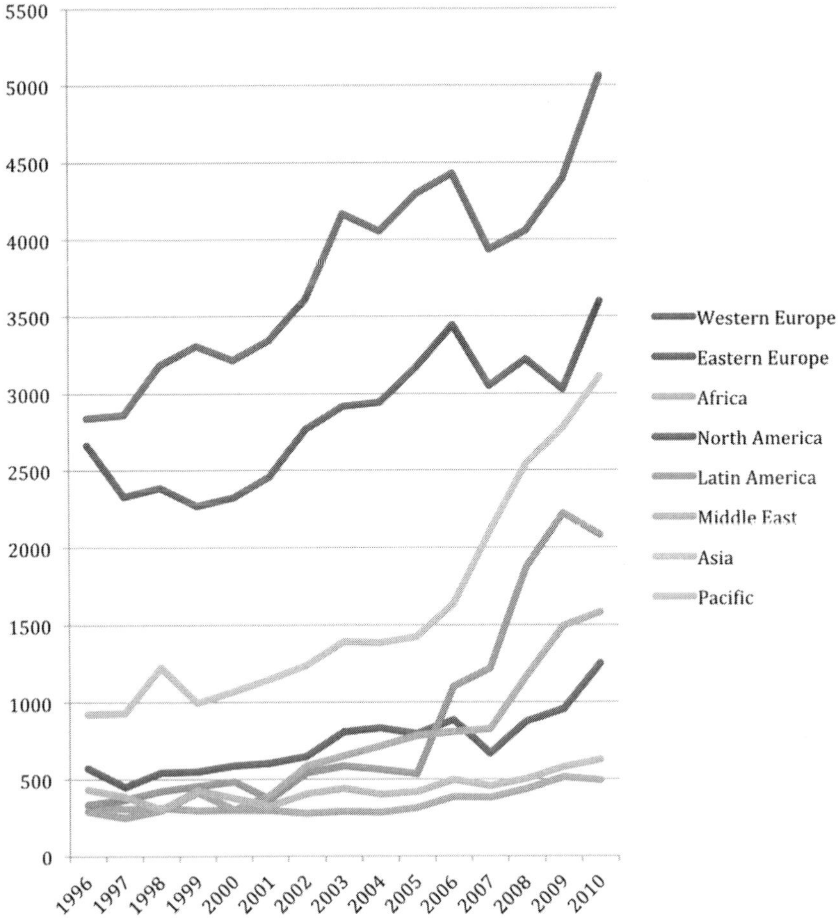

Figure 1. Regional Increases In The Number Of Veterinary Articles From 1996 To 2010. Data were obtained from SCImago [18].

Species-specific output and relation to demographic and economic variables

During the combined period of 1996–2010, 50,756 articles (31.1% of veterinary articles worldwide) were published in species-specific journals; of these, 15,587 (50.4%) were in FA journals, 15,715 (30.9%) were in SA journals, and 11,454 (22.5%) were in EQ journals (Table 4). The remaining 112,494 articles (68.9% of veterinary articles worldwide) were published in MISC journals. Western Europe and North America together published 64.0% of FA, 80.1% of SA, and 77.3% of EQ articles in species-specific journals in 1996–2010. Except for Asia, which contributed to 11% of world output in FA journals, other regions contributed <10% of articles in species-specific categories. The US had the most veterinary articles in all categories, with 5548 in FA journals (3.2-fold more than 2nd-ranked Germany; 23.5% of FA articles), 5957 in SA journals (4.8-fold more than 2nd-ranked UK; 37.9% of SA articles), 3506 in EQ journals (2.6-fold more than 2nd-ranked UK; 30.6% of EQ articles), and 22,957 in MISC journals (2.0-fold more than 2nd-ranked India; 20.4% of MISC articles)". Articles in species-specific journals increased between 1996 and 2010 in all regions except Africa (Figure 2). The number of articles in 1996 and 2010 respectively was 1139 and 2421 in FA journals; 774 and 1280 in SA journals; 713 and 1063 in EQ journals; and 6361 and 13,390 in MISC journals. The UK had the largest net increase in SA articles (33 to 113), with high relative increases in Brazil, China, Poland, Taiwan, and Turkey. The UK also had the largest increase in EQ articles (84 to 124), with high relative increases in Brazil, China, and Taiwan. Germany had the largest increase in FA articles (59 to 193), with high relative increases in China, Ethiopia, Iran, and Turkey.

Table 4. Regional Publication Output In Veterinary Species-Specific Journals (1996-2010)*

Region	Food Animal			Small Animal			Equine		
	No. articles	% of world	Top country	No. articles	% of world	Top country	No. articles	% of world	Top country
Western Europe	8451	36.1	Germany	5451	34.7	UK	4716	41.2	UK
Eastern Europe	756	3.2	Poland	468	3.0	Poland	281	2.5	Poland
Africa	1492	6.4	Nigeria	97	0.6	So. Africa	72	0.6	So. Africa
North America	6645	27.6	US	7144	45.5	US	4144	36.2	US
Latin America	1383	5.9	Brazil	553	3.5	Brazil	570	5.0	Brazil
Middle East	1380	5.9	Iran	368	2.3	Turkey	227	2.0	Turkey
Asia	2581	11.1	India	853	5.4	Japan	982	8.6	Japan
Pacific	899	3.8	Australia	781	5.0	Australia	462	4.0	Australia
World	23,587	100	US	15,715	100	US	11,454	100	US

*Data are from Scopus, obtained via SCImago [18]. Top country is based on number of articles published.

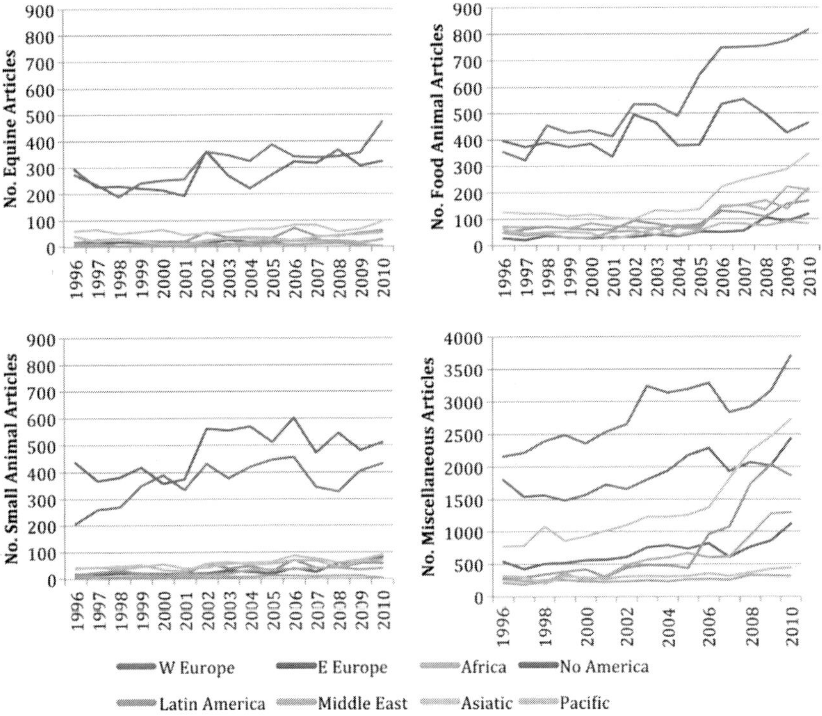

Figure 2. Regional increases in the number of articles published in equine, food animal, small animal, and miscellaneous veterinary journals from 1996 to 2010. Data were obtained from SCImago [18].

As a percentage of regional output, North America had the highest proportion of SA and EQ articles and Africa had the highest proportion of FA articles (Figure 3A). Within countries, the proportion of SA articles was highest in the US, France, and Australia (Figure 3B); the proportion of EQ articles was highest in Sweden; and the proportion of FA articles was highest in Denmark and the Netherlands. Less than 15% of articles from Turkey, India, Brazil and Poland were published in species-specific journals in the period 1996–2010.

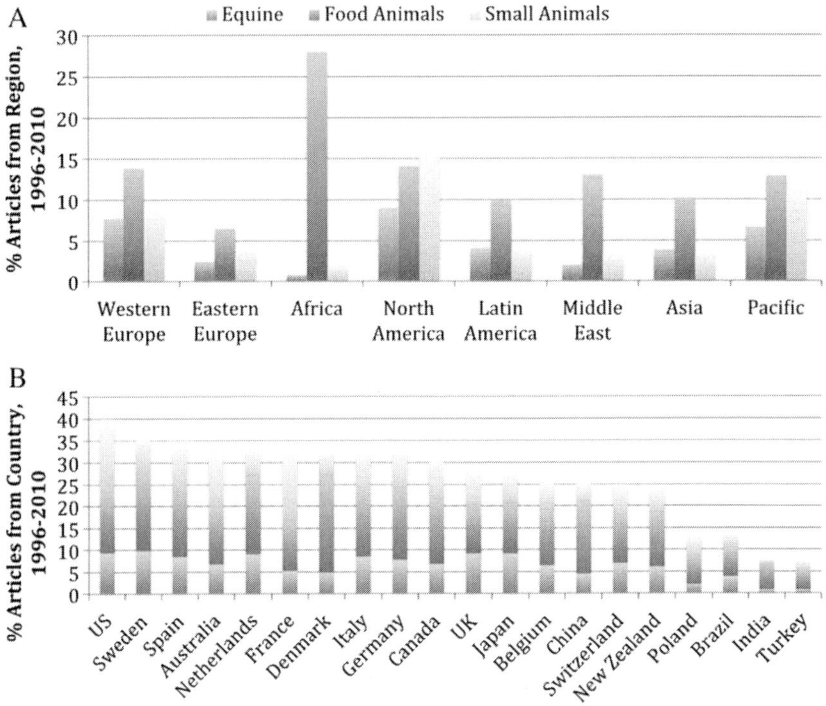

Figure 3. Percentage of articles in species-specific veterinary journals within each region (a) and within the top 20 countries for species-specific article output (B). Data were obtained from SCImago [18].

Demographic, economic, and veterinary faculty data were tabulated for the 193 countries that published veterinary articles from 1996–2010 (Table 5). Africa and Asia had the largest rural populations and the fewest articles per million people. When article output was normalized for GDP, the Pacific region ranked highest and the Middle East ranked lowest. Principal component analysis yielded three distinct clusters (Figure 4). Strong correlation ($P<0.0001$) was found between GDP, number of articles (total and species-specific), and number of veterinary faculties (component 1). A cluster consisting of agricultural output, % rural population, and % articles in FA journals was clearly and inversely separated from a cluster consisting of GNI per capita, % urban population, % R&D, and % articles in SA and EQ journals (component 2).

Table 5. Demographic and economic data (2010) by region, for countries that published veterinary articles from 1996 to 2010*

Variable	Western Europe	Eastern Europe	Africa	North America	Latin America	Middle East	Asia	Pacific†
No. countries	25	22	46	2	42	20	26	2
No. veterinary faculties	60	85	29	33	174	55	165	8
Total population (in millions)	413.5	337.3	842.6	343.5	588.4	454.0	3798.8	26.7
GDP, current US$ (in billions)	15.9	3.1	1.1	16.0	5.2	2.8	1.1	1.3
GNI per capita, current US$ (median)	42,970	7295	750	45,300	6115	6570	1870	37,775
Agriculture, value added, % of GDP (median)	1.8	6.4	23.9	1.5	6.3	7.3	16.2	4.0
R&D, % of GDP (median)§	1.8	0.7	0.2	2.4	0.4	0.6	0.2	1.8
% Rural population (median)	22.4	37.6	61.7	18.6	33.3	28.1	63.1	12.1
% Urban population (median)	77.6	62.5	38.4	81.5	66.7	71.9	36.9	88.0
Articles per million population (median)	133.4	7.3	3.4	160.0	8.9	9.7	2.4	329.7
Articles per million GDP (median)	3.4	1.7	3.3	3.4	1.2	1.0	1.1	8.9

*Data for veterinary faculties were obtained from the American Veterinary Medical Association [25] and the World Veterinary Association [26]. Data for population, gross domestic product (*GDP*), gross national income (*GNI*), agriculture value added, research and development (*R&D*), and % rural and urban populations were obtained from the World Bank [24]. Number of articles was for 1996-2010 from the Scopus database, obtained via SCImago [18].
†Data are for Australia and New Zealand.
§The number of countries for which R&D data were available: Western Europe (19), Eastern Europe (19), Africa (16), North America (2), Latin America (15), Middle East (8), Asia (13), Pacific (2).

The number of veterinary faculties was highest in Latin America, followed closely by Asia (Table 5). When normalized to population, Eastern Europe and Latin America had the most veterinary faculties, but publications per veterinary faculty were among the fewest (Figure 5). Between 1996 and 2010, the Netherlands, Denmark, and Sweden had the most articles per veterinary faculty (4457, 2443, and 2333, respectively), whereas, Russia, the Philippines, and Peru had the least articles per veterinary faculty (0.3, 5.7, and 13.4, respectively).

Within-country co-citations between veterinary medicine and other disciplines

Co-citation maps were available for 93 countries for one or more years between 1998–99 and 2009–10. In all regions, veterinary articles were linked by co-citations primarily with Agriculture & Biological Sciences (ABS) journals (Figure 6A). Several Eastern European countries had co-citation links with Medicine and basic sciences; in the Middle East, Hong Kong and Singapore, veterinary articles were often linked with Medicine (Table 6). Co-citation links shifted from ABS to Medicine in Switzerland and the US between 2003 and 2005. Co-citations shifted from ABS to Biochemistry, Genetics, and Molecular Biology (BGM) in China, Taiwan, and South Korea between 2004 and 2006.

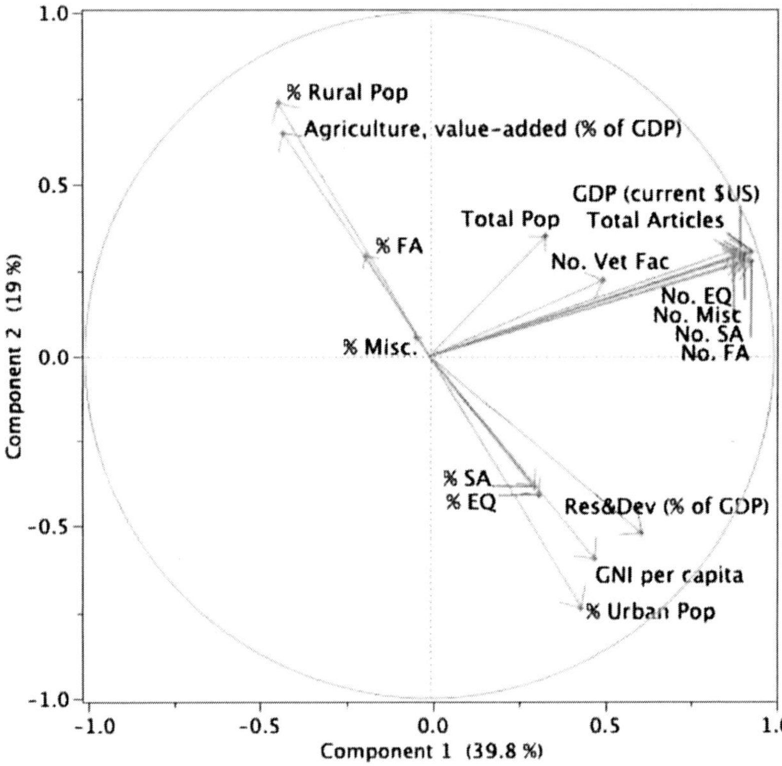

Figure 4. PRINCIPAL COMPONENT ANALYSIS (CORRELATION CIRCLE) OF VETERINARY ARTICLE OUTPUT AND DEMOGRAPHIC AND ECONOMIC VARIABLES. The first principal component (horizontal axis) accounts for 39.8% of variation and comprises the highly correlated cluster of gross domestic product (GDP) and the number of articles (total, small animal [SA], equine [EQ], food animal [FA], miscellaneous [Misc]). The second principal component (vertical axis) accounts for 19% of variation and comprises the opposing clusters of agriculture value-added and % rural population (Pop.) versus gross national income (GNI) per capita, research and development (Res & Dev), and % urban Pop. Variables close to the center are less strongly correlated with variables far from the center in the same cluster because some information is carried on other axes. %SA, %EQ, %FA, and %Misc refer to the proportions of articles in species-specific journals. Bibliographic data [18] from 1996–2010 were used in the analysis; demographic and economic data were primarily from 2010 [24].

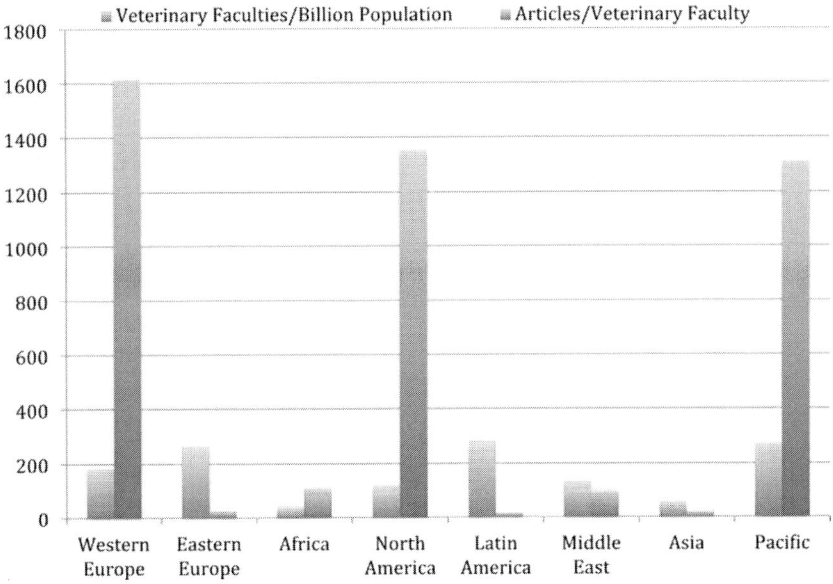

Figure 5. MEDIAN NUMBER OF VETERINARY FACULTIES (NORMALIZED TO POPULATION) AND MEDIAN ARTICLE OUTPUT (1996–2010) PER FACULTY, BY REGION. Only countries with veterinary faculties were included in each region: Western Europe (n = 16), Eastern Europe (n = 21), Africa (n = 14), North America (n = 2), Latin America (n = 22), Middle East (n = 14), Asia (n = 21), and Pacific (n = 2).

Within the Veterinary subject area, articles in FA journals were linked by co-citations almost exclusively with Animal Science & Zoology (ASZ) journals (within ABS); articles in EQ journals were linked with ASZ and other veterinary journals (data not shown). Articles in SA journals were linked with ASZ and Medicine, and, in the US, Western Europe (especially Switzerland), South Africa, and Australia, with MISC veterinary articles. In Asia, SA and EQ articles were linked with BGM. In Singapore, SA articles were linked with a variety of medical disciplines, including radiology & ultrasound, genetics, and biotechnology. Articles in MISC journals were linked by co-citations with Immunology, Parasitology, and Microbiology in all regions except North America and the Pacific (Figure 6B). In Japan, veterinary MISC articles were also linked with Marketing.

Figure 6. FREQUENCY OF CO-CITATION LINKS BETWEEN VETERINARY MEDICINE AND OTHER DISCIPLINES, BY REGION. (**A**) Frequency of co-citation links between veterinary journals and journals in other subject areas. (**B**) Frequency of co-citation links between Miscellaneous veterinary journals and journals in other subject categories. Annual co-citation maps were available for up to 12 years (1998–2010) for 93 countries in Western Europe (n = 19), Eastern Europe (n = 20), Africa (n = 8), North America (n = 2), Latin America (n = 11), Middle East (n = 15), Asia (n = 16), and the Pacific (n = 2).

Table 6. Changes in subject co-citation links with veterinary journals over time in selected countries*

Country	1998-99	1999-00	2000-01	2001-02	2002-03	2003-04	2004-05	2005-06	2006-07	2007-08	2008-09	2009-10
United States	ABS	ABS	ABS	ABS	ABS	ABS	ABS	**MED**	**MED**	**MED**	ABS	**MED**
Switzerland	ABS	ABS	ABS	ABS	ABS	**MED**	**MED**	**MED**	**MED**	**MED**	ABS	**MED**
Austria	ABS	ABS	ABS	**MED**	ABS	ABS	**MED**	**MED**	ABS	ABS	ABS	ABS
Lithuania	ABS	ABS	ABS	**IMM**	**IMM**	ABS	ABS	ABS	ABS	ABS	ABS	ABS
Bulgaria	ABS	ABS	**IMM**	ABS	**IMM**	**MED**	**MED**	ABS	ABS	ABS	ABS	ABS
Serbia	–	–	–	–	–	–	–	–	**MED**	**BGM**	ABS	ABS
Russian Fed.	ABS	ABS	ABS	ABS	**BGM**	**MED**	**MED**	**BGM**	ABS	**BGM**	ABS	ABS
Romania	ABS	ABS	**BGM**	**BGM**	ABS	**BGM**	**BGM**	ABS	ABS	**MED**	ABS	ABS
Estonia	ABS	ABS	**BGM**	ABS	ABS	**BGM**	**BGM**	ABS	ABS	ABS	ABS	ABS
UAE	–	–	–	–	**MED**	**MED**	ABS	**MED**	**MED**	ABS	ABS	ABS
Kuwait	ABS	ABS	ABS	**MED**	**MED**	**MED**	**IMM**	ABS	ABS	**MED**	ABS	ABS
Algeria	–	–	ABS	ABS	ABS	**IMM**	ABS	**IMM**	ABS	ABS	ABS	ABS
China	ABS	ABS	ABS	ABS	ABS	ABS	ABS	ABS	**BGM**	**BGM**	**BGM**	**BGM**
South Korea	ABS	ABS	ABS	ABS	ABS	ABS	**BGM**	**BGM**	**BGM**	**BGM**	**BGM**	ABS
Taiwan	ABS	ABS	ABS	ABS	ABS	ABS	ABS	ABS	**BGM**	**BGM**	**BGM**	**BGM**
Hong Kong	**IMM**	ABS	**BGM**	**MED**	**MED**	**MED**	**MED**	**MED**	**MED**	**MED**	**MED**	**MED**
Singapore	ABS	**IMM**	**IMM**	**IMM**	ABS	ABS	**BGM**	**BGM**	ABS	**MED**	**MED**	**MED**
Thailand	**IMM**	ABS	ABS	ABS	**IMM**	ABS	ABS	ABS	ABS	ABS	ABS	ABS
Malaysia	**IMM**	**IMM**	ABS	ABS	ABS	ABS	ABS	ABS	ABS	ABS	ABS	ABS

*Subject area abbreviations are bolded for those years where co-citation links occurred between veterinary medicine and non-agricultural journal disciplines. Only countries with 50 or more veterinary articles between 1998 and 2010 and two or more years of co-citation links with non-agricultural journals are shown. *ABS* Agricultural & Biological Sciences, *MED* Medicine, *IMM* Immunology and Microbiology, *BGM* Biochemistry, Genetics, & Molecular Biology. Dashes indicate data were not available. Data were obtained from co-citation maps generated for each year and country via SCImago [18].

DISCUSSION

We used bibliographic data to provide a quantitative image of the current and evolving status of veterinary research capacity, species specialization, and interdisciplinary affiliations in regions and countries around the world between 1996 and 2011. Veterinary research productivity, closely associated with GDP, was strongly dominated by Western Europe and North America, but has increased rapidly in Asia, Latin America, and to a lesser extent in the Middle East and Eastern Europe. Western Europe and North America also led strongly in species specialization based on article output in FA, SA, and EQ journals. Reliance on the Scopus species-specific journal designations was a limitation, in that the majority of veterinary journals (in the MISC category) as well as other medical, animal science, and basic bioscience journals also publish species-specific veterinary articles. Because of this, the true magnitude and geographic distribution of species-specialized research remains uncertain; however, trends in this study were supported by principal component analysis, in which species emphasis aligned with opposing demographic and economic factors. Co-citation links supported strong alignment between veterinary medicine and agriculture in most countries, and also indicated distinct and evolving interdisciplinary links with medicine and basic sciences. Overall, these results provide a unique picture of the veterinary profession and its research capacity globally, and impel additional investigation into species-specific research output as an indicator of evolving education and workforce needs and priorities.

The Scopus database was a logical one to use for this study because it indexes the highest percentage of core veterinary journals [17], indexes geographically diverse journals [16], and to our knowledge is the only index to categorize veterinary journals according to species; data also are freely accessible through the SCImago portal, making it easy for readers to get detailed bibliographic data for their country, region, and discipline [18]. Compared with other indexes, Scopus is limited to more recent articles (beginning in 1995–96), however, this time period was relevant to our study and was balanced by the validity and large size of the Scopus co-citation journal database [21, 27]. Although veterinary journals published in Western Europe, North America, and the Pacific included those with the highest impact factors, SJRs, and H indexes, the country of publication does not necessarily reflect the country of authors or affiliated societies, since many international journals are clustered within major publishers in Western Europe. Thus, although not all veterinary research is published in veterinary journals, Scopus provided a good cross-section of small to large journals of varying impact factors and national origin that reflected a wide range of veterinary research publications.

Western Europe and North America, led by the US and UK, accounted for more than 60% of veterinary articles and more than 70% of citations during the past 14 years, a regional dominance that has been described previously for other biomedical disciplines [14, 15]. An earlier limited analysis of highly cited articles in veterinary journals in 2002 and 2003 found that 51% were from the US and UK [28]. In our study, notable growth in Italy and Spain has opened up a widening gap in research productivity between Western Europe and North America, and Asia is on a trajectory to exceed North America in veterinary publications in the near future. Indeed, Asia, with strong growth in India and China, surpassed North America in 2007 in the number of articles published in MISC veterinary journals. The increase in articles from Asia, Latin America, and the Middle East also likely reflected, in part, the addition of journals (and hence articles) from these regions to the Scopus database during the study period. Concomitant with growth in Asia, Latin America (Brazil), and the Middle East (Turkey, Iran), US contribution to world veterinary research output in veterinary medicine fell by nearly 7% from 1996 to 2010, supporting a trend noted in the 1990s and in other disciplines [29, 30]. Although the decrease in the US and

Germany was relative, decreases in other countries were absolute, and continued change is likely in the future.

Similar to bibliographic studies in other disciplines (which used different journal indexes and slightly different world regions), research output in the present study was strongly related to GDP [14, 15, 31]. The Pacific region had especially high article output relative to both GDP and population, as well as high quality articles, based on citation metrics. Notably, Africa produced about the same number of articles as Western Europe, North America, and the Pacific when article output was normalized to GDP. A possible explanation for this is the high average level of international collaboration (52%) for articles from Africa, suggesting that increasing international collaborations could be a fruitful way of enhancing veterinary research productivity in other regions, especially Eastern Europe and the Middle East. Uruguay and the UAE also had strong international collaboration, and ranked higher (based on H index) than other countries in their regions that produced more articles.

Despite increasing rates of publication, the quality and scientific impact of articles from Asia, Latin America, the Middle East, and Eastern Europe remain lower than in other regions, based on fewer citations, more self-citations and uncited articles, and lower citation metrics (impact factor, SJR, and H-index), parameters used as surrogates for research quality [32, 33]. India, for example, published nearly as many articles as the UK during the study period, but had only about one-tenth of the citations. Language and accessibility of articles can affect citations, although a higher number of journals from these regions were open access. The H-index increases the longer articles are available to be cited, which favors well-established journals and countries with strong research infrastructures. Although citation-based metrics are commonly used to measure scientific impact, usage-based metrics (eg, online usage logs, social network analysis) may be just as important [34] and were not evaluated in the present study.

Another factor affecting regional differences in veterinary article output and quality appears to be the number of veterinary faculties relative to population, especially in Eastern Europe and Latin America, where limited resources are spread among a larger number of faculties than in other regions, making it more difficult to achieve critical mass and infrastructure for research. Although veterinary research is done not only in veterinary faculties, faculties are one indicator of a country's capacity and structure for supporting veterinary research and education.

Countries with high total article output also tended to have high numbers of species-specific articles, and differences in urbanization and economic development correlated with species specialization within regions. The US had the highest overall rate of specialization (nearly 40% of US articles were in species-specific journals) and led the world in articles in SA-specific journals, followed by France and Western Europe in general. This SA (and to a lesser extent EQ) predominance was closely aligned with GNI and urbanization, consistent with expanding demand for companion animal specialty services in metropolitan areas, and the alignment of academic programs to meet research and training priorities for these species [2]. Notable differences in companion animal research output likely reflected differences in development between countries within a region. In Asia, for example, Japan ranked first in article output in SA and EQ journals, whereas India led in articles in FA journals. Similarly, South Africa ranked first in article output in SA and EQ journals, whereas Nigeria led in articles in FA journals. Increases in veterinary articles over time in most regions occurred in MISC and FA journals, but notable increases in SA and EQ journal articles were seen in Brazil, China, Poland, Taiwan, and Turkey. As previously noted, an important limitation of the Scopus classification of journals is that many species-specific articles are published in MISC journals (which comprised the largest group), as well as in journals in related subject areas. In addition, a few species-specific journals appeared to be misclassified (see Appendix). Thus, although the results of our study were consistent with what might be expected relative to demographic and economic indicators, additional studies are warranted to evaluate species-specific publications at the article rather than the journal level.

The strong co-citation affiliation between agriculture and veterinary journals, especially FA journals, was consistent with the importance of livestock production and disease in much of the world. Co-citations represent intellectual relationships between journals and have been used to study the specialty structure of science through the pattern of linkages, to monitor the development of scientific fields, and to assess developing interrelationships [27]. We noted several interesting patterns, including links between veterinary medicine and BGM in China, Taiwan, and South Korea (and some Eastern European countries), and links between veterinary medicine and medicine in several other countries. These patterns could be related to differences in culture, veterinary research infrastructure (relative to medicine and basic sciences), and geographic priorities, and warrant further investigation as potential indicators of shifts in species or interdisciplinary focus.

In the US and Switzerland, co-citation links provided tangible evidence of a recent shift in primary research affiliation from agriculture to medicine, although there could be more than one reason for this. The high output of articles in SA-specific journals, and parallels between SA and human medical practice (ie, residency/specialty training), make SA specialization a plausible explanation. If so, other countries in Western Europe with strong SA specialization, such as the UK and France, would be expected to undergo a similar shift in the near future. North America and Western Europe also had a high frequency of co-citation links between SA and MISC veterinary journals, the latter of which include veterinary specialty journals in subdisciplines like internal medicine, cardiology, dermatology, and ophthalmology. Increased laboratory animal, public health, and zoonotic disease research, such as that promoted by the One Health initiative [2, 3], could also have contributed to co-citations with medicine in these countries. Again, a more detailed study with granularity at the article level is warranted to further assess the strength of co-citation affiliations and the evolving structure of veterinary research.

CONCLUSIONS

In conclusion, bibliographic data documented geographic trends in research output and quality, species specialization, and interdisciplinary affiliations, and their relationship with demographic and economic factors that underlie many of the changes affecting the veterinary profession. Although conclusions about specialization based solely on species-specific journals are limited, it is clear that veterinary research output in the US and Western Europe has expanded notably in companion animal specialty areas, while research output in most regions of the world remains strongly focused on agriculture and food animal production. An emerging relevance gap along species lines could affect the ability of wealthy countries to contribute to veterinary training and research in developing countries, underlining concerns about the profession's ability to meet global animal health and food security needs [1, 2, 3, 4, 7]. Facilitating international collaboration and focusing resources on fewer veterinary faculties also could help strengthen the quantity and quality of veterinary research and enable all regions to compete effectively. Future bibliographic studies that monitor and investigate evolving trends in veterinary and species-specific publications could improve our ability to understand and balance veterinary research and education needs globally.

COMPETING INTERESTS

The authors declare that they have no competing interests.

AUTHORS' CONTRIBUTIONS

MMC defined the scope and hypothesis of the study, contributed to study design and data interpretation, acquired and analyzed the data, and drafted the manuscript. AM inspired the study, contributed to study design and data interpretation, and critically revised the manuscript. All authors read and approved the final manuscript.

REFERENCES

1. Willis NG, Monroe FA, Potworowski JA, Halbert G, Evans BR, Smith JE, Andrews KJ, Spring L, Bradbrook A: Envisioning the future of veterinary medical education: the association of American veterinary medical colleges foresight project, final report. J Vet Med Educ. 2007, 34: 1-41.
2. National Academy of Sciences: Workforce needs in veterinary medicine.Washington DC: National Academy Press; 2012.
3. US National Research Council: Animal Health at the Crossroads: Preventing, Detecting,and Diagnosing Animal Diseases. Washington DC: National Academy Press; 2005.
4. Marshak R: Veterinary schools and the profession: a search for bearings in the new century. J Am Vet Med Assoc. 2005, 227: 1234-1238.
5. Larson RI: Food animal veterinary medicine: leading a changing profession. J Vet Med Educ. 2004, 31: 341-345.
6. USDA Economic Research Service: Farm employment losses outstrip job gains in farm-related industries in some nonmetro areas. Rural Conditions and Trends. 2000, 10: 22-26.
7. Narver HL: Demographics, moral orientation, and veterinary shortages in food animal and laboratory animal medicine. J Am Vet Med Assoc. 2007, 230: 1798-1804.
8. Rucker M: Gender changes and the future of our profession. J Vet Med Educ. 2002, 29: 63-65.
9. Prasse KW, Heider LE, Maccabe AT: Envisioning the future of veterinary medicine and the imperative for change in veterinary medical education. J Am Vet Med Assoc. 2007, 231: 1340-1342.
10. Lumeij JT, Herrtage ME: Veterinary specialization in Europe. J Vet Med Educ. 2006, 33: 176-179.
11. Chieffo C, Kelly AM, Ferguson J: Trends in gender, employment, salary, and debt of graduates of US veterinary medical schools and colleges. J Am Vet Med Assoc. 2008, 233: 910-917.
12. American Veterinary Medicine Association Communications Division: Employment, starting salaries, and educational indebtedness of year-2011 graduates of US veterinary medical colleges. J Am Vet Med Assoc. 2011, 239: 953-957.
13. Market research statistics: U.S. Veterinarians. American Veterinary Medical Association. 2011, http://www.avma.org/KB/Resources/Statistics/Pages/Market-research-statistics-US-veterinarians-2011.aspx.
14. Soteriades ES, Rosmarakis ES, Paraschakis K, Falagas ME: Research contribution of different world regions in the top 50 biomedical journals (1995–2002). FASEB J. 2005, 20: 29-34.
15. Bliziotis IA, Paraschakis K, Vergidis PI, Karavasiou AI, Falagas ME: Worldwide trends in quantity and quality of published articles in the field of infectious diseases. BMC Inf Dis. 2005, 5: 16-23.

16. deMoya-Anegón F, Chinchilla-Rodríguez Z, Vargas-Quesada B, Corera-Alvarez E, Muñoz FJ, González-Molina A, Herrero-Solana V: Coverage analysis of Scopus: a journal metric approach. Scientometrics. 2007, 73: 53-78.
17. Grindley DJC, Brennan ML, Dean RS: Searching the veterinary literature: a comparison of the coverage of veterinary journals by nine bilbiographic databases. J Vet Med Educ. 2012, 39: 404-412.
18. SCImago. http://www.scimagojr.com.
19. Hailat N: Present status and vision for veterinary higher education in the Arab world. J Vet Med Educ. 2005, 32: 151-155.
20. González-Pereira B, Guerrero-Bote VP, Moya-Anegón F: A new approach to the metric of journals' scientific prestige: the SJR indicator. J Informetrics. 2010, 4: 379-391.
21. Falagas ME, Pitsouni EI, Malietzis GA, Pappas G: Comparison of , Scopus, Web of Science, and : strengths and weaknesses. FASEB J. 2008, 22: 338-342.
22. JCR Science Edition Veterinary Sciences. ISI Journal Citation Reports. New York: Thomson-Reuters 2011.
23. DOAJ Directory of Open Access Journals. http://www.doaj.org.
24. World Bank. http://data.worldbank.org.
25. American Veterinary Medical Association. http://www.avma.org/ professionald evelopment/education/foreign/pages/ECFVG-world-colleges-colleges.aspx.
26. World Veterinary Association. http://www.worldvet.org.
27. Leydesdorff L, de Moya-Anegón F, Guerrero-Bote VP: Journal maps on the basis of Scopus data: a comparison with the journal citation reports of the ISI. J Am Soc Infor Sci Technol. 2010, 61: 352-369.
28. Elsinghorst TA: Analysis of the 96 most often cited articles published in veterinary journals in 2002 and 2003. Vet Q. 2005, 27: 183-189.
29. Falagas ME, Papastamatiaki PA, Bliziotis IA: A bibliometric analysis of research productivity in parasitology by different world regions during a 9-year period (1995–2003). BMC Inf Dis. 2006, 6: 56-61.
30. Rahman M, Fukui T: A decline in the U.S. share of research articles. N Engl J Med. 2002, 347: 1211-1212.
31. Hefler L, Tempfer C, Kainz C: Geography of biomedical publications in the European union, 1990–98. Lancet. 1856, 1999: 353.
32. Weale AR, Bailey M, Lear PA: The level of non-citation of articles within a journal as a measure of quality: a comparison to the impact factor. BMC Med Res Methodol. 2004, 4: 14-21.
33. Falagas ME, Kouranos VD, Arencibia-Jorge R, Karageorgopoulos DE: Comparison of SCImago journal rank indicator with journal impact factor. FASEB J. 2008, 22: 2623-2628.
34. Bollen J, Van de Sompel H, Hagberg A, Chute R: A principal component analysis of 39 scientific impact measures. PLoS One. 2009, 4: e6022-

CHAPTER 20

Fluorescence-Guided Surgery of Liver Metastasis in Orthotopic Nude-Mouse Models

Takashi Murakami

Anti Cancer, Inc., San Diego, California, United States of America

ABSTRACT

We report here the development of fluorescence-guided surgery of liver metastasis. HT29 human colon cancer cells expressing green fluorescent protein (GFP) were initially injected in the spleen of nude mice. Three weeks later, established liver metastases were harvested and implanted on the left lobe of the liver in other nude mice in order to make an orthotopic liver metastasis model. Fourteen mice with a single liver metastasis were randomized into bright-light surgery (BLS) or fluorescence-guided surgery (FGS) groups. Seven mice were treated with BLS, seven were treated with FGS. Three weeks after implantation, the left lobe of the liver with a single metastasis was exposed through a median abdominal incision. BLS was performed under white light. FGS was performed using a hand-held portable fluorescence imaging system (Dino-Lite). Post-surgical residual tumor fluorescence was visualized with the OV100 Small Animal Imaging System. Residual tumor fluorescence after BLS was clearly visualized at high magnification with the OV100. In contrast, residual tumor fluorescence after FGS was not detected even at high magnification with the OV100. These results demonstrate the feasibility of FGS for liver metastasis.

INTRODUCTION

The ability of the surgeon to accurately visualize tumor margins is essential at the time of surgery and is of particular importance for resection of metastatic disease, especially in the liver [1].

A variety of labeling compounds have been used for fluorescence-guided surgery in the clinic. Sentinel lymph nodes in breast cancer patients were labeled by a near-infrared (NIR) fluorescing dye indocyanine [2]. However, indocyanine does not specifically label tumor cells. The metabolite 5-aminolevulinic acid, a precursor of hemoglobin, labels porphyrins in malignant glioma for FGS [3], which significantly improved outcome.

Folate was coupled to fluorescein isothiocyanate (FITC) for targeting folate receptor–α (FR-α) in ovarian cancers. Under fluorescence-guided surgery, tumor deposits less than 1 mm in size could be visualized and resected [4].

Monoclonal antibodies directed against cancer antigen 19–9 (CA19-9) or carcino-embryonic antigen (CEA) were conjugated to a green fluorophore and delivered intravenously into nude mice with orthotopic human pancreatic or colon tumors [5]. The tumors become fluorescent and were resected under fluorescence-guidance improved outcome [5–10].

Kishimoto *et al.* [11] labeled tumors with green fluorescence protein (GFP) using a telomerase-dependent adenovirus (OBP-401) that expresses the *gfp* gene only in cancer cells, which, in contrast to normal cells, express the telomerase enzyme. The labeled tumors could also be resected under fluorescence guidance. Tumors that recurred after fluorescence-guided surgery maintained GFP expression [12]. Since the cancer cells stably express GFP, detection of cancer recurrence and metastasis is possible with OBP-401 GFP labeling, and is not possible with nongenetic probes.

The present report demonstrates the feasibility of FGS of liver metastasis labeled with GFP in orthotopic mouse models.

MATERIALS AND METHODS

Ethics Statement

All animal studies were conducted with an AntiCancer Institutional Animal Care and Use Committee (IACUC)-protocol specifically approved for this study and in accordance with the principals and procedures outlined in the National Institute of Health Guide for the Care and Use of Animals under Assurance Number A3873-1. In order to minimize any suffering of the animals, anesthesia and analgesics were used for all surgical experiments. Animals were anesthetized by intramuscular injection of a 0.02 ml solution of 20 mg/kg ketamine, 15.2 mg/kg xylazine, and 0.48 mg/kg acepromazine maleate. The response of animals during surgery was monitored to ensure adequate depth of anesthesia. Ibuprofen (7.5 mg/kg orally in drinking water every 24 hours for 7 days post-surgery) was used in order to provide analgesia post-operatively in the surgically-treated animals. The animals were observed on a daily basis and humanely sacrificed by CO_2 inhalation when they met the following humane endpoint criteria: prostration, skin lesions, significant body weight loss, difficulty breathing, epistaxis, rotational motion and body temperature drop. The use of animals was necessary to develop fluorescence-guided surgery of liver metastasis. Animals were housed with no more than 5 per cage. Animals were housed in a barrier facility on a high efficiency particulate arrestance (HEPA)-filtered rack under standard conditions of 12-hour light/dark cycles. The animals were fed an autoclaved laboratory rodent diet (S1 ARRIVE Checklist).

Cell Line

The human colon cancer cell line HT-29 [10], obtained from the American Type Culture Collection (Rockville, MD) was maintained in DMEM (Irvine Scientific, Irvine, CA) supplemented with heat-inactivated 10% fetal bovine serum (FBS) (Gemini Biologic Products, Calabasas, CA), 2 mM glutamine, 100 units/ml penicillin, 100 μg/ml streptomycin, and 0.25 μg/ml amphotericin B (Life Technologies, Inc., Grand Island, NY). The cells were incubated at 37°C in 5% CO_2.

Mice

Athymic *nu/nu* nude mice (AntiCancer Inc., San Diego, CA), 4–6 weeks old, were used in this study. Mice were kept in a barrier facility under HEPA filtration (as noted above). Mice were fed with an autoclaved laboratory rodent diet. All mouse surgical procedures and imaging were performed with the animals anesthetized by subcutaneous injection of the ketamine mixture described above. All animal studies were conducted with an AntiCancer Institutional Animal Care and Use Committee (IACUC)-protocol specifically approved for this study and in accordance with the principals and procedures outlined in the National Institute of Health Guide for the Care and Use of Animals under Assurance Number A3873-1.

Establishment of GFP-labeled HT29

For GFP gene transfection, 25% confluent HT-29 cells were incubated with a 1:1 mixture of GFP retroviral supernatants of PT67 packaging cells and RPMI 1640 cell culture medium (GIBCO Life Technologies, New. York, NY) containing 10% FBS (Gemini Biological Products) for 72 h. Fresh medium was replenished at this time. Cells were harvested by trypsin-EDTA 72 h after transduction and subcultured at a ratio of 1:15 into medium, which contained 200 μg/ml G418 to select for high GFP expression, since the GFP retrovirus also contained the neomycin-resistance gene. The level of G418 was increased to 400 μg/ml stepwise. Clones stably expressing GFP were isolated with cloning cylinders (Bel-Art Products, Pequannock, NJ) with the use of trypsin-EDTA. The high-GFP-expressing cells were then amplified and transferred by conventional culture methods. High GFP-expression clones were then isolated in the absence of G418 for 10 passages to select for stable expression of GFP [13–15].

Initial Establishment of Liver Metastasis

HT-29-GFP cells were harvested by trypsinization and washed twice with serum-free medium. Cells (5×10^5 in 50 μl serum-free medium with 50% Matrigel) were injected into the superior and inferior pole of the spleen in mice (Fig 1A). Three week after injection, liver metastasis was confirmed by laparotomy (Fig 1C).

Surgical orthotopic Implantation (SOI) of Liver Metastasis

Liver metastases were resected and cut into 8 mm^3 blocks. A small 6- to 8-mm midline incision was made in other nude mice. The left lobe of the liver was exposed through this incision, and a single tumor fragment (3-mm^3) was orthotopically implanted to the left lobe of the liver (Fig 1B). The left lobe of the liver was returned to the abdominal cavity, and the incision was closed in one layer using 6–0 nylon surgical sutures (Ethilon; Ethicon Inc., NJ, USA).

Bright-Light and Fluorescence-Guided Surgery of Liver Metastases

Four weeks after SOI of HT-29-GFP to the liver, the liver metastasis was exposed and imaged preoperatively with the OV100 Small Animal Imaging System (Olympus, Tokyo, Japan) [16] at a magnification of 0.14x. Fourteen mice underwent surgery: Fluorescence-guided surgery (FGS) in 7 mice and bright-light surgery (BLS) in 7 mice (Fig 2A and 2B). FGS was performed using the Dino-Lite imaging system (AM4113TGFBW Dino-Lite Premier; AnMo Electronics Corporation, New Taiwan). The surgical resection bed was imaged with the Olympus OV100 at a magnification of 0.14x or 0.56x to detect microscopic residual cancer. Residual tumor area was analyzed with ImageJ v1.49f (National Institutes of Health). The incision was closed in one layer using 6–0 nylon surgical sutures.

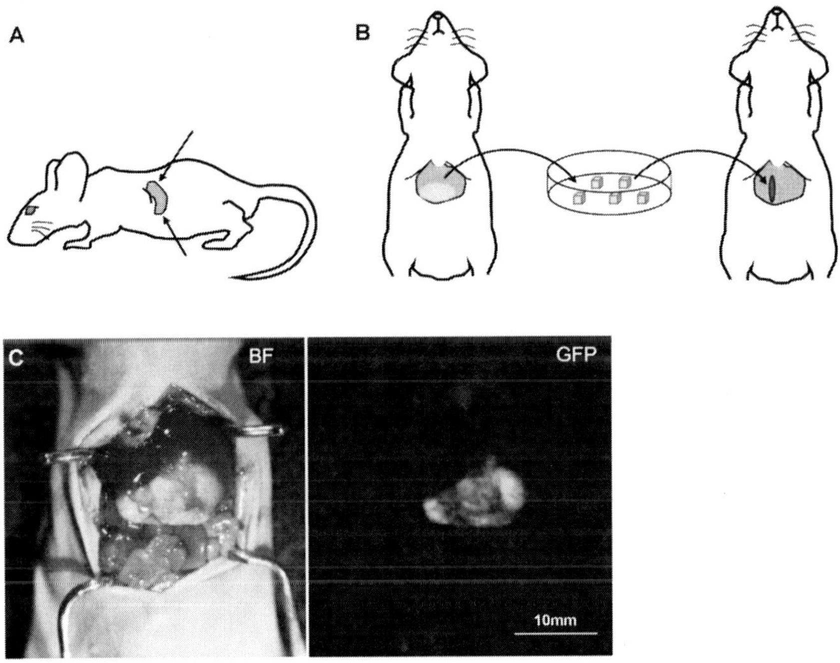

Figure 1. HT29-GFP colon cancer experimental liver metastasis.

(A) HT-29-GFP cells (5×10^5 in 50 μl with 50% Matrigel) were injected into the superior pole and inferior pole of the spleen (arrows), respectively. **(B)** Three weeks after injection, liver metastasis was confirmed by laparotomy, which was resected and cut into 3-mm^3 blocks. Each tumor fragment was implanted by surgical orthotopic implantation (SOI) in the left lobe of the liver on other nude mice. **(C)** Representative images of liver metastasis established after spleen injection. The large tumor in the liver strongly expressed GFP. (GFP, green fluorescent protein; BF, bright field)

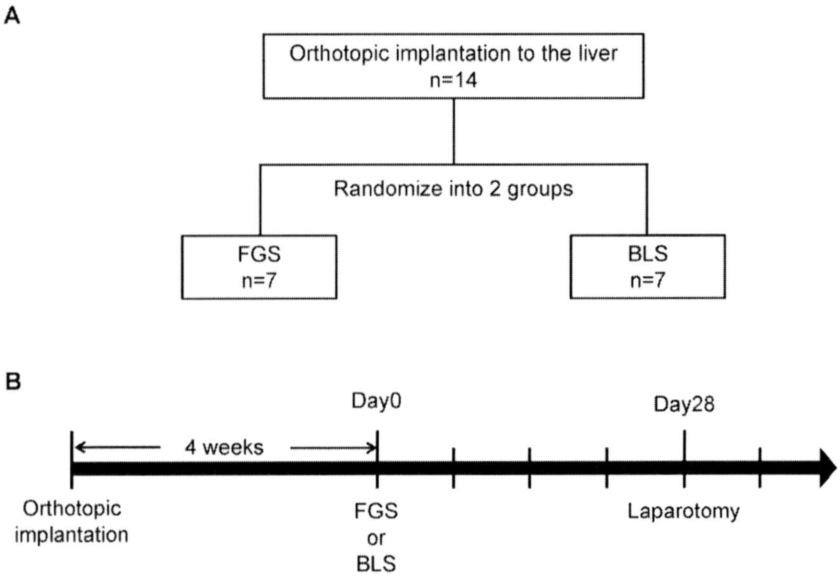

Figure 2. Schematic diagram of the experimental design.

(A) Fourteen mice were randomized into 2 groups; FGS: n = 7, BLS: n = 7. **(B)** Timeline from orthotopic implantation. Four weeks after implantation, all mice were treated with FGS or BLS. Twenty-eight days after the surgery, all mice underwent laparotomy to detect GFP for evaluation of recurrence. After the laparotomy, follow-up examination for survival was continued. (FGS, fluorescence-guided surgery; BLS, bright-light surgery).

Imaging of Tumor Progression

To evaluate tumor fluorescence around the resected site in the liver, laparotomy was performed in all mice at the 28th post-operative day (Fig 2B). Tumor fluorescence was detected with the OV100 and analyzed with ImageJ.

Tumor histology

Resected fresh tumor samples from either BLS or FGS were fixed in 10% formalin, and then embedded in paraffin. The tissue sections were deparaffinized in xylene and rehydrated in an ethanol series. Hematoxylin and eosin (H & E) staining was performed according to standard protocols. The margin between the tumor and the liver tissue was evaluated with a BHS System microscope (Olympus).

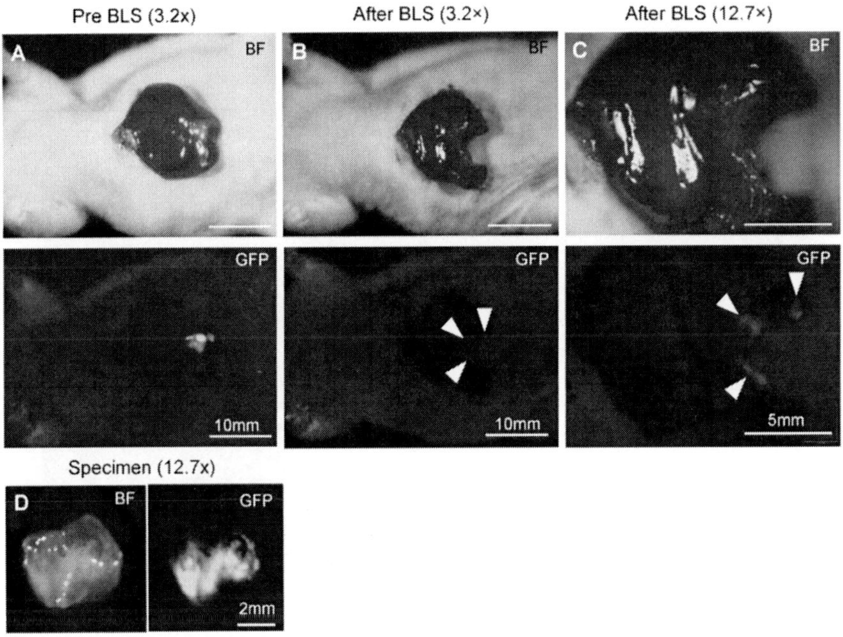

Figure 3. Pre-operative and post-operative images from the orthotopic liver metastasis model treated with BLS.

(A)—(C) Upper panels show bright field images, and lower panels are images of tumor fluorescence obtained with the OV100. At low magnification, residual tumor fluorescence was marginally detected. **(B)** However, at high magnification, residual tumor fluorescence was clearly visualized (arrows) **(C)**. Arrowheads show residual tumor fluorescence in **B** and **C**. **(D)** Resected specimen. Magnifications are indicated above the columns.

Statistical Analysis

SPSS statistics version 21.0 was used for all statistical analyses (IBM, New York City, NY, USA). Residual tumor area is expressed as mean ± SD. Significant differences for continuous variables were determined using the Mann-Whitney U test. A probability value of $P \leq 0.05$ was considered statistically significant.

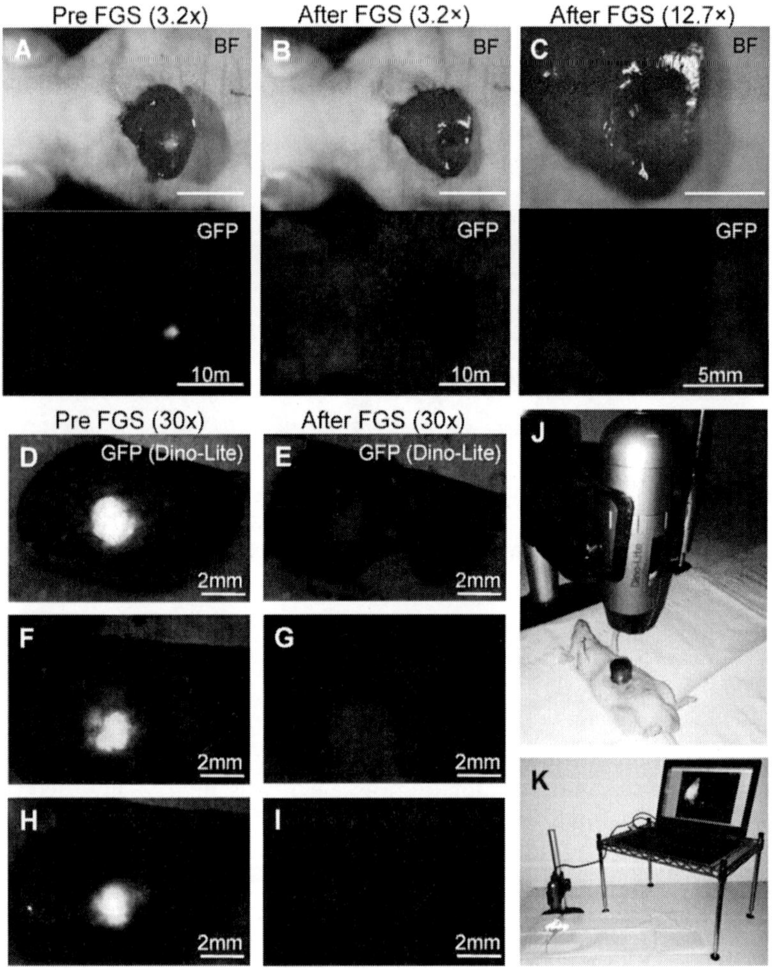

Figure 4. Pre-operative and post-operative images from the orthotopic liver metastasis model treated with FGS.

(A)—(C) Upper panels show bright field images, and lower panels are images of tumor fluorescence obtained with the OV100. Residual tumor fluorescence could not be detected even at high magnification **(C)**. **(D,F,H)** Pre-FGS tumor fluorescence was clearly visualized with the Dino-Lite imaging system. **(E,G,I)** Dino-Lite imaging showed no evidence of tumor after FGS. **(J-K)** Dino-Lite settings. **(J)** After exposing the left lobe of the liver, the mouse was put under the Dino-Lite. **(K)** Connection between the Dino-Lite and computer. Tumor fluorescence was imaged on the monitor during FGS. Magnifications are indicated above the columns.

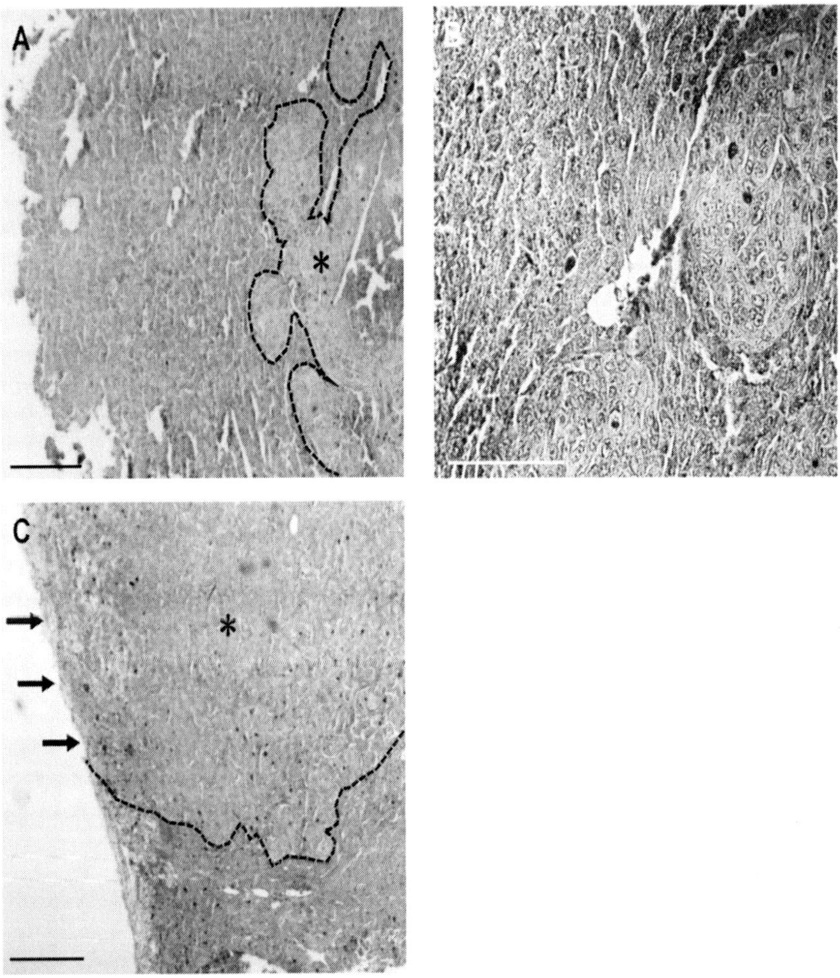

Figure 5. Histological tumor margin of resected specimen.

(A)-(C) H&E staining of resected specimen. **(A)** In the mouse treated with FGS, viable cancer tissue (marked by an asterisk) is surrounded by normal liver tissues. **(B)** High magnification of **(A)**. **(C)** In the BLS-treated mouse, viable cancer cells are visible along the resection line. Arrows show residual cancer tissue. Dashed lines separate viable cancer and normal liver tissue. Scale bars: 200 μm. https://doi.org/10.1371/journal.pone.0138752.g005

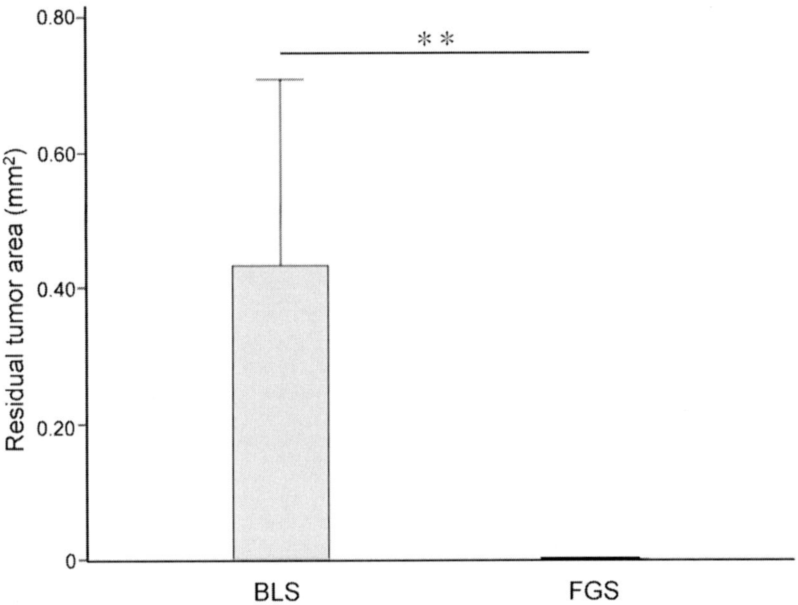

Figure 6. Bar graph of residual tumor area after surgery.

No residual tumor was detected in the FGS group. Residual tumor area after BLS was relatively large. Error bar shows SD. $^{**}P<0.01$.

RESULTS AND DISCUSSION

FGS significantly reduces residual liver metastasis

Single liver metastases were clearly detected preoperatively with the OV100 (Figs 3A and 4A). In the BLS group, residual tumor fluorescence was marginally detected at a magnification of 0.14×. However, at a magnification of 0.56×, residual cancer was clearly visualized (Fig 3B and 3C). By contrast, residual tumor fluorescence could not be detected even at a magnification of 0.56× in the FGS group (Fig 4B and 4C). Tumor fluorescence was also clearly visualized before FGS with the Dino-Lite at a magnification of 30× (Fig 4D). Dino-Lite imaging showed no evidence of residual cancer after FGS (Fig 4E). All resected specimens exhibited strong GFP expression (Fig 3D). Histologically, viable cancer was detected on the resection line in the BLS group, but not in the FGS group which had clear margins (Fig 5), which is consistent with the fluorescence data. Residual tumor area after BLS was significantly larger than after FGS (0.34 ± 0.28 mm^2 and 0 mm^2, respectively; $P = 0.003$; Fig 6).

FGS significantly decreases recurrent tumor area

Laparotomy was performed at 28 postoperative day. Recurrent tumor fluorescence was visualized next to the resected site in the liver in the BLS group, but not in the FGS group (Fig 7). In the FGS group, only autofluorescence was detected.

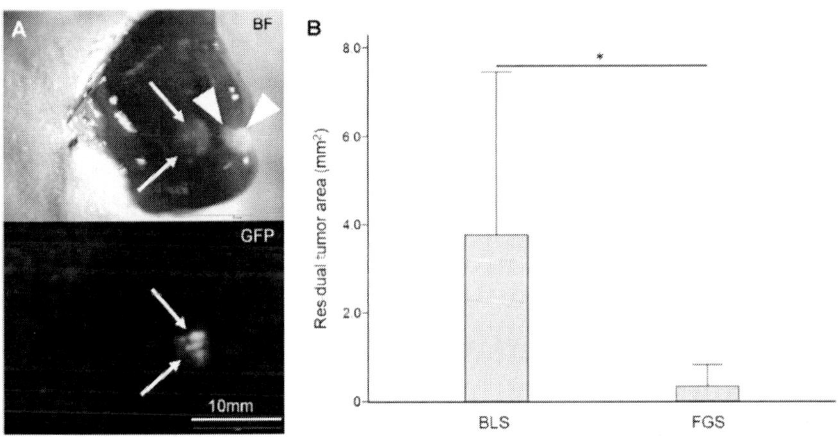

Figure 7. Evaluation of tumor fluorescence at day 28 after surgery.

(A) Upper panel shows the bright field image, and lower panel shows the GFP tumor fluorescence image obtained with the OV100 at a magnification of 0.56. Laparotomy was performed at the 28th postoperative day. Bright field image shows tumor in the resection site in the liver (arrows). Strong GFP fluorescence from the tumor is seen in the lower panel. Arrows show recurrent tumor in the resection site. Arrowheads show operative scar on the liver. **(B)** The GFP tumor fluorescence area was significantly larger in the BLS group compared to the FGS group, where only autofluorescence was detected. Error bars show SD. *$P<0.05$.

The FGS procedure for liver metastasis described in the present report had short- and long-term advantages over BLS in that most if not all of the metastasis was resected by FGS compared to BLS (Figs 3–6), and this advantage was maintained even at 28 days post-surgery, where recurrence was greatly reduced by FGS compared to BLS (Fig 7). Future studies will devise curative strategies for FGS of liver metastasis.

CONCLUSIONS

The results of present study suggest that FGS has clinical potential of reducing residual cancer in patients who have liver metastasis resected. Clinical application strategies could include labeling the liver metastasis with a tumor-specific monoclonal antibody such as anti-carcinoembryonic antigen (anti-CEA) conjugated to a fluorophore [5–10] or labeled by a telomerase-dependent cancer-specific adenovirus that contains the GFP gene, OBP-401 [11, 12, 17–19]. The labeled-antibody strategy is probably feasible in the near future. The viral-labeling strategy will require more safety studies. However, the parent virus of OBP-401, OBP-301, has proven safe in a clinical trial [20]. However, OBP-401 will also have safety tested in a clinical trial.

DEDICATION

This paper is dedicated to the memory of A. R. Moossa, M.D.

AUTHOR CONTRIBUTIONS

Conceived and designed the experiments: TM RMH. Performed the experiments: TM YH YZ. Analyzed the data: TM YH YZ TC KT MB IE RMH. Contributed reagents/materials/analysis tools: RMH. Wrote the paper: TM RMH.

REFERENCES

1. Bouvet M, Hoffman RM (2011) Glowing tumors make for better detection and resection. Sci Transl Med 3:110fs10. pmid:22116932
2. Troyan SL, Kianzad V, Gibbs-Strauss SL, Gioux S, Matsui A, Oketokoun R, et al. (2009) The FLARE intraoperative near-infrared fluorescence imaging system: A first-in-human clinical trial in breast cancer sentinel lymph node mapping. Ann Surg Oncol 16:2943–2952. pmid:19582506
3. Stummer W, Pichlmeier U, Meinel T, Wiestler OD, Zanella F, Reulen HJ, et al. (2006) Fluorescence-guided surgery with 5-aminolevulinic acid for resection of malignant glioma: A randomised controlled multicentre phase III trial. Lancet Oncol 7:392–401. pmid:16648043
4. van Dam GM, Themelis G, Crane LM, Harlaar NJ, Pleijhuis RG, Kelder W, et al. (2011) Intraoperative tumor-specific fluorescence imaging in ovarian cancer by folate receptor-α targeting: First inhuman results. Nat Med 17:1315–1319. pmid:21926976

5. McElroy M, Kaushal S, Luiken G, Moossa AR, Hoffman RM, Bouvet M (2008) Imaging of primary and metastatic pancreatic cancer using a fluorophore-conjugated anti-CA19-9 antibody for surgical navigation. World J Surg 32:1057–1066. pmid:18264829
6. Metildi CA, Kaushal S, Snyder CS, Hoffman RM, Bouvet M (2013) Fluorescence-guided surgery of human colon cancer increases complete resection resulting in cures in an orthotopic nude mouse model. J Surg Res 179:87–93. pmid:23079571
7. Kaushal S, McElroy MK, Luiken GA, Talamini MA, Moossa AR, Hoffman RM, et al. (2008) Fluorophore-conjugated anti-CEA antibody for the intraoperative imaging of pancreatic and colorectal cancer. J Gastrointest Surg 12:1938–1950. pmid:18665430
8. Metildi CA, Kaushal S, Luiken GA, Talamini MA, Hoffman RM, Bouvet M. (2014) Fluorescently-labeled chimeric anti-CEA antibody improves detection and resection of human colon cancer in a patient-derived orthotopic xenograft (PDOX) nude mouse model. J Surg Oncol 109:451–458. pmid:24249594
9. Hiroshima Y, Maawy A, Metildi CA, Zhang Y, Uehara F, Miwa S, et al. (2014) Successful fluorescence-guided surgery on human colon cancer patient-derived orthotopic xenograft mouse models using a fluorophore-conjugated anti-CEA antibody and a portable imaging system. J Laparoendosc Adv Surg Tech A 24:241–247. pmid:24494971
10. Maawy A, Hiroshima Y, Zhang Y, Luiken GA, Hoffman RM, Bouvet M (2014) Polyethylene glycol (PEG) conjugations of chimeric anti-carcinoembryonic antigen (CEA) labeled with near infrared (NIR) dyes enhances imaging of liver metastases in a nude-mouse model of human colon cancer. PLoS One 9: e97965. pmid:24859320
11. Kishimoto H, Zhao M, Hayashi K, Urata Y, Tanaka N, Fujiwara T, et al. (2009) In vivo internal tumor illumination by telomerase-dependent adenoviral GFP for precise surgical navigation. Proc Natl Acad Sci USA 106:14514–14517. pmid:19706537
12. Kishimoto H, Aki R, Urata Y, Bouvet M, Momiyama M, Tanaka N, et al. (2011) Tumor-selective adenoviral-mediated GFP genetic labeling of human cancer in the live mouse reports future recurrence after resection. Cell Cycle 10:2737–2741. pmid:21785265
13. Hoffman RM, Yang M (2006) Subcellular imaging in the live mouse. Nature Protoc 1:775–782.
14. Hoffman RM, Yang M (2006) Color-coded fluorescence imaging of tumor-host interactions. Nature Protoc 1:928–935.
15. Hoffman RM, Yang M (2006) Whole-body imaging with fluorescent proteins. Nature Protoc 1:1429–1438.
16. Yamauchi K, Yang M, Jiang P, Xu M, Yamamoto N, Tsuchiya H, et al. (2006) Development of real-time subcellular dynamic multicolor imaging of cancer-cell trafficking in live mice with a variable-magnification whole-mouse imaging system. Cancer Res 66:4208–4214. pmid:16618743
17. Yano S, Miwa S, Kishimoto H, Uehara F, Tazawa H, Toneri M, et al. (2015) Targeting tumors with a killer-reporter adenovirus for curative fluorescence-guided surgery of soft-tissue sarcoma. Oncotarget 6:13133–13148. pmid:26033451
18. Yano S, Miwa S, Kishimoto H, Toneri M, Hiroshima Y, Yamamoto M, et al. (2015) Experimental curative fluorescence-guided surgery of highly invasive glioblastoma multiforme selectively labeled with a killer-reporter adenovirus. Molecular Therapy 23:1182–1188. pmid:25896244
19. Yano S, Hiroshima Y, Maawy A, Kishimoto H, Suetsugu A, Miwa S, et al. (2015) Color-coding cancer and stromal cells with genetic reporters in a patient-derived orthotopic xenograft (PDOX) model of pancreatic cancer enhances fluorescence-guided surgery. Cancer Gene Therapy 22:344–350. pmid:26088297
20. Nemunaitis J, Tong AW, Nemunaitis M, Senzer N, Phadke AP, Bedell C, et al. (2010). A phase I study of telomerase-specific replication competent oncolytic adenovirus (telomelysin) for various solid tumors. Mol Ther 18:429–434. pmid:19935775

CHAPTER 21

Fire Ignition During Laser Surgery in Pet Rodents

Tommaso Collarile[1] , Nicola Di Girolamo[1*], Giordano Nardini[2] , Ivano Antonio Ciraci[1] and Paolo Selleri[1]

[1] *Clinica per Animali Esotici, Centro Veterinario Specialistico, Via Sandro Giovannini 53, Rome, Italy.*

[2] *Clinica Veterinaria Modena Sud, Spilamberto, MO, Italy.*

ABSTRACT

Background

Laser surgery is an attractive alternative to other means of section device in terms of tissue inflammation and interaction, which has been extensively used in human and veterinary medicine. Although accidental ignition during laser surgeries is sporadically reported in human medical literature, to the authors' knowledge this is the first report regarding laser-dependent fire ignition during surgery in veterinary medicine.

Case presentation

Two rodents, a 13-month old, 27-gram, male pet mouse (*Mus musculus*) and a 1-year old, female Russian hamster (*Phodopus sungorus*), underwent surgical removal of masses with diode laser. During the surgical procedures fires ignited from the face masks. The mouse presented severe burns on the head and both forelimbs, it was hospitalized and approximately 2 months after surgery burns were resolved. The hamster presented severe burns on the face and the proximal regions of the body. At 72 hours from the accident the hamster was euthanized.

Conclusion

The present report suggests that fire ignition is a potential life-threatening complication of laser surgery in non-intubated rodents maintained under volatile anesthesia. High oxygen concentrations, the presence of combustible, and the narrowness of the surgical field with the face mask during laser surgery on rodents are risk factors for fire ignition.

Keywords : Laser Rodent Pet Surgery Fire Ignition Face mask Burn

BACKGROUND

Laser surgery is an attractive alternative to other means of section device in terms of tissue inflammation and interaction [1], which has been extensively used in human [2, 3, 4, 5] and veterinary medicine [6, 7, 8].

In laboratory rodents, laser surgery promoted rapid postoperative healing [9] and allowed a significant reduction in tumor recurrence and mortality rates [10]. Several experienced surgeons propose the use of laser surgery in pet rodents [11, 12].

Indeed, several articles on the complications associated with laser surgery have been published [13, 14, 15]. Among other complications (e.g. accidental burns, edema of mucus membranes and airway obstruction in larynx surgery), ignition is probably the most dangerous one [16, 17].

Although some authors reported fire ignition as a risk associated with laser procedures [18], to the authors' knowledge specific reports of laser-dependent fire ignition in veterinary medicine are lacking. The objective of this report is to describe two cases of fire ignition observed during laser surgery in non-intubated rodents under volatile anesthesia.

CASE PRESENTATION

Case 1: Mouse

A 13-month old, 27-gram, male pet mouse (*Mus musculus*) was presented for the evaluation of a growing mass located in the right ear pinna. A thorough physical examination was performed and no other abnormalities were found. Due to the need of anesthesia to perform any additional diagnostic evaluation, direct removal of the mass by use of surgical laser was proposed. Anesthesia was induced with 8% sevoflurane (Sevorane, Abbott, UK) administered via an induction chamber. Anesthetic maintenance was facilitated with 4.5-5% sevoflurane and 0.5 litre/minute oxygen delivered via a small, modified, face mask (Figure 1). The mouse was laid supine and retained spontaneous ventilation throughout surgery. Surgical scrub was performed with chlorhexidine (Clorexyderm, ICFpet, Palazzo Pignano, Italy) and sterile saline. A diode laser (Veilure S9, Lasering S.r.l., Modena, Italy) transduced by an optic fiber was used in contact [19], continuous mode at 3 W to excise the mass. During surgical excision of the mass an accidental fire ignited from the face mask. Emergency procedures for free fire were adopted: the primary surgeon moved immediately the mouse away from the face mask. The anesthetist turned off the oxygen and the volatile anesthesia. The assistant surgeon poured water on the mouse. These procedures were executed almost at the same time. Thus in few seconds the fire was extinguished.

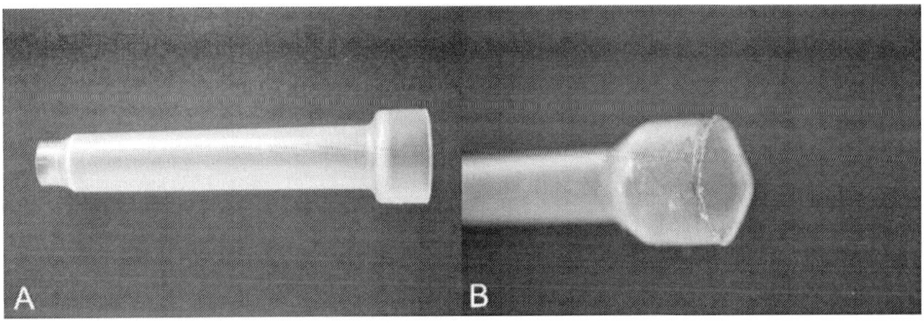

Figure 1. FACE MASK EMPLOYED BY THE AUTHORS FOR MAINTENANCE OF VOLATILE ANESTHESIA IN RODENTS. The face mask in figure A, due to the absence of a diaphragm that permits to achieve a seal on the animal's nose or neck, did not prevent leakage of oxygen and anesthetic gasses. In figure B it is depicted the face mask used for anesthesia of the mouse after fire ignition. The use of tight-fitting face masks providing a hermetical seal seems necessary during laser surgery under volatile anesthesia in non-intubated rodents.

Nevertheless, the mouse presented severe burns on the head and both forelimbs (Figure 2A). The animal was hospitalized and standard medical measures for burns were administered. In the days following the surgery the mouse body weight dropped to 21 grams. Mouse's clinical conditions slowly improved and four weeks following the accident it was discharged (Figure 2B). The mouse gradually returned to its initial weight and approximately 2 months after surgery the burns were resolved (Figure 2C).

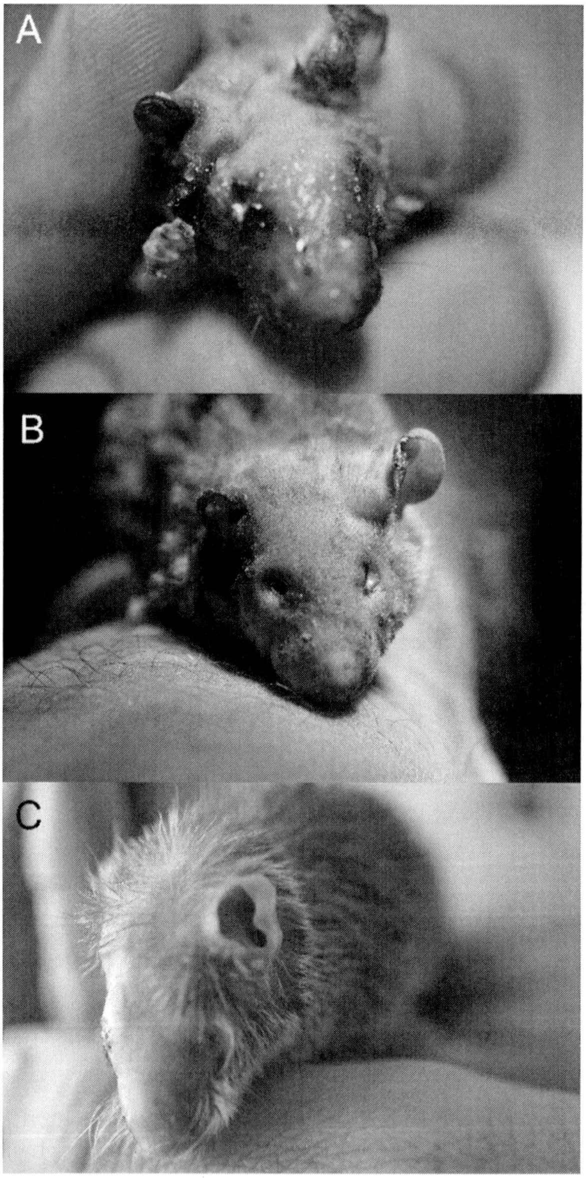

Figure 2. CLINICAL PROGRESSION OF BURN SECONDARY TO LASER-DEPENDENT IGNITION OF ANESTHETIC GASSES IN A PET MOUSE. The mouse at day 5 (A), 30 (B) and 60 (C) after fire ignition.

Case 2: Hamster

A 1-year old, female Russian hamster (*Phodopus sungorus*) was presented for eyelid swelling and lameness. Physical examination revealed a firm, encapsulated mass located in the inner surface of the left upper eyelid and a round, pedunculated, 1-mm mass dorsally to the first finger of left forelimb. Surgical removal of both masses was elected. Anesthesia was induced with 5% isoflurane (Isoba, Schering-Plough, UK) administered via an induction chamber. After induction the hamster was maintained on 2.5-3% isoflurane with a modified face-mask. Trichotomy and surgical scrub with chlorhexidine (Clorexyderm, ICFpet, Palazzo Pignano, Italy) were performed around the two masses. The hamster's eye was covered with moist gauze sponges in order to protect the corneas from damage by exposure to the laser beam. A 400 micron transducer of the diode laser (Veilure S9, Lasering S.r.l., Modena, Italy) was used to excise the palpebral mass. The transducer was settled on 1.5 W, non-continuous mode (On: 10 milliseconds, Off: 50 milliseconds). The mass on the eyelid was removed without adverse reactions during the procedure. Diode laser was then used in continuous mode, 3.5 W, to excise the mass on the left forelimb. After few seconds from the beginning of the procedure, some sparks were seen originating from the laser beam and fire ignite from the face mask. Although the fire was immediately extinguished, the hamster presented severe burns on the face and the proximal regions of the body. Left eye vision was impaired due to corneal burn (Figure 3). The hamster was hospitalized and aggressive supportive therapy was initiated. Nevertheless there was a rapid worsening of its clinical status. At 72 hours from the accident the hamster was humanely euthanized.

DISCUSSION

The present report underline that ignition is a potential life-threatening complication of laser surgery in non-intubated rodents under volatile anesthesia. Although accidental ignition during laser surgeries is sporadically reported in human medical literature [17, 20, 21, 22, 23], to the author's best knowledge this is the first report in which laser-secondary ignition is described in veterinary literature.

For any fire to start, three elements must be present: heat or an ignition source, fuel and an oxidizer. The three elements constitute the so-called "fire triad" or "fire triangle" [24, 25]. An oxidizer is a substance that gains electrons in a red-ox chemical reaction. The most common oxidizers in the operating room are oxygen and nitrous oxide [25]. The most common

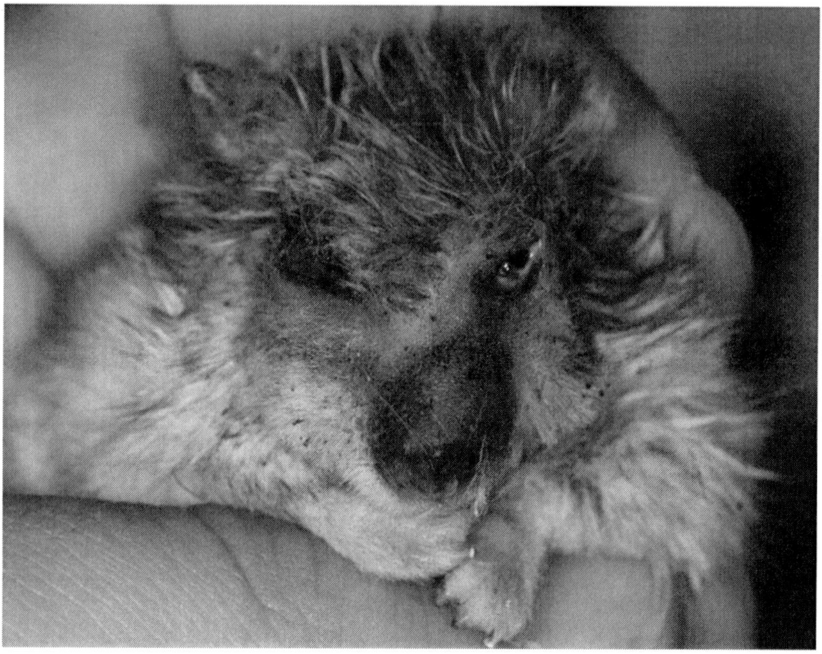

Figure 3. Photograph Of A Pet Hamster After That Fire Ignited During Laser Surgery**.** Notice the spread of the lesion along the face.

ignition sources cited in operating room fires seem to be the electrosurgical unit [24], although also lasers are another common ignition source. As previously stated heat can replace the ignition source: the heat produced by the fiberoptic light source was responsible of operating room fire [26]. Fuels can be from alcohol based solutions, to any cloth-containing or paper-containing materials on or around the patient, including abdominal gasses [24].

Indeed, an oxygen-rich atmosphere, flammable materials, and ignition sources are virtually ubiquitous in the modern-day operating room [24].

In human medical literature, laser-secondary ignitions are mainly reported during upper airways surgeries [20], especially laryngeal [17, 22, 23] and tracheal [21] surgeries. On 20,000 laryngeal laser-surgical procedures, Sesterhenn and colleagues [27] reported 15 cases (0.075%) of tube fires. The mechanism of combustion is the initial penetration of the endotracheal tube by the laser and then the ignition of the tube. This is facilitated by the heat produced by the laser, the flow of oxygen, inflammable anesthetic agents or combustible materials [16, 23, 28]. We suppose that in the present cases the elements composing the "fire triangle" (oxidizer/inflammable substance/ignition source) [25] were: 100 per cent oxygen, sevoflurane or isoflurane and diode laser at 3–3.5 W continuous pulsation, respectively. Decrease or removal of any element of the fire triangle could have prevented fire ignition.

In a recent study 15 rats underwent resection of a 3-mm transverse area of the anterior tongue by use of a carbon dioxide laser [29]. The rats in the study were not exposed to inhalant anesthesia or to oxygen. It is not unexpected that no fire ignition consequent to the procedure was reported in that study. This finding corroborates the role that oxygen and inhalant gasses play in fire ignition during laser surgery.

In the present work we presented two cases in which the oxygen was administered at 100 per cent concentration. Oxygen concentrations of 50 and 75 per cent did not alter the time to ignition of surgical patties exposed to laser beam. Indeed the time to ignition felt significantly when the oxygen concentration was further increased to 100 per cent [30]. A previous study already demonstrated that surgical drapes present a lower time to ignition when exposed to higher oxygen concentration [31]. Considering such findings, ventilation with 100 per cent oxygen should be avoided when laser surgeries are performed under face mask ventilation. An alternative to 100 per cent oxygen ventilation may be the ventilation with a lower oxygen concentration (e.g. 30-60-80 per cent), as it has been suggested in human medicine to prevent atelectasis [32, 33, 34]. Although lower oxygen concentrations may decrease the risk of fire ignition, specific reports investigating its impact on rodent surgery outcome are necessary before its application in clinical practice.

Intubation in rodents is, in most cases, not easily performed; therefore anesthesia is often maintained with a tight-fitting face mask connected to the breathing system [35, 36]. As suggested by Dave and Mahaffey [37] "*The use of face masks and nasal cannulae should be avoided as there is always some leakage around these devices*". In the present reports the face mask employed (Figure 1) did not prevent leakage of oxygen and anesthetic gasses. Face masks similar to those employed by the authors are often used during surgeries of small exotic animals [35, 38, 39]. The use of tight-fitting face masks providing a hermetical seal seems necessary during laser surgery under volatile anesthesia in rodents. Anesthetic gas leakage from standard rodent non-rebreathing circuits has been demonstrated [40]. The use of modified face masks significantly reduced the volume of gas leakage. The masks were modified by addition of a latex diaphragm to the conical mask attached to the Mapleson E circuit [40]. Nevertheless, before laser can be safely employed on non-intubated animals under volatile anesthesia, specific trials focusing on the real efficacy of face masks in preventing gas leakage are necessary.

In the two cases described topical scrub was not performed with alcohol due to its intrinsic inflammable properties. Unfortunately, also most commercial chlorhexidine solutions contain alcohol, and their flammability is in direct proportion to their alcohol concentration [24]. Although some case reports identifying alcohol-based skin preparations as a fuel source in surgical fires have been published in human medical literature [41, 42, 43], a recent study was unable to demonstrate flammability of isopropyl alcohol exposed to electrofulguration in experimental settings [44]. Conversely, when materials in the study were poured with chlorhexidine gluconate under saturated and damp conditions produced a spark and associated mild to moderate char [44]. Only water-based prep solutions such as betadine contain no alcohol and, therefore, can be considered truly nonflammable [24]. It should be also mentioned that the use of alcohol in surgeries of rodents is discouraged, as it could promote hypothermia [11, 45].

CONCLUSIONS

Although in human medical literature most cases of fire ignition seems to be managed without vital damage of the patient [22, 46], in the present cases, catastrophic, life-threatening injuries were suffered from both animal patients.

Fire ignition is a potential complication during laser surgeries in rodents due to the narrowness of surgical field with oxygen outflow and anesthetic gasses. Besides paying appropriate care when specific regions (e.g. airways and oral cavity [7, 8]) are target of laser surgery, the surgeon should always be aware of fire ignition when laser surgeries are performed on non-intubated animals under volatile anesthesia.

Consent

Consent was obtained from the owners of the animals for publication of this case report and any accompanying images.

Competing interests

The authors declare that they have no competing interests

Authors' contributions

TC and GN were the surgeons of the mouse and the hamster, respectively. ND was the assistant surgeon of the first case and drafted the manuscript. IC was the anesthetist of the first case and took care of the mouse during its recovery. PS conceived the report and helped to draft the manuscript. All authors read and approved the final manuscript.

REFERENCES

1. Dunié-Mérigot A, Bouvy B, Poncet C: Comparative use of CO2 laser, diode laser and monopolar electrocautery for resection of the soft palate in dogs with brachycephalic airway obstructive syndrome. Vet Rec. 2010, 167: 700-704.
2. Hall RR, Hill DW, Beach AD: A carbon dioxide surgical laser. Ann R Coll Surg Engl. 1971, 48: 181-188.
3. Bellina JH: The use of the carbon dioxide laser in the management of condyloma acuminatum with eight-year follow-up. Am J Obstet Gynecol. 1983, 147: 375-378.
4. Rau BK, Harikrishnan KM, Krishna S: Endoscopic laser ablation of duodenal carcinoids: a new treatment modality. J Clin Laser Med Surg. 1995, 13: 37-38.
5. Puthussery FJ, Shekar K, Gulati A, Downie IP: Use of carbon dioxide laser in lingual frenectomy. Br J Oral Maxillofac Surg. 2011, 49: 580-581.
6. Mison MB, Steficek B, Lavagnino M, Teunissen BD, Hauptman JG, Walshaw R: Comparison of the effects of the CO2 surgical laser and conventional surgical techniques on healing and wound tensile strength of skin flaps in the dog. Vet Surg. 2003, 32: 153-160.
7. Kovács K, Jakab C, Szász AM: Laser-assisted removal of a feline eosinophilic granuloma from the back of the tongue. Acta Vet Hung. 2009, 57: 417-426.
8. Taney K, Smith MM: Resection of mast cell tumor of the lip in a dog. J Vet Dent. 2009, 26: 28-34.
9. Wang Z, Devaiah AK, Feng L, Dasai U, Shapira G, Weisberg O, Torres DS, Shapshay SM: Fiber-guided CO2 laser surgery in an animal model. Photomed Laser Surg. 2006, 24: 646-650.
10. Peled I, Shohat B, Gassner S, Kaplan I: Excision of epithelial tumors: CO2 laser versus conventional methods. Cancer Lett. 1976, 2: 41-45.
11. Capello V: Common surgical procedures in pet rodents. J Exot Pet Med. 2011, 20: 294-307.
12. Bennett RA: Soft tissue surgery. Ferrets, Rabbits, and Rodents Clinical Medicine and Surgery. Edited by: Quesenberry KE, Carpenter JW. 2012, St. Louis: Elsevier Saunders, 373-391. 3
13. Meyers A: Complications of CO2 laser surgery of the larynx. Ann Otol Rhinol Laryngol. 1981, 90: 132-134.
14. Fried MP: Complications of CO2 laser surgery of the larynx. Laryngoscope. 1983, 93: 275-278.
15. Vilaseca-González I, Bernal-Sprekelsen M, Blanch-Alejandro JL, Moragas-Lluis M: Complications in transoral CO2 laser surgery for carcinoma of the larynx and hypopharynx. Head Neck. 2003, 25: 382-388.
16. Handa KK, Bhalla AP, Arora A: Fire during the use of Nd-Yag laser. Int J Pediatr Otorhinolaryngol. 2001, 60: 239-242.
17. Chou AK, Tan PH, Yang LC, Sun GC, Hsieh SW: Carbon dioxide laser induced airway fire during larynx surgery: case report. Chang Gung Med J. 2001, 24: 393-398.

18. Berger N, Eeg PH: Veterinary laser surgery. 2006, Iowa: Blackwell Publishing Ames
19. Hernandez-Divers S: Diode laser surgery: Principles and application in exotic animals. Semin Avian Exot Pet Med. 2002, 11: 208-220.
20. Snow JC, Norton ML, Saluja TS, Estanislao AF: Fire hazard during CO2 laser microsurgery on the larynx and trachea. Anesth Analg. 1976, 55: 146-147.
21. Komatsu T, Kaji R, Okazaki S, Miyawaki I, Ishihara K, Takahashi Y: Endotracheal tube ignition during the intratracheal laser treatment. Asian Cardiovasc Thorac Ann. 2008, 16: 49-51.
22. Santos P, Ayuso A, Luis M, Martínez G, Sala X: Airway ignition during CO2 laser laryngeal surgery and high frequency jet ventilation. Eur J Anaesthesiol. 2000, 17: 204-207.
23. Wang HM, Lee KW, Tsai CJ, Lu IC, Kuo WR: Tracheostomy tube ignition during microlaryngeal surgery using diode laser: a case report. Kaohsiung J Med Sci. 2006, 22: 199-202.
24. Rinder CS: Fire safety in the operating room. Curr Opinion Anaesthesiol. 2008, 21: 790-795. 10.1097/ACO.0b013e328318693a.
25. Blazquez E, Thorn C: Fires and explosions. Anaesth Intensive Care Med. 2010, 11: 455-457.
26. Williams DM, Littwin S, Patterson AJ, Brock-Utne JG: Fiberoptic light source induced surgical fires: the contribution of forced-air warming blankets. Acta Anesthesiol Scand. 2006, 50: 505-508.
27. Sesterhenn AM, Dunne AA, Braulke D, Lippert BM, Folz BJ, Werner JA: Value of endotracheal tube safety in laryngeal laser surgery. Lasers Surg Med. 2003, 32: 384-390.
28. Walker P, Temperley A, Thelfo S, Hazelgrove A: Avoidance of laser ignition of endotracheal tubes by wrapping in aluminium foil tape. Anaesth Intensive Care. 2004, 32: 108-112.
29. Shires CB, Saputra JM, King L, Thompson JW, Heck DH, Sebelik ME, Boughter JD: Histopathological and postoperative behavioral comparison of rodent oral tongue resection: fiber-enabled co2 laser versus electrocautery. Otolaryngol Head Neck Surg. In press
30. Dhar V, Young K, Nouraei SA, Sandhu GS, Tatla T, Farrell R, Vaughan D, Sesterhenn AM, Dunne AA: Impact of oxygen concentration and laser power on occurrence of intraluminal fires during shared-airway surgery: an investigation. J Laryngol Otol. 2008, 122: 1335-1338.
31. Wolf GL, Sidebotham GW, Lazard JL, Charchaflieh JG: Laser ignition of surgical drape materials in air, 50% oxygen, and 95% oxygen. Anesthesiology. 2004, 100: 1167-1171.
32. Rothen HU, Sporre B, Engberg G, Wegenius G, Reber A, Hedenstierna G: Prevention of atelectasis during general anaesthesia. Lancet. 1995, 345: 1387-1391.
33. Edmark L, Kostova-Aherdan K, Enlund M, Hedenstierna G: Optimal oxygen concentration during induction of general anesthesia. Anesthesiology. 2003, 98: 28-33.
34. Edmark L, Auner U, Enlund M, Ostberg E, Hedenstierna G: Oxygen concentration and characteristics of progressive atelectasis formation during anaesthesia. Acta Anaesthesiol Scand. 2011, 55: 75-81.
35. Longley LA: Anaesthesia of Exotic Pets. 2008, Philadelphia: Saunders Elsevier
36. Wenger S: Anesthesia and analgesia in rabbits and rodents. J Exot Pet Med. 2012, 21: 7-16.
37. Dave R, Mahaffey PJ: The control of fire hazard during cutaneous laser therapy. Lasers Med Sci. 2002, 17: 6-8.
38. Redrobe S: Soft tissue surgery of rabbits and rodents. Semin Avian Exot Pet Med. 2002, 11: 231-245.
39. Smith JC, Bolon B: Isoflurane leakage from non-rebreathing rodent anaesthesia circuits: comparison of emissions from conventional and modified ports. Lab Anim. 2006, 40: 200-209.

40. Hawkins MG, Pascoe PJ: Anesthesia, analgesia, and sedation of small mammals. Ferrets, Rabbits, and Rodents Clinical Medicine and Surgery. Edited by: Quesenberry KE, Carpenter JW. 2012, St. Louis: Elsevier Saunders, 429-451. 3
41. Batra S, Gupta R: Alcohol based surgical prep solution and the risk of fire in the operating room: a case report. Patient Saf Surg. 2008, 2:
42. Roy S, Smith LP: What does it take to start an oropharyngeal fire? Oxygen requirements to start fires in the operating room. Int J Pediatr Otorhinolaryngol. 2011, 75: 227-230.
43. Patel R, Chavda KD, Hukkeri S: Surgical field fire and skin burns caused by alcohol-based skin preparation. J Emerg Trauma Shock. 2010, 3: 305-
44. Arefiev K, Warycha M, Whiting D, Alam M: Flammability of topical preparations and surgical dressings in cutaneous and laser surgery: A controlled simulation study. J Am Acad Dermatol. In press
45. Bennett RA, Mullen HS: Soft tissue surgery. Ferrets, Rabbits and Rodents: Clinical Medicine and Surgery. Edited by: Quesenberry KE, Carpenter JW. 2004, St. Louis: Elsevier Saunders, 316-328. 2
46. Munksgaard AB, Bonde J, Valentin N: Ignition of the endotracheal tube during CO2 laser therapy of upper airway disorders. Ugeskr Laeger. 1805, 1991: 153

Index